CHARACTERIZATION OF ORGANIC THIN FILMS

MATERIALS CHARACTERIZATION SERIES

Surfaces, Interfaces, Thin Films

Series Editors: C. Richard Brundle and Charles A. Evans, Jr.

Series Titles

Encyclopedia of Materials Characterization, C. Richard Brundle, Charles A. Evans, Jr., and Shaun Wilson

Characterization of Metals and Alloys, Paul H. Holloway and P. N. Vaidyanathan

Characterization of Ceramics, Ronald E. Loehman

Characterization of Polymers, Ned J. Chou, Steven P. Kowalczyk, Ravi Saraf, and Ho-Ming Tong

Characterization in Silicon Processing, Yale Strausser

Characterization in Compound Semiconductor Processing, Yale Strausser

Characterization of Integrated Circuit Packaging Materials, Thomas M. Moore and Robert G. McKenna

Characterization of Catalytic Materials, Israel E. Wachs

Characterization of Composite Materials, Hatsuo Ishida

Characterization of Optical Materials, Gregory J. Exarhos

Characterization of Tribological Materials, William A. Glaeser

Characterization of Organic Thin Films, Abraham Ulman

CHARACTERIZATION OF ORGANIC THIN FILMS

EDITOR

Abraham Ulman

MANAGING EDITOR

Lee E. Fitzpatrick

BUTTERWORTH-HEINEMANN
Boston London Oxford Singapore Sydney Toronto Wellington

MANNING
Greenwich

This book was acquired, developed, and produced by Manning Publications Co.
Design: Christopher Simon

Recognizing the importance of preserving what has been written, it is the policy of Butterworth-Heinemann and of Manning to have the books they publish printed on acid-free paper, and we exert our best efforts to that end.

Library of Congress Cataloging–in–Publication Data
Characterization of organic thin films / editor, Abraham Ulman.
p. cm. — (Materials characterization series)
Includes bibliographical references and index.
ISBN 0-7506-9467-X
1. Organic compounds. 2. Thin films. I. Ulman, Abraham, 1946– .
II. Series.
QC176.9.073C48 1994 94–26831
530.4′175—dc20 CIP

Butterworth-Heinemann
80 Montvale Avenue
Stoneham, MA 02180

Manning Publications Co.
3 Lewis Street
Greenwich, CT 06830

10 9 8 7 6 5 4 3 2 1

Printed in the United States of America

Contents

PART II: ANALYSIS OF FILM AND SURFACE PROPERTIES

SPECTROSCOPIC ELLIPSOMETRY

INFRARED SPECTROSCOPY IN THE CHARACTERIZATION OF ORGANIC THIN FILMS

RAMAN SPECTROSCOPIC CHARACTERIZATION OF ORGANIC THIN FILMS

Preface to Series

This Materials Characterization Series attempts to address the needs of the practical materials user, with an emphasis on the newer areas of surface, interface, and thin film microcharacterization. The Series is composed of the leading volume, *Encyclopedia of Materials Characterization*, and a set of about 10 subsequent volumes concentrating on characterization of individual materials classes.

In the *Encyclopedia*, 50 brief articles (each 10 to 18 pages in length) are presented in a standard format designed for ease of reader access, with straightforward technique descriptions and examples of their practical use. In addition to the articles, there are one-page summaries for every technique, introductory summaries to groupings of related techniques, a complete glossary of acronyms, and a tabular comparison of the major features of all 50 techniques.

The 10 volumes in the Series on characterization of particular materials classes include volumes on silicon processing, metals and alloys, catalytic materials, integrated circuit packaging, etc. Characterization is approached from the materials user's point of view. Thus, in general, the format is based on properties, processing steps, materials classification, etc., rather than on a technique. The emphasis of all volumes is on surfaces, interfaces, and thin films, but the emphasis varies depending on the relative importance of these areas for the materials class concerned. Appendixes in each volume reproduce the relevant one-page summaries from the *Encyclopedia* and provide longer summaries for any techniques referred to that are not covered in the *Encyclopedia*.

The concept for the Series came from discussion with Marjan Bace of Manning Publications Company. A gap exists between the way materials characterization is often presented and the needs of a large segment of the audience—the materials user, process engineer, manager, or student. In our experience, when, at the end of talks or courses on analytical techniques, a question is asked on how a particular material (or processing) characterization problem can be addressed the answer often is that the speaker is "an expert on the technique, not the materials aspects, and does not have experience with that particular situation." This Series is an attempt to bridge this gap by approaching characterization problems from the side of the materials user rather than from that of the analytical techniques expert.

We would like to thank Marjan Bace for putting forward the original concept, Shaun Wilson of Charles Evans and Associates and Yale Strausser of Surface Science Laboratories for help in further defining the Series, and the Editors of all the individual volumes for their efforts to produce practical, materials user based volumes.

C. R. Brundle *C. A. Evans, Jr.*

Preface

Materials science is at the center of academic and industrial research today. In the past ten years, it has become apparent that the way materials scientists operate should change, and that a design approach must be used in the preparation of new materials. This is best represented by the research done in the area of organic thin films, where a useful property is identified a priori, an appropriate molecule designed and synthesized, and the corresponding film prepared.

Materials scientists, physicists, chemists, and even biologists are interested in both amorphous (spin-coated polymer) and highly organized (Langmuir–Blodgett and self-assembled) organic films because of their relevance to science and technology. For example, spin-coated polymer films that contain polar aromatic molecules may have applications in electro-optic devices. Monolayers on piezoelectric crystals may serve as chemical and biological sensors, and self-assembled monolayers, due to their dense and stable structure, have potential application in corrosion prevention, wear protection, and more. The ability to tailor both head and tail groups of the constituent molecules makes self-assembled monolayers ideally suited for a more fundamental understanding of phenomena affected by competing intermolecular, molecular–substrate, and molecule–solvent interactions, such as ordering and growth, wetting, adhesion, lubrication, and corrosion.

While scientists and engineers typically have the background needed to understand the subtleties of the molecular material important for their applications, they may not have the training to extract all the information from the analysis of organic thin films. Often, analysts have the opposite strengths and weaknesses. This volume, *Characterization of Organic Thin Films*, together with the others in the Characterization Series published by Manning Publications and Butterworth-Heinemann, are intended to rectify this situation. This volume does not emphasize the characteristics of the different techniques—that is accomplished in the lead volume in the Series, *Encyclopedia of Materials Characterization*. Instead, a case study approach is used in most chapters to illustrate how important problems in organic thin films can be resolved using a given analytical technique.

In arranging this volume, we have not followed the general pattern used in other volumes of the series. We have decided to organize it according to different analytical techniques for a couple of reasons. The first is that by writing a chapter on only a specific material, using different analytical techniques as examples, a serious discussion on any of the techniques cannot be developed, and an understanding of the uniqueness, strengths, and weaknesses of a particular technique cannot be achieved. Helping readers achieve such an understanding is my first goal. The second reason is that if the book is structured according to materials, readers will probably choose to read the one or two chapters closely related to their work, and will not be introduced to other types of materials. By taking the analytical techniques approach, on the other hand, a range of important structural issues are presented, and at the same time the reader gains an understanding of interpretation schemes. The reader also is exposed to different types of materials through the examples used. The purpose of the book clearly is not to make readers experts, but to get them started should they desire to become experts, and to plant ideas so they can question an associate who is an expert analyst. Providing a broad perspective on organic thin films is my second goal.

We begin this volume with introductory chapters on Langmuir–Blodgett and self-assembled films. We then turn to discussions of both their surface (interfacial) and bulk properties, as studied using different analytical techniques. The techniques discussed are ellipsometry, Fourier transform infrared (FTIR) spectroscopy, Raman spectroscopy, surface potential measurements, X-ray diffraction measurements, high resolution electron energy loss spectroscopy (HREELS), wetting and surface energy, secondary ion mass spectrometry (SIMS), X-ray photoelectron (XPS) and Auger electron spectroscopies, and optical second harmonic generation (SHG).

A number of techniques have not been included in this volume, and the reader is referred to the complementary volume *Encyclopedia of Materials Characterization.* The development of analytical tools for the study of organic thin films has been dramatic in the past decade. Using such tools it has become possible to get structural information at the molecular level and thus relate structure to properties. The fundamental understanding of structure–properties relationships makes molecular engineering of advanced materials possible and opens new opportunities in material science and molecular manufacturing.

Abraham Ulman

Contributors

David L. Allara The Pennsylvania State University University Park, PA	Spectroscopic Ellipsometry; Infrared Spectroscopy in the Characterization of Organic Thin Films
Robert W. Collins The Pennsylvania State University University Park, PA	Spectroscopic Ellipsometry
Robert M. Corn University of Wisconsin-Madison Madison, WI	Molecular Orientation in Thin Films as Probed by Optical Second Harmonic Generation
Lawrence H. Dubois AT&T Bell Laboratories Murray Hill, NJ	High Resolution EELS Studies of Organic Thin Films and Surfaces
P. Dutta Northwestern University Evanston, IL	X-Ray Diffraction
James F. Elman Eastman Kodak Company Rochester, NY	X-Ray Photoelectron Spectroscopy of Organic Thin Films
Stephen D. Evans The University of Leeds Leeds, U.K.	Surface Potential
George L. Gaines, Jr. Rensselaer Polytechnic Institute Troy, NY	Langmuir–Blodgett Films
Joseph A. Gardella, Jr. State University of New York at Buffalo Buffalo, NY	Secondary Ion Mass Spectrometry as Applied to Thin Organic and Polymeric Films
Daniel A. Higgins University of Wisconsin-Madison Madison, WI	Molecular Orientation in Thin Films as Probed by Optical Second Harmonic Generation
Robert W. Johnson, Jr. State University of New York at Buffalo Buffalo, NY	Secondary Ion Mass Spectrometry as Applied to Thin Organic and Polymeric Films
Yeon-Taik Kim Los Alamos National Laboratory Los Alamos, NM	Spectroscopic Ellipsometry

D. Li University of Alberta Edmonton, Alberta	Wetting
Jian-Xin Li State University of New York at Buffalo Buffalo, NY	Secondary Ion Mass Spectrometry as Applied to Thin Organic and Polymeric Films
Yiwei Lu The Pennsylvania State University University Park, PA	Spectroscopic Ellipsometry
A. W. Neumann University of Toronto Toronto, Ontario	Wetting
Jeanne E. Pemberton The University of Arizona Tucson, AZ	Raman Spectroscopic Characterization of Organic Thin Films
Jianou Shi The Pennsylvania State University University Park, PA	Spectroscopic Ellipsometry
Abraham Ulman Polytechnic University Brooklyn, NY	Self-Assembled Monolayers; X-Ray Photoelectron Spectroscopy of Organic Thin Films

Part I

Preparation and Materials

1

Langmuir–Blodgett Films

GEORGE L. GAINES, JR.

Contents

1.1 Introduction

In 1919, Katharine Blodgett and Irving Langmuir, working in the General Electric Research Laboratory in Schenectady, New York, showed that monomolecular films of such insoluble substances as long-chain fatty acids floating on water could be transferred to solid substrates (e.g., glass or metal plates) by a simple dipping procedure.[1] Some 15 years later, Blodgett announced the discovery that monolayer transfers could be accomplished sequentially to form built-up multilayer films—the structures now universally referred to as Langmuir–Blodgett (or L–B) films.[2–4] The technique, as Blodgett first developed it, involved spreading the monolayer on water and compressing it to some appropriate surface pressure where the molecules were close packed; while a constant pressure was maintained, the solid substrate, held vertically, was dipped in and out of the water surface through the film. Figure 1.1 depicts Blodgett's apparatus for carrying out the process. In the most common type of transfer—producing "Y films," in Langmuir's notation—a layer is deposited in both the insertion and withdrawal parts of the cycle (no transfer occurs on the first insertion of a hydrophilic substrate, whereas a hydrophobic substrate does accumulate film on the first insertion; hence, for both, a Y film is composed of

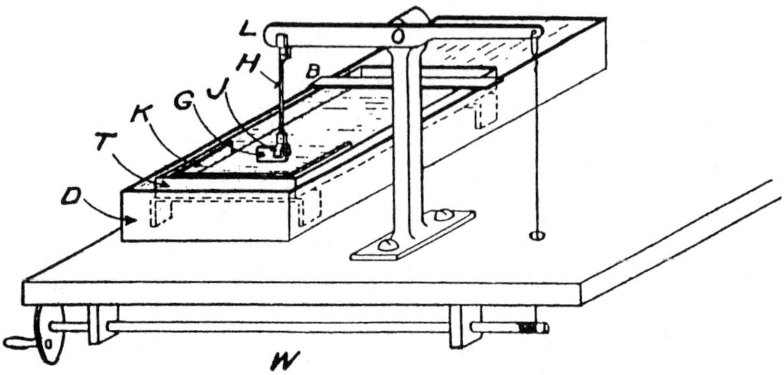

Figure 1.1 **Blodgett's apparatus for layer deposition. *D* = thermostat bath; *T* = trough; *K* = glass strips, which prevent benzene spreading solution from reaching waxed trough edges; *G* = glass slide substrate; *J* = clamp to hold substrate; *H* = connecting rod; *L* = lever arm; *W* = windlass (hand operated); and *B* = sweeping barrier. Not shown is the silk thread arrangement to separate the spread monolayer from the "piston oil" to maintain constant pressure. (Reproduced from *J. Amer. Chem. Soc.*, Vol. 57, p. 1008, 1935.)**

head-to-head and tail-to-tail bilayers) (see Figure 1.2). Alternatively, transfer may occur on insertion only ("X films") or withdrawal only ("Z films"). Langmuir and Blodgett and subsequent workers have developed alternative methods, such as the horizontal touching method, in which the substrate is lowered onto the film in a horizontal position and excess monolayer is then swept away from the water surface before removing the solid plate. A variety of techniques have also been used to maintain constant surface pressure. Blodgett devised and extensively employed "piston oils," which are substances that, when applied as a small droplet to the surface, spread to maintain a constant equilibrium spreading pressure. Typical of these is oleic acid, which at room temperature on neutral or acidic aqueous subphases produces a pressure of ~30 dyn/cm. Alternatives include various mechanical arrangements.

The first multilayers studied by Blodgett and Langmuir were composed of long-chain saturated fatty acids—especially palmitic (C_{16}), stearic (C_{18}), and arachidic (C_{20})—spread on slightly alkaline aqueous solutions containing low concentrations of divalent cations, such as calcium, barium, or cadmium—hence, the fatty acids were partially converted to the divalent ion soap. Among Blodgett's most stable, reproducible, and extensively studied films were layers of barium stearate and cadmium arachidate.

Blodgett took advantage of optical interference effects to estimate the thickness of the built-up films (21.5, 24.4, and 27.4 Å per layer, respectively, for the C_{16}, C_{18}, and C_{20} fatty-acid calcium soaps), and this in turn led to her development of the first practical application of L–B films, nonreflecting glass.[5–7] Though soap films

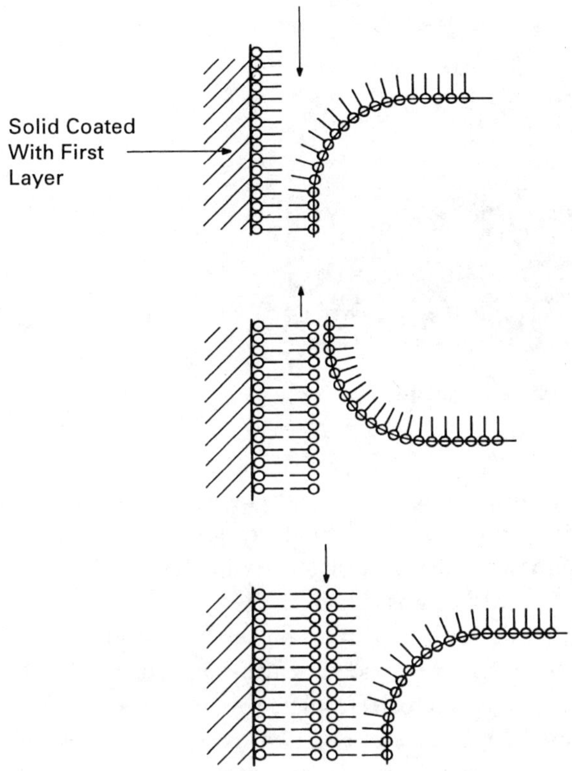

Figure 1.2 Schematic of Y-layer deposition.

are mechanically too fragile for most applications, she demonstrated that reflection could be eliminated by applying a coating of the proper thickness and refractive index, which laid the foundation for the subsequent development of practical coatings. As she showed, such coatings must have a thickness of one-quarter of the wavelength of light and have a refractive index which is the geometric mean between that of glass and air. With L–B soap films, the thickness requirement is easily met by depositing a certain number of layers of the proper length chain. Blodgett discovered that the refractive index could be adjusted by building the layers from an aqueous subphase of the proper pH and then soaking the film in a solvent which dissolves free fatty acid but not soap. Since the degree to which the fatty acid is converted to the divalent ion soap depends on the pH of the bath, this process (which she called "skeletonization") permits a very sensitive control of refractive index.

Langmuir and his co-workers devised other ways to take advantage of the optical properties of L–B films for practical applications. For example, they showed that treatment of fatty-acid soap films with solutions of large polyvalent ions, such as aluminum or thorium, rendered them strongly adsorbent for various other materials, such as proteins, from solution(s). Such "conditioned" multilayers, if they were

Figure 1.3 The Blodgett barium stearate step gauge.

in the right thickness range for interference in the visible light range, then exhibited obvious color changes when adsorption occurred.[8, 9] Blodgett also developed the "step gauge"—which contained films of different thicknesses in the visible light range—as a color comparator for other thin films (Figure 1.3).[10]

Between 1940 and the late 1970s, interest in L–B films for practical applications waned, but in the past 15 years, there has been a resurgence. Sparked largely by potential new applications in optics and microelectronics, this new interest has spawned a succession of international conferences[11] and two recent books.[12–14] It has also led to the invention and development of new devices aimed at the rapid and continuous deposition of L–B films, which might be required for practical large-scale commercial applications (Figure 1.4). It is still true, however, that many variables affecting the formation of L–B films are poorly understood, and the "recipes" in the literature must be followed very carefully to fabricate films of high quality.[12–15]

In this chapter, I review with a broad brush the substances which have been fabricated into L–B films and some of their possible applications. It must be emphasized that this review is far from complete: the twelfth collective *Index of Chemical Abstracts* contains nearly 1000 entries under the heading "Film, Langmuir–Blodgett." For more details, the interested reader should refer to material there or in References 8–10.

1.2 L–B Films of Long-Chain Compounds

Fatty Acids

As already noted, the long-chain fatty acids (usually partially converted to divalent ion soaps) were the earliest and still most extensively studied compounds for L–B film formation. The most stable films of saturated straight-chain fatty acids appear to be formed of cadmium arachidate, optimally deposited from the acetate buffer system originally suggested by Blodgett. With 2.5×10^{-4} M CdCl in the subphase

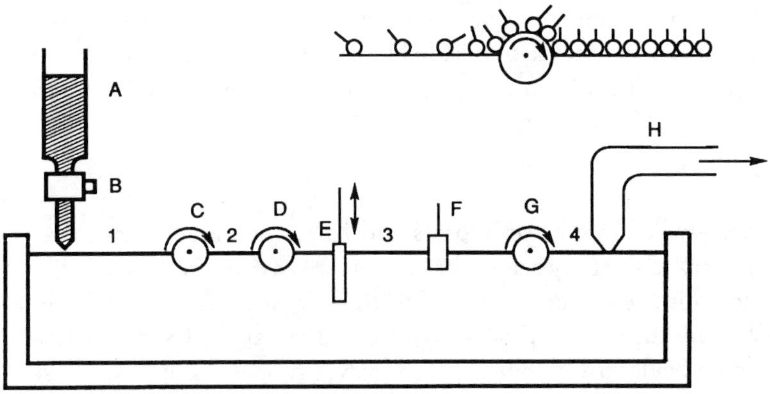

Figure 1.4 **Schematic of a device for the continuous spreading and compression of monolayer.[14] Spreading is accomplished on area 1 by dripping spreading solution from reservoir A, with the rate of delivery controlled by valve B. Rollers C and D compress the film (shown in the inset—upper right) to areas 2 and 3. The surface pressure in area 3 is sensed by Wilhelmy plate F while deposition of the L–B film on substrate E is occurring. Excess monolayer can be moved by roller G to area 4 and then removed by aspirator H. With the addition of an appropriate drive with feed and uptake reels and submerged pulleys, it would be possible to pass a beltlike substrate into and out of the trough and accomplish continuous deposition as well.**

at 20 °C, the arachidate system goes from pure free acid to completely saponified over the pH range of 15.1 5.9.[16] Such films have been extensively used as spacer layers, especially in the elegant studies of Kuhn, Möbius, and their co-workers.[17, 18] It has been shown that the presence of Cd^{2+} in arachidic acid films changes the structure from monoclinic, with chains at 25° to the normal,[19] to orthorhombic, with near-vertical chains.[20] Recently, atomic force microscopy (AFM) has been used to demonstrate the presence and structure of 30–200 μm size domains in L–B films of stearic acid deposited from a poly(ethyleneimine)-containing aqueous subphase; presumably this polymeric anion is incorporated in the film.[21] Another ion, Mn^{2+}, has been incorporated into multilayers to produce "literally two-dimensional magnets."[22] Binks has written an excellent recent review of L–B film formation by fatty acids and amines.[23]

In addition to saturated fatty acids, compounds containing double and triple bonds in the hydrocarbon chain have now been extensively studied, primarily because of their potential application as very high-resolution electron beam resists for microelectronic fabrication. One such compound carefully studied by Barraud and his colleagues at Gif-sur-Yvette is ω-tricosenoic acid, the 23-carbon fatty acid with a terminal double bond in the chain. L–B films of this compound have been shown to be useful as electron beam resists with resolutions of better than 500 Å.[24, 25] Other acids, containing triple bonds, are the diacetylenes (Figure 1.5), which are of interest as polymerizable resists and because they exhibit interesting nonlinear optical properties. Such nonlinear responses, which may be produced when materials are subjected to the very intense optical fields available from high-power lasers, have potential application in switching, optical communications, and information processing.[26]

Amines

Long-chain amines—which are organic bases and hence provide a class of compounds complementary to the fatty acids—have been far less extensively studied. The probable reason for this (not recognized until 1982) is that monolayers of amines interact with CO_2 in the air to produce amine carbamates, $RNH_2 + CO_2 \rightarrow RNHCOOH$, which leads to variability in monolayer properties unless it is controlled.[27] It is also true that longer chains are required to prevent dissolution of the amine monolayer on acidic subphases, as compared to the fatty acids; *n*-docosylamine appears to be an optimal candidate for L–B film formation. It has also been demonstrated that the best films are formed when divalent anions are used to produce the amine salts, and that if pH is adjusted to provide partial salt formation, skeletonization and conditioning reactions analogous to those useful with fatty-acid soap L–B films are available.[27]

Other Long-Chain Compounds

A wide variety of other long-chain compounds—including alcohols, esters (including glycerides and hence biological lipids), nitriles, and phosphorus and sulfur derivatives—have been studied as monolayers on water. L–B multilayer deposition of a variety of esters was demonstrated not long after Blodgett's original work; Stenhagen reported in 1938 the deposition of methyl and ethyl stearates and octadecyl acetate (up to 750 Y layers), as well as the glycerides tripalmitin and α,α'- and α,β-dipalmitins. He also reported that he could not satisfactorily deposit α-monopalmitin or esters with short chains longer than C_2, such as cetyl propionate.[28]

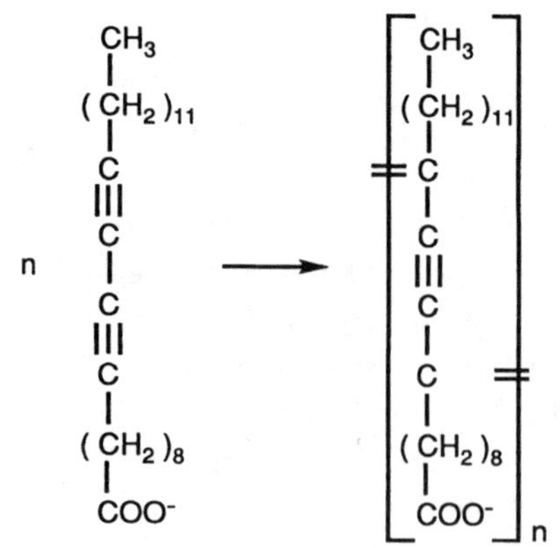

Figure 1.5 A diacetylene fatty acid and the mode of its polymerization.[26]

Subsequently, the fact that the molecules in ester multilayers undergo a crystallization process was recognized,[29] and this process has now been studied by both electron microscopy[30] and infrared spectroscopy.[31]

Mixed monolayers, such as methyl esters plus fatty-acid soaps, have also been reported to form good L–B films; the same report also indicated that pure fatty alcohols could not be built into satisfactory multilayers.[32] It has also been reported that nonadecane nitrile, $C_{18}H_{37}CN$, forms good L–B films from a 2.5×10^{-4} M $CdCl_2$ subphase, but not from pure water.[33]

Because of their relevance as constituents of biological membranes, both natural and synthetic phospholipids have been extensively studied as L–B films. A review of some of this work has been given by Swart,[34] while an interesting report on the surface spectroscopy of phospholipid L–B films typifies recent work.[35]

In addition to the use of conventional hydrocarbon chain compounds, a few reports of L–B film formation by fluorocarbon chain compounds have appeared. Apparently, in the first demonstration of the fabrication of a perfluorinated L–B film, which was achieved in 1986, $C_{10}F_{21}COOH$ was deposited from a subphase containing aluminum ions.[36] More recently, Takahara et al. built multilayers of a two-chain fluorinated surfactant from a subphase containing potassium poly(styrene sulfonate),[37] and Naselli et al. built up layers of $CF_3(CF_2)_7(CH_2)_{10}COOH$ from a $CdCl_2$ solution subphase.[38] It has also been discovered that monolayers are formed, and L–B films can be built, from certain semifluorinated alkanes, $F(CF_2)_m(CH_2)_nH$, even though these molecules do not contain any conventional polar functional groups.[39]

1.3 Cyclic Compounds and Chromophores

Naselli et al.[40] fabricated L–B films of *p*-tetradecyl benzoic acid

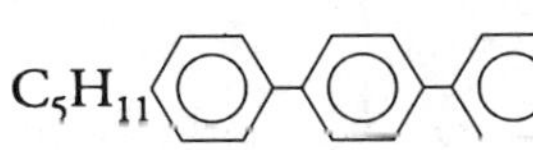

from a Cd^{2+}-containing subphase and found that though chain melting occurred at ~60 °C, head group (aromatic ring) orientation was maintained to >125 °C, and only a temperature of >150 °C led to irreversible damage. A liquid crystal forming terphenyl derivative

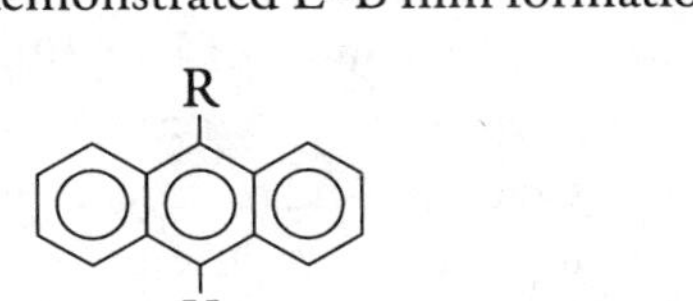

can be built up as Z-type multilayers.[33, 41] Work at ICI and Durham University demonstrated L–B film formation from a series of anthracene derivatives of the form

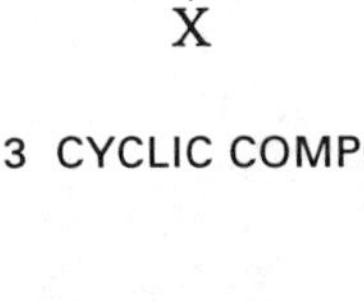

where the R are alkyl chains from 4 to 12 carbon atoms long, and the X are various polar functional groups.[42, 43] When $R = C_4H_9$ and $X = -CH_2CH_2COOH$, excellent films up to 500 layers thick were prepared, and X-ray diffraction showed that the $-C_4H_9$ chains were interdigitated.

N-Docosyl pyridinium TCNQ

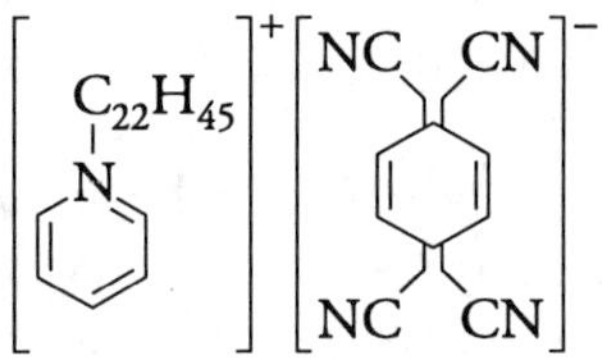

forms L–B films, which are highly conductive if doped with iodine.[44]

Cyanine (and merocyanine and hemicyanine) dyes have been extensively studied in L–B films, especially by Kuhn, Möbius, and their co-workers.[17, 45] Their elegant demonstrations of energy- and electron-transfer in different molecular assemblies involving such dyes have led to Kuhn's concept of "supramolecular machines."[46]

A merocyanine

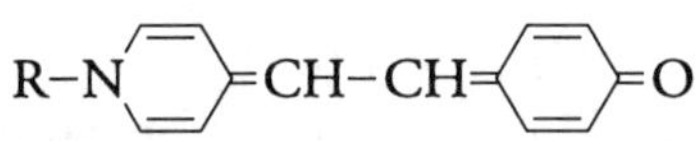

(where R is a long alkyl chain), first synthesized by the author,[47] produced the first L–B films observed to produce optical second-harmonic generation, as long as the film had a noncentrosymmetric structure.[48] Other noncentrosymmetric L–B films have been shown to exhibit pyroelectric activity.[49]

Many porphyrins and phthalocyanines form good L–B films. Langmuir and Schaefer built up chlorophyll multilayers in 1937,[50] which have recently been of interest because of their utility as a model for the natural photosynthetic system.[51] Films that are very stable mechanically and thermally can be built from substituted phthalocyanines such as dichlorosilicon tetra-*t*-butylphthalocyanine.[52] Certain phthalocyanine L–B films promise to be sensitive gas detectors.[53]

The very interesting compound, buckminsterfullerene (C_{60}), can be built into L–B films if it is spread as a mixed monolayer with *n*-octadecanol.[54]

1.4 Polymers and Proteins

Langmuir and his colleagues extensively studied protein monolayers and multilayers in the 1930s.[55] L–B films of synthetic preformed polymers have been studied more recently. Though a large number of studies of spread monolayers of polymers have been reported over the past 65 years,[56] the first detailed accounts of L–B film deposition have appeared only since 1982.[57] Among the types of polymer which have been demonstrated to form good L–B films are those having long hydrocarbon side chains, such as poly(octadecyl acrylate) or the corresponding methacrylate.[58] However, this structural feature is not required, and it has long been known that stable monolayers may be formed of polymers composed of small repeating monomer units which if not polymerized would be water soluble. Examples include some polypeptides[59] and materials such as poly(methyl methacrylate); we found that a syndiotactic PMMA on water at room temperature could be built into good Y-films, with an ellipsometric thickness (9.1 Å per layer) similar to the thickness calculated from the Π–A curve and considerably smaller than that of conventional fatty-acid soap films (20–30 Å per layer).[56]

Many polymer monolayers seem too rigid to permit multilayer formation by conventional L–B dipping. Some cellulose derivatives with this characteristic have, however, been deposited by a horizontal lifting technique.[60] At least two research groups have been concerned with synthetic methods to make polymers whose monolayers will be sufficiently flexible to permit good L–B deposition. Tredgold and co-workers have studied various copolymers including maleic anhydride as one constituent.[57, 61] Ringsdorf and his colleagues have constructed monomer molecules in which the polymerizable group is attached to the head group by a flexible hydrophilic chain, a typical grouping being $R{-}(CH_2CH_2O)_4{-}OC{-}C(CH_3){=}CH_2$. This material is then spread on the water surface and polymerization initiated; the flexibility of the $(CH_2CH_2{-}O)_n$ group permits polymerization to go to completion, and the resulting polymer undergoes satisfactory L–B deposition.[62]

1.5 Polymerization In Situ

As noted in the preceding paragraph, it is possible to spread a monomer and then induce polymerization. Another possibility is to build up an L–B film of monomers and then polymerize the built-up multilayer. A variety of studies have been reported utilizing one or the other of these approaches. One of the earliest examples of multilayer polymerization utilized vinyl stearate,[63] which could be polymerized on the water surface. However, the films were too rigid to be built up. The monomer, on the other hand, could be deposited as X-layers, which then were polymerized in situ with UV or gamma radiation. Such polymerization reactions proceed by a free-radical mechanism which can be quenched by oxygen; polymerization in an inert atmosphere is therefore required. A similar polymerization was of multilayers of cadmium octadecylfumarate. When the L–B film was built of alternating layers of the octadecylfumarate and stearic acid, it was found that the polymerization rate was the same as that observed in the absence of the intervening inert layers. Complete sheets of polymer were therefore being produced with no need for interaction between adjacent layers[64]

Besides addition reactions involving double bonds, there are several examples of condensation polymerizations, often involving soluble reactants in the subphase combining with a spread monolayer. A monolayer of stearaldehyde spread on a subphase containing polyfunctional amines was shown to increase in viscosity, presumably as a result of the condensation reaction, as early as 1941.[65] More recently, the same reaction has been studied in more detail with direct verification of its nature.[66, 67]

As previously discussed in "Fatty Acids" in Section 1.2, compounds such as ω-tricosenoic acid and the diacetylene fatty acids have been studied for their application as high-sensitivity resists for microelectronic fabrication. Another interesting compound in this category is 2-octadecylacrylic acid, $C_{18}H_{37}C(=CH_2)COOH$; L–B films of the monomer can be polymerized by UV or electron-beam irradiation. Higher doses of the electron beam will cause depolymerization; so these films can be used as either positive or negative resists.[68] In addition to diacetylene fatty acids, diacetylene alcohols, such as $C_{12}H_{25}{-}C{\equiv}C{-}C{\equiv}C{-}CH_2OH$, can give good quality L–B films (an exception to the usual observation that long-chain alcohol layers are difficult to build up).[68] Both these and the fatty acid films, when polymerized, are very tough and resistant to solvents. A drawback to their use as resists, however, is that the monolayers as spread are not single crystals, but consist of an array of two-dimensional domains, usually with dimensions in the range of 1–300 μm. These domains persist in built-up films and on polymerization, and polymerization tends to continue to the edge of a domain; on development, the domain either is or is not dissolved, so the resolution is limited to the domain size.[69, 70] For integrated optics applications, on the other hand, the largest possible domain size is desired. It has been reported that a horizontal electric field of 10^4 V/m above the water surface can increase the domain size to the order of 1 mm.[71]

1.6 Alternating Films (Superlattices)

For certain applications, it is desirable to fabricate L–B films with layers having different compositions. Most commonly, alternating layers (–ABABAB–) are desired, but it is conceivable that (especially in the context of "supramolecular machines") any sequence such as –ABCDE– might be required. One case in which alternating layers may be required is the fabrication of devices for nonlinear optics applications involving second-order processes, such as second-harmonic generation. For such processes to be allowed within a bulk material, it is necessary that there be no inversion symmetry of the optically active molecules. Typical examples of structures fabricated to avoid this problem are L–B films containing layers of the merocyanine dye (previously mentioned in Section 1.3) or a closely related hemicyanine dye alternating with layers of ω-tricosenoic acid.[72] For monolayers that deposit as either X- or Z-films, such fabrication presents no problem, the solid substrate simply being dipped into separate monolayer troughs in the desired order. For materials which undergo Y-type deposition, however, a different technique is required. One method is to dip the substrate into a monolayer-covered surface and, while it is held immersed in the subphase, sweep the monolayer off, and spread and compress a film of different composition. The substrate is then withdrawn through the second film. Such a procedure is obviously very tedious if more than a few layers are to be built up, and several research groups have designed dipping devices in which the substrate can be manipulated beneath the subphase surface so that it can be immersed through one monolayer and withdrawn through a different

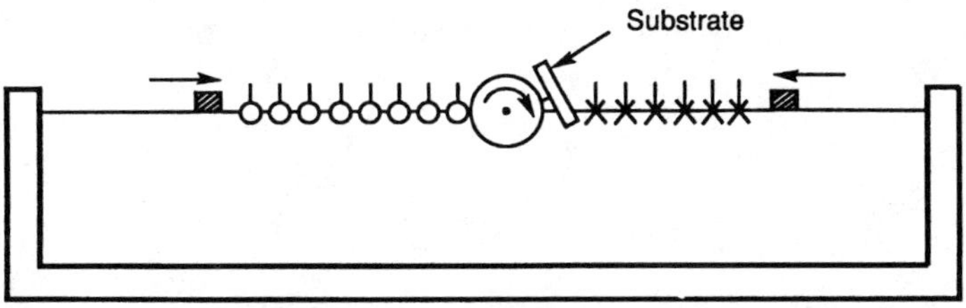

Figure 1.6 **Schematic of one class of devices for depositing alternating (XOXOXO) layers.**

film. Figure 1.6 shows the basic operation of one class of such devices, in which the substrate is carried under a barrier that separates the two monolayers. In another approach, the substrate is held beneath the liquid surface while the monolayers are transported by the movement of appropriate barriers. Some of these alternate-layer dipping devices are now available commercially.

1.7 Potential Applications

Mention has already been made of various practical applications of L–B films, from the early Blodgett and Langmuir methods and devices based on their optical properties to the more recent interest in very high-resolution photo or electron-beam resists for microelectronics. A very large number of other applications have been suggested. Recent books by Roberts[12] and Ulman[13] contain substantial chapters on applications—the former containing 422 references and the latter 351. I have space here to mention only a few possible applications.

In addition to Blodgett's step gauge, one of the few applications which had early commercial application was the use of heavy-metal soap L–B films as diffraction gratings for soft X rays.[73] In the X-ray wavelength range of 1.5–15 mm, lead stearate multilayers are excellent gratings.

Many people believe that the next applications to achieve commercial viability will be in the realm of optics. Successful waveguide measurements in fatty-acid soap films illuminated with He–Ne lasers were reported in 1977,[74] and there is considerable interest in using such films in integrated optics. As already noted, L–B assembly may possibly be used to obtain the noncentrosymmetric arrays required for second-order nonlinear optical effects (second-harmonic generation and sum-and-difference frequency generation). At the present time, commercial devices applying these effects (most Nd–YAG lasers have second-harmonic-generating crystals, so that they can deliver both the fundamental frequency at 1064 nm and also green light at 532 nm) are inorganic materials, such as lithium niobate. It is known, however, that many organic materials have higher intrinsic nonlinear susceptibilities than these inorganics. Noncentrosymmetry, in some cases, can be achieved through crystal growth (as it is with the inorganics), but it may be easier, and in some cases only possible, through L–B methods. It has been demonstrated that in

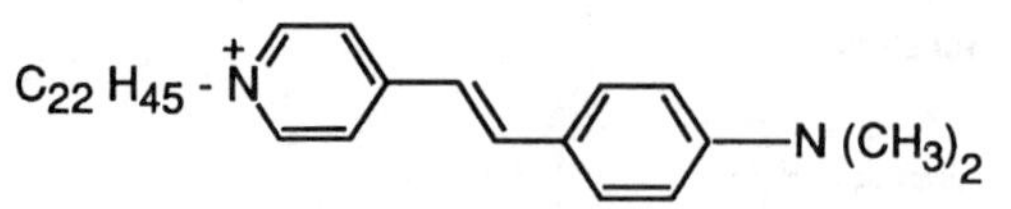

HEMICYANINE

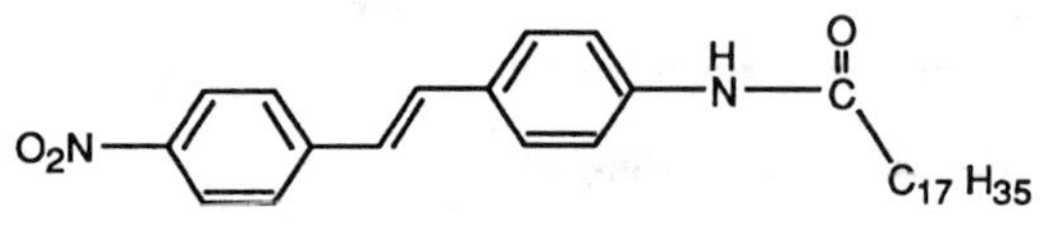

NITROSTILBENE

Figure 1.7 The hemicyanine and nitrostilbene molecules whose alternating-layer L–B films give superadditive second-harmonic generation.[75]

certain cases, alternate-layer structures may possess higher second-order susceptibilities than can be expected from the sum of the individual, separate monolayers. Thus, an alternate-layer structure of a hemicyanine and a nitrostilbene (Figure 1.7) possesses a second-harmonic generation coefficient approximately five times the average value measured for the separate layers, and this corresponds to about fifty times the susceptibility of lithium niobate.[75, 76]

Another promising area requiring noncentrosymmetric layers is that of piezoelectric or pyroelectric elements—films that develop a charge across the layer in response to either pressure (piezo-) or heating (pyro-). Novak and Myagkov[77] measured the charge developed in a capacitor containing an L–B film (X-type) of 4-nitro-4′-*n*-octadecylazobenzene when the film was extended in the longitudinal direction (stress applied in the transverse direction). As in the case of nonlinear optical elements, it may be desirable to build in asymmetry by alternate-layer fabrication. Such has been used by at least two research groups to produce films of various acid/amine combinations, with resulting pyroelectric response.[78, 79]

The most common practical applications of piezoelectric elements are as sound detectors, as in hydrophones or acoustic testing, and for pyroelectrics, as thermal imagers. These may be considered within the class of sensors, and L–B films have also been suggested for other sensor applications. As previously noted, Langmuir patented the application of a "conditioned" stearate film as an optical detector of substances adsorbed from solution.[9] This involved a simple interference color change on adsorption. More recently, other optical effects, such as surface plasmon resonance, have been applied.[80] (Surface plasma waves are electromagnetic waves at the interface between a metal and a dielectric whose quanta are called surface plasmons. They can be excited by light, and resonance can be observed by changing the angle of incidence and monitoring the light reflected from the metal [usually silver]/L–B film interface.)

Perhaps even more interesting than optical-detection techniques are electrical means of detection. One such is the detection of electron acceptors in the gas phase, such as NO_2 and halogens, by films of phthalocyanimes or porphyrins. The L–B film is built on a substrate bearing two planar electrodes separated by a gap, and the dc conductivity is found to be altered when charge carriers are introduced if the gas is adsorbed. The earliest example of such a sensor was described by Baker et al.[81] Such devices are highly sensitive but unfortunately not very selective, because all powerful electron acceptors give essentially the same response.

Though earlier attempts to use L–B films to alter the selectivity of membranes for gas permeation were unsuccessful,[82] more recently this has been found possible. Regen's group at Lehigh have found that certain macrocyclic (and hence "porous") surfactants can be built as L–B films exhibiting high permselectivity (e.g., He/N_2 permeation ration of 17, compared with the value 2.6 expected on the basis of Graham's law).[83] The combination of high permeation rate obtainable with very thin and permeable support films and high selectivity provided by the L–B film coating may make such composite membranes attractive for many separations.

It was a lubrication problem (friction in meter bearings) that led to the discovery of the L–B film process.[4] It is appropriate, therefore, that we finish this brief and incomplete discussion of possible applications by noting a recent attempt to apply L–B films as lubricants. In high-density magnetic recording systems, it is essential that the spacing between the magnetic layer and the recording head be very thin, typically no more than 10–20 nm. Since momentary contact between head and tape cannot be completely prevented, it is desirable that the tape carry a lubricant layer to prevent high friction and wear during such contacts. Seto et al.[84] found that with a bare cobalt metal layer, the head–tape friction coefficient rose to a value greater than 0.5 after only a few transits of the tape over the head. When an L–B film consisting of seven layers of barium stearate was applied to the cobalt, the friction coefficient remained at the relatively low value of 0.22 for 100 transits.

Acknowledgment

The preparation of this chapter was supported in part by the Office of Naval Research under Grant NOOO14-91-J-1690.

References

1 I. Langmuir. *Trans. Faraday Soc.* **15**, 62, 1920.

2 K. B. Blodgett. *J. Am. Chem. Soc.* **56**, 495, 1934.

3 K. B. Blodgett. *J. Am. Chem. Soc.* **57**, 1007, 1935.

4 G. L. Gaines, Jr. *Thin Solid Films.* **99**, ix, 1983. A history of the development of L–B films.

5 K. B. Blodgett. *Science.* **89**, 60, 1939.

6 K. B. Blodgett. *Phys. Rev.* **55**, 391, 1939.

7 K. B. Blodgett. U.S. Patent 2,220,861, 1940.

8 I. Langmuir and V. J. Schaefer. *J. Am. Chem. Soc.* **59**, 1406 and 1762, 1937.

9 I. Langmuir. U.S. Patent 2,232,539, 1941.

10 K. B. Blodgett. U.S. Patent 2,587,282, 1952.

11 Proceedings of the First through Fifth International Conferences on L–B Films have appeared in *Thin Solid Films*, Vols. 99 (1983), 132–134 (1985), 159–160 (1988), 178–180 (1989), and 210–211 (1992). The sixth conference (renamed "Sixth International Conference on Organized Molecular Films" to recognize films prepared by other than the classical L–B method) was held in July 1993.

12 *Langmuir-Blodgett Films.* (G. Roberts, Ed.) Plenum, New York, 1990.

13 A. Ulman. *An Introduction to Ultrathin Organic Films, from Langmuir–Blodgett to Self-Assembly.* Academic Press, San Diego, 1991.

14 A. Barraud and M. Vandevyver. *Thin Solid Films.* **99**, 221, 1983.

15 G. L. Gaines, Jr. *Thin Solid Films.* **68**, 1, 1980.

16 G. L. Gaines, Jr. *J. Colloid Interface Sci.* **59**, 438, 1977.

17 H. Kuhn, D. Möbius, and H. Bucher. In *Physical Methods of Chemistry.* Vol. 1, Part 3B. (A. Weissberger and B. Rossiter, Eds.) Wiley, New York, 1972, p. 577.

18 D. Möbius. In *Langmuir-Blodgett Films.* (G. Roberts, Ed.) Plenum, New York, 1990, chapt. 5.

19 P. A. Chollet. *Thin Solid Films.* **52**, 343, 1978.

20 J. F. Rabolt, F. C. Burns, W. E. Schlotter, and J. D. Swalen. *J. Chem. Phys.* **78**, 946, 1983.

21 L. F. Chi, M. Anders, H. Fuchs, R. R. Johnston, and H. Ringsdorf. *Science.* **259**, 213, 1993.

22 M. Pomerantz, F. H. Dacol, and A. Segmuller. *Phys. Rev. Lett.* **40**, 246, 1978.

23 B. P. Binks. *Advan. Colloid Interface Sci.* **34**, 343, 1991.

24 A. Barraud, C. Rosilio, and A. Ruaudel-Teixier. *J. Colloid Interface Sci.* **62**, 509, 1977.

25 A. Barraud. *Thin Solid Films.* **99**, 317, 1983.

26 F. Kajzar and J. Messier. *Thin Solid Films.* **132**, 1, 1985.

27 G. L. Gaines, Jr. *Nature.* **298**, 544, 1982.

28 E. Stenhagen. *Trans. Faraday Soc.* **34**, 1328, 1938.

29 G. Knott, J. H. Schulman, and A. F. Wells. *Proc. Roy. Soc.* **A176**, 534, 1940.

30 F. Kopp, V. P. Fringeli, K. Muhlethaler, and H. H. Gunthard. *Biophys. Struct. Mechanism.* **1**, 75, 1975.

31 F. Kopp, V. P. Fringeli, K. Muhlethaler, and H. H. Gunthard. *Z. Naturforsch.* **30C**, 711, 1975.

32 E. P. Honig, J. H. T. Hengst, and D. den Engelsen. *J. Colloid Interface Sci.* **45**, 92, 1973.

33 M. F. Daniel, O. C. Lettington, and S. M. Small. *Molec. Cryst. Liq. Cryst.* **96**, 373, 1983.

34 R. M. Swart. In *Langmuir-Blodgett Films.* (G. Roberts, Ed.) Plenum, New York, 1990, pp. 287–292.

35 R. Linton, V. Guariseo, J. J. Lee, B. Hagenhoff, and A. Benninghoven. *Thin Solid Films.* **211**, 565, 1991.

36 N. Nakahara, S. Miyata, T. T. Wang, and S. Tasaka. *Thin Solid Films.* **141**, 165, 1986.

37 A Takahara, N. Morotomi, S. Hiraoka, N. Higashi, T. Kunitake, and T. Kajiyama. *Macromolecules.* **22**, 617, 1989.

38 C. Naselli, J. D. Swalen, and J. F. Rabolt. *J. Chem. Phys.* **90**, 3855, 1989.

39 G. L. Gaines, Jr. *Langmuir.* **7**, 3054, 1991.

40 C. Naselli, J. P. Rabe, J. F. Rabolt, and J. D. Swalen. *Thin Solid Films.* **134**, 173, 1985.

41 M. F. Daniel, O. C. Lettington, and S. M. Small. *Thin Solid Films.* **99**, 61, 1983.

42 P. S. Vincett, W. A. Barlow, F. T. Boyle, J. A. Finney, and G. G. Roberts. *Thin Solid Films.* **60**, 265, 1979.

43 P. S. Vincett and W. A. Barlow. *Thin Solid Films.* **71**, 305, 1980.

44 A. Ruaudel-Teixier, M. Vandevyver, and A. Barraud. *Mol. Cryst. Liq. Cryst.* **120**, 319, 1985.

45 D. Möbius. In *Langmuir-Blodgett Films.* (G. Roberts, Ed.) Plenum, New York, 1990, chapt. 5.

46 H. Kuhn. *Thin Solid Films.* **178**, 1, 1989.

47 G. L. Gaines, Jr. *Anal. Chem.* **48**, 450, 1976.

48 I. R. Girling, N. A. Cade, P. V. Kolinsky, and C. M. Montgomery. *Electron. Lett.* **21**, 169, 1985.

49 G. W. Smith, M. F. Daniel, J. W. Barton, and W. Ratcliffe. *Thin Solid Films.* **132**, 125, 1985.

50 I. Langmuir and V. J. Schaefer. *J. Amer. Chem. Soc.* **59**, 2075, 1937.

51 G. Munger, R. M. Leblanc, B. Zelent, A. G. Volkov, M. I. Gugeshashvili, J. Gallant, H. Tajmir-Riahi, and J. Aghion. *Thin Solid Films.* **211**, 739, 1991.

52 Y. L. Hua, G. G. Roberts, M. M. Ahmad, M. C. Petty, M. Hanack, and M. Rein. *Phil. Mag.* **B53**, 105, 1986.

53 S. Baker, G. G. Roberts, and M. C. Petty. *IEEE Proc.* **130**, 260, 1983.

54 J. Milliken, D. D. Dominguez, H. H. Nelson, and W. R. Barger. *Chem. Mater.* **4**, 252, 1992.

55 I. Langmuir and V. J. Schaefer. *Chem. Rev.* **24**, 181, 1939.

56 G. L. Gaines, Jr. *Langmuir.* **7**, 834, 1991.

57 R. H. Tredgold and C. S. Winter. *J. Phys. D.* **15**, L55, 1982.

58 S. J. Mumby, J. D. Swalen, and J. F. Rabolt. *Macromolecules.* **19**, 1054, 1986.

59 C. S. Winter and R. H. Tredgold. *Thin Solid Films.* **123**, L1, 1985.

60 T. Kawaguchi, N. Nakahara, and K. Fukuda. *J. Colloid Interface Sci.* **104**, 290, 1985.

61 C. S. Winter, R. H. Tredgold, A. J. Vickers, E. Khoshdel, and P. Hodge. *Thin Solid Films.* **134**, 49, 1985.

62 R. Elbert, A. Laschewsky, and H. Ringsdorf. *J. Am. Chem. Soc.* **107**, 4134, 1985.

63 A. Cemel, T. Fort, Jr., and J. B. Lando. *J. Poly. Sci.* **10**, 2061, 1972.

64 D. Naegele, J. B. Lando, and H. Ringsdorf. *Macromolecules.* **10**, 1339, 1977.

65 S. Bresler, M. Judin, and D. Talmud. *Acta Physicochim, URSS.* **14**, 71, 1941.

66 S. J. Valenty. *Macromolecules.* **11**, 1221, 1978.

67 S. J. Valenty. *Advan. Chem. Series.* **184**, 69, 1980.

68 G. Fariss, J. Lando, and S. Rickert. *Thin Solid Films.* **99**, 305, 1983.

69 B. Tieke, H. J. Graf, G. Wegner, B. Naegele, H. Ringsdorf, A. Banerjie, D. Day, and J. B. Lando. *Colloid Polym. Sci.* **255**, 521, 1977.

70 D. Day and H. Ringsdorf. *J. Poly. Sci., Polym. Lett. Ed.* **16**, 205, 1978.

71 F. Grunfeld and C. W. Pitt. *Thin Solid Films.* **99**, 249, 1983.

72 I. R. Girling, N. A. Cade, P. V. Kolinsky, J. D. Earls, G. H. Cross, and I. R. Peterson. *Thin Solid Films.* **132**, 101, 1985.

73 B. L. Henke. *Advan. X-Ray Anal.* **7**, 460, 1964.

74 C. W. Pitt and L. M. Walpita. *Electron. Lett.* **12**, 479, 1977.

75 D. B. Neal, M. C. Petty, G. G. Roberts, M. M. Ahmad, W. J. Feast, I. R. Girling, N. A. Cade, P. V. Kolinsky, and I. R. Peterson. *Electron. Lett.* **22**, 460, 1986.

76 D. B. Neal, M. C. Petty, G. G. Roberts, M. M. Ahmad, W. J. Feast, I. R. Girling, N. A. Cade, P. V. Kolinsky, and I. R. Peterson. *Proc. 6th IEEE Conf. on Applications of Ferroelectrics.* **1**, 89, 1986.

77 V. R. Novak and I. V. Myagkov. *Sov. Tech. Phys. Lett.* **11**, 159, 1985.

78 G. W. Smith, N. M. Ratcliffe, S. J. Roser, and M. F. Daniel. *Thin Solid Films.* **151**, 9, 1987.

79 G. A. Jones, M. C. Petty, and G. G Roberts. *Thin Solid Films.* **160**, 117, 1988.

80 C. A. Brown, F. C. Burns, W. Knoll, J. D. Swalen, and A. Fischer. *J. Phys. Chem.* **87**, 3616, 1983.

81 S. Baker, G. G. Roberts, and M. C. Petty. *Proc. IEEE.* **130**, 260, 1983.

82 G. L. Gaines, Jr., and W. J. Ward III. *J. Colloid Interface Sci.* **60**, 210, 1977.

83 M. Conner, V. Janout, and S. L. Regen. *J. Am. Chem. Soc.* **115**, 1178, 1993.

84 J. Seto, T. Nagai, C. Ishimoto, and H. Watanabe. *Thin Solid Films.* **134**, 101, 1985.

2

Self-Assembled Monolayers

ABRAHAM ULMAN

Contents

2.1 Introduction

Self-assembled monolayers (SAMs) have been intensively studied in the past few years because of their relevance in science and technology.[1] Due to their dense and stable structure, SAMs have potential application in corrosion prevention, wear protection, biosensing devices, etc. The ability to tailor both the head and tail groups of the constituent molecules makes them ideally suited for a more fundamental understanding of phenomena affected by competing intermolecular, molecular–substrate, and molecule–solvent interactions, such as ordering and growth, wetting, adhesion, lubrication, and corrosion.

SAMs are formed by the immersion of an appropriate substrate into a solution of an active surfactant (Figure 2.1). The molecular order in these two-dimensional systems is produced by a *spontaneous* chemical synthesis at the interface as the system approaches equilibrium.[1] In contrast to Langmuir–Blodgett (L–B) monolayers, which for significant bonding require substrate pretreatment and are often metastable, SAMs exhibit more uniform and stable behavior.

2.2 Monolayers of Fatty Acids

Carboxylic acid derivatives self-assemble on surfaces by an acid–base (e.g., on glass, Al_2O_3, and Ag_2O surfaces).[1] Spontaneous adsorption of long-chain *n*-alkanoic acids ($C_nH_{2n+1}COOH$) has been studied in the past few years. It was found that the time required to reach a complete monolayer increased with decreasing concentration, and that it took a much higher concentration to produce good monolayers on glass than on Al_2O_3 (10^{-2} versus 10^{-4} M). This concentration dependence is in agreement with the expected affinity of the COOH groups to the Al_2O_3 and glass surfaces, and to the expected surface concentrations of the salt forming oxides in the two substrates.

Recently, Tao showed that the chemisorption of alkanoic acids on amorphous metal oxide surfaces is not unique. On AgO, the carboxylate two oxygen atoms bind to the surface nearly symmetrically, and ordered monolayers are formed with a chain tilt angle from the surface normal of 15° to 25°.[2] On surfaces of CuO and Al_2O_3, the carboxylate binds asymmetrically to the surface, displaying tilt angles close to 0°. This is a classical case of order determined by chemisorption potential (the symmetry is determined by interchain interactions). In a later study, Tao et al. investigated the self-assembly of biphenyl- and naphthyl-containing alkanoic acids.[3] They demonstrated that a biphenyl or naphthyl group causes a higher tilt of molecular chains in carboxylic acid monolayers on silver and a correlation of wetting properties and chain tilt. The different behavior observed on silver and copper has been attributed to site specificity, binding geometry, and binding strength, among other things. The structure of the monolayers studied is a balance of various interactions, as described above. Hence, chemisorption mode, interchain interactions, intrachain energy, and lattice spacing determine the equilibrium structure of the assembly.

2.3 Monolayers of Organosilicon Derivatives

SAMs of alkyltrichlorosilanes require hydroxylated surfaces as substrates (glass, SiO_2/Si, Al_2O_3, etc.), because the formation of the SAM essentially is an in situ

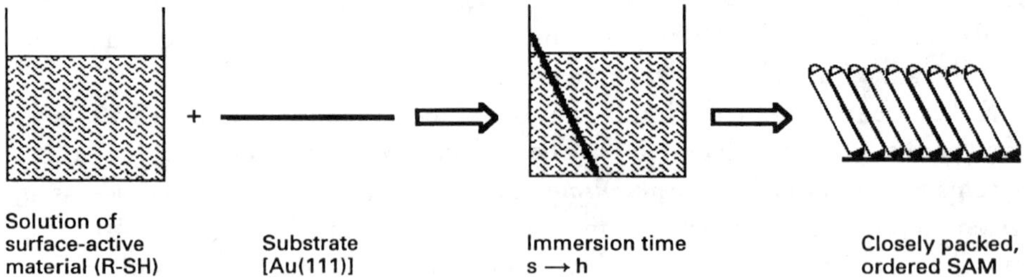

Figure 2.1 **Self-assembled monolayers are formed by simply immersing a substrate into a solution of the surface-active material.**

formation of polysiloxane, which is connected to surface silanol groups (–Si–OH) via Si–O–Si bonds.[1] These monolayers show remarkable stability. For example, octadecyltrichlorosilane (OTS) monolayers on SiO_2 can be washed with a 1% detergent solution and hot tap water and with organic solvents without any detectable damage; they were found to be stable in 0.1 M HCl (at room temperature) and in hexadecane (at 75 °C) for two days. The unique stability of SAMs of silane derivatives makes them ideal materials for surface modification and functionalization applications, as adhesion promoters, boundary lubricants, etc.

The Si···Si spacing (~4.4 Å) is determined by the formation of a polysiloxane chain. This spacing leaves little free volume, and the chains are locked perpendicular to the surface. During adsorption, Si–Cl bonds react with the OH groups present on the surface of the substrate and with trace water, which is adsorbed on the surface, to form a network of Si–O–Si bonds (Figure 2.2). X-ray photoelectron spectroscopy (XPS) studies confirm the complete surface reaction of the $-SiCl_3$ groups.[4] The kinetics of the formation of long-chain alkyltrichlorosilane monolayers was studied by Wasserman et al., showing that a completion of monolayer formation is achieved within ~40 min.[5]

The construction of multilayers from alkyltrichlorosilane derivatives requires that the monolayer surface be modified to a hydroxylated one. Such surfaces can be prepared by a chemical reaction and the conversion of a nonpolar terminal group to a hydroxyl group. Examples of such reactions are the reduction of a surface ester group, the hydrolysis of a protected surface hydroxy group, the hydroboration–oxidation of a terminal double bond, and the conversion of a terminal bromide using silver chemistry. Once a subsequent monolayer is adsorbed on the "activated" monolayer, multilayer films may be built by a repetition of this process. Multilayers built by self-assembly are noncentrosymmetric; hence, they may be utilized in second-order nonlinear optics applications.

Multilayer films of ~0.1 μm thickness by the self-assembly of methyl 23-trichlorosilyl-tricosanoate ($H_3CO_2C-(CH_2)_{22}-SiCl_3$, MTST) on silicon substrates were

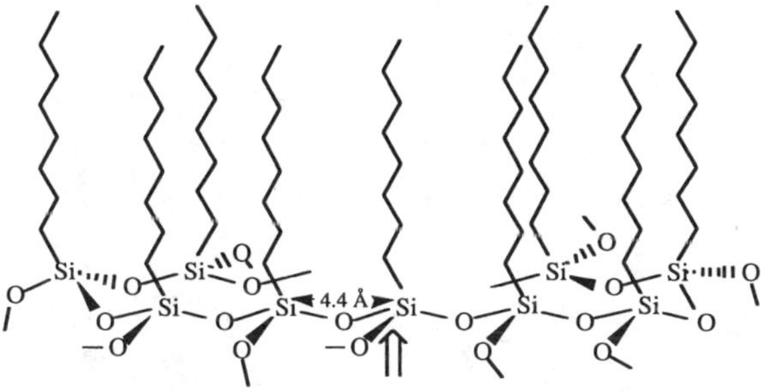

Figure 2.2 A schematic description of a polysiloxane at the monolayer–substrate surface. The arrow points to an equatorial Si–O bond that can be connected either to another polysiloxane chain or to the surface.

prepared by Ulman et al.[6] A linear relationship between the film thickness and the layer number showed a slope of 35 Å/layer (Figure 2.3).

2.4 Monolayers of Alkanethiolates on Metal and Semiconductor Surfaces

Nuzzo and Allara published the first paper in this area, showing that dialkyldisulfides (RS–SR) form oriented monolayers on gold surfaces.[7] Later it was found that sulfur compounds coordinate very strongly to gold,[1] silver,[1] copper,[1] platinum,[1] and gallium arsenide surfaces.[8] However, the most thoroughly studied and best-characterized SAMs are those formed by thiols adsorbed from solution on a Au(111) substrate, mainly because gold does not have a stable oxide, making it easy to handle in ambient conditions.

Studies of alkanethiolate monolayers on Au(111) show that at relatively dilute solutions (10^{-3} M), two distinct adsorption kinetics can be observed: (1) a very fast step taking a few minutes, by which time the contact angles are close to their limiting values and the thickness to about 80–90% of its maximum and (2) a slow step taking several hours, at the end of which time the thickness and contact angles reach their final values.[9] It also was found that the kinetics is faster for longer alkyl chains, probably due to the increased van der Waals interactions.[9] Studies also showed pronounced differences between the short ($n < 9$) and long ($n > 9$) alkanethiolates:[10] the first, a fast one, describing the chemisorption process and the second, a slow one, describing a two-dimensional crystallization process.[11, 12]

The chemisorption of alkanethiols on gold gives the Au(I) thiolate (RS^-) species by the reaction

$$RS{-}H + Au_n^0 \rightarrow RS^-Au^+ \bullet Au_n^0 + \tfrac{1}{2}H_2$$

The electron transfer from the gold surface to the thiol sulfur atoms weakens the S–H bond and results in its dissociation. On Au(111) surfaces, the S–H bond is

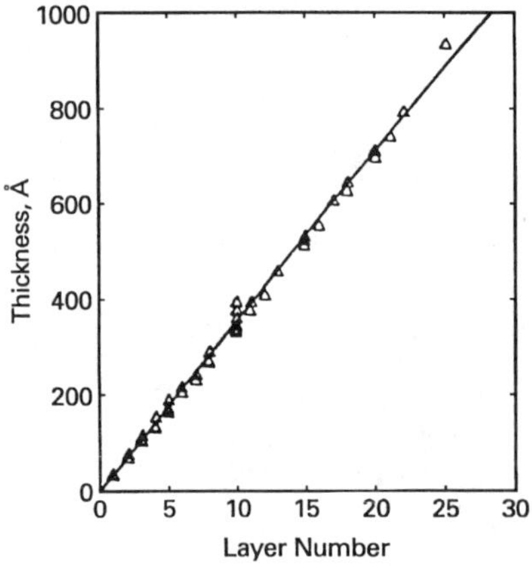

Figure 2.3 **Film thickness versus layer number. (From Tillman et al.[6] © 1988, American Chemical Society.)**

cleaved at the on-top site, and the thiolate then moves to its final equilibrium position at the hollow site. The combination of hydrogen atoms at the metal surface to yield H_2 molecules is an important exothermic step in the overall chemisorption energetics. That the adsorbing species is the thiolate (RS^-) has been shown by XPS,[13] Fourier transform infrared (FTIR) spectroscopy,[14–17] Fourier transform mass spectrometry,[18] electrochemistry,[19] and Raman spectroscopy.[20, 21] The bonding of the thiolate group to the gold surface is very strong (homolytic bond strength is approximately 44 kcal/mol).[22]

Electron diffraction studies (both high-energy[23, 24] and low-energy[25]) of monolayers of alkanethiolates on Au(111) surfaces show that the symmetry of sulfur atoms is hexagonal, with an S···S spacing of 4.97 Å and a calculated area *per* molecule of 21.4 Å^2. Early helium diffraction[26] and atomic force microscopy (AFM)[27] studies confirmed that the structure formed by docosanethiol on Au(111) is commensurate with the underlying gold lattice and is a simple $\sqrt{3} \times \sqrt{3}R$ 30° overlayer (Figure 2.4). Recent X-ray and helium diffraction studies show evidence of a rectangular primitive unit mesh with dimensions of 8.68 × 10.02 Å containing four chains, equivalent to a $c(4 \times 2)$ superlattice.[28] A careful scanning tunneling microscopy (STM) study confirmed this observation.[29]

At the hollow site of Au(111), the sulfur charge is −0.4 e,[30] whereas at the on-top site this charge is about −0.7 e.[30, 31] Since S–H bond cleavage occurs at the on-top site,[32] if this cleavage is the rate-determining step, the adsorption rate should be faster in polar solvents. However, if the migration of a thiolate from the on-top site to the hollow site is the slow step, the reaction should be faster in a nonpolar solvent. Recent second-harmonic generation studies showed that while the rate constant in ethanol is 1.3×10^6 cm^3/mol·s, it is 4.7×10^6 cm^3/mol·s in hexane,[33] suggesting that the S–H bond cleavage is not the rate-determining step. The migration of thiolates between hollow sites, essential for healing defects, occurs either through the on-top or the bridge sites, and should be sensitive to the dielectric constant. Indeed, ethanol has been found to consistently yield highly ordered monolayers.[9] Note that the thiolate is chemically bonded to *one* gold atom at the on-top site, forming a neutral gold thiolate molecule (RS–Au). This may desorb

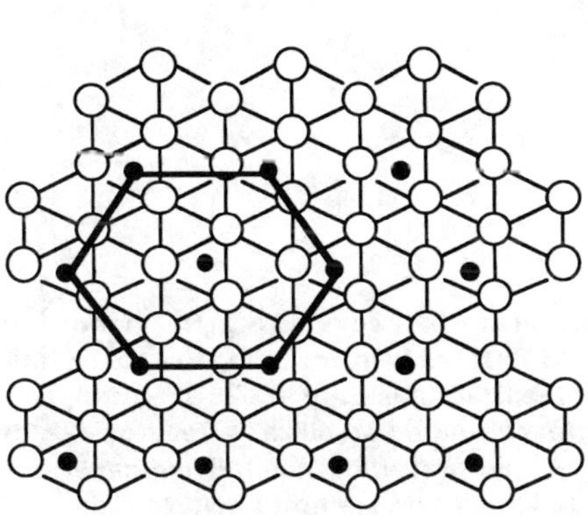

Figure 2.4 Hexagonal coverage scheme for Au(111) (Scheme I, *top*) and for Ag(111) (Scheme II, *bottom*). The S···S distance in Scheme II is 4.41 Å.

before the thiolate moves to the hollow site, thus leaving a defect. Recent STM studies suggest that some of the pinholes observed in monolayers on Au(111) may be the result of such an etching process.[34]

Alkanethiolates have two binding modes at the Au(111) hollow site, one with a bend angle around the sulfur of 180° (*sp*) and the other of 104° (sp^3), the latter being more stable by 0.41 kcal/mol.[30] Thus, packing requirements may dictate the final surface–S–C angle. Many studies have suggested that this angle in monolayers on Au(111) surfaces must be tetrahedral.[14–17] However, recent studies on triphenylthiolate (C_6H_5–C_6H_4–C_6H_4–S^-) monolayers on Au(111) suggest a tilt angle of ~6° from the surface normal,[35] providing the first evidence for two chemisorption modes.

Molecular mechanics (MM) energy minimization indicates that the two modes lead to monolayers that exhibit different types of packing arrangements (Figure 2.5) but are comparable in their ground-state energies. (The monolayers resulting from the sp^3 mode are more stable by 0.6 kcal/mol, although one should consider the monolayers to have the same ground-state energy given the accuracy of these calculations.)[30] Therefore, monolayers may consist of two different chemisorption modes ordered in different domains—simultaneously coexisting homogeneous clusters—each characterized by a different conformer in its unit cell. Annealing may result either in a unique chemisorption mode or in a much more complex structure, as the different modes are redistributed.

The energy barrier between the two chemisorption modes is very small (2.5 kcal/mol),[30] suggesting that the thiolate may easily cross from one of these minima to the other, thus providing a facile annealing mechanism (Figure 2.6). This predicts that the process of changing tilt direction should occur well below the melting point of the monolayer and should be chain-length dependent. Recent X-ray data show a narrowing of the diffraction peak when monolayers of alkanethiolates on Au(111) were annealed.[36] The development of larger domain size was the apparent result of the heating and cooling. Thus, close packing and the high ordering of

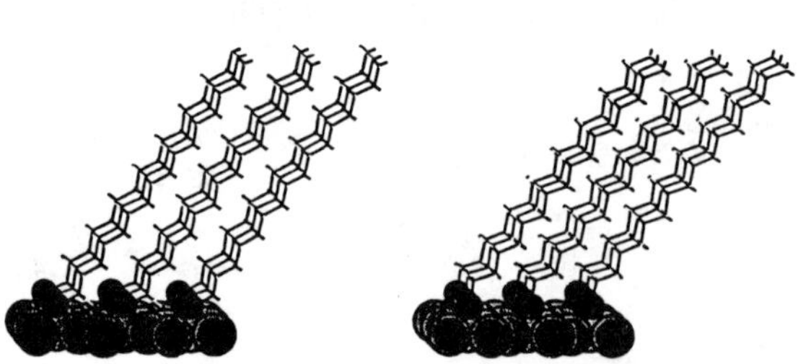

Figure 2.5 A view of the tilt in a nine-molecule section of a fully covered $SC_{16}H_{33}$ monolayer on Au(111) minimized with a modified MM2 force field, including the sp^3 chemisorption parameters (*left:* calculated stabilization energy –23.29 kcal/mol) and the *sp* chemisorption parameters (*right:* calculated stabilization energy –22.25 kcal/mol). The gold atoms are shown in light gray with their full van der Waals radii. The sulfur atoms are shown in dark gray and are not drawn to scale.

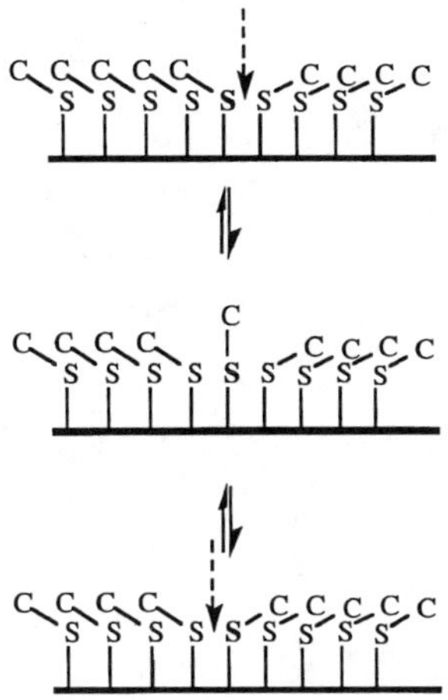

Figure 2.6 **A possible scenario for annealing alkanethiolate monolayers. The hybridization change results in the motion of only one chain. The arrow represents the grain boundary. Note that after the molecule has moved, the grain boundary moved from the right side of that molecule to the left side.**

alkanethiolates on Au(111) apparently result from the relatively easy two-dimensional "recrystallization" process.

2.5 Self-Assembled Monolayers Containing Aromatic Groups

A model of a $CH_3(CH_2)_8–C_6H_4–O–(CH_2)_{11}–SiCl_3$ monolayer on SiO_2/Si suggests that for alkyl chains perpendicular to the surface, the phenyl ring has to exhibit a 35° tilt from the surface normal (Figure 2.7). However, infrared spectroscopy studies suggest a much smaller angle (~20°) and an average tilt of ~15° for the chains.[37] This is a result of the mismatch of the cross-sectional area of the phenyl ring and the alkyl chains. Because the packing in a monolayer cannot be any closer than the spacing of the largest moiety and because the cross-sectional area depends upon tilt and orientation, a clockwise rotation of 15° in the plane of the figure will result in a better cross-sectional area match between the different molecular parts. Thus, as the cross-sectional area of the alkyl chains increases from 19.6 to 20.7 Å², that of the phenyl ring decreases from 25.1 to 24.8 Å². No

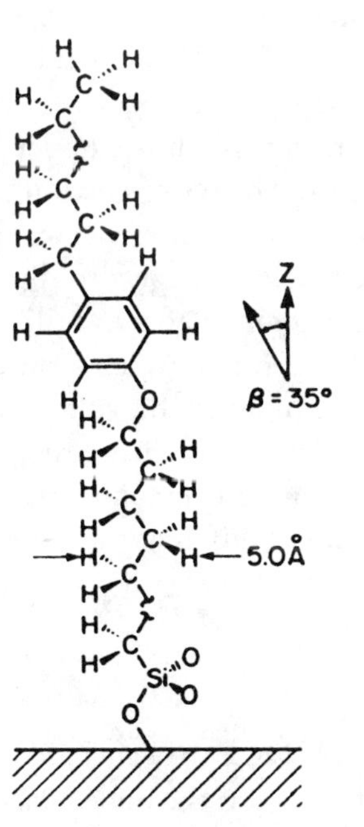

Figure 2.7 **Representation of $CH_3(CH_2)_8–C_6H_4–O–(CH_2)_{11}–SiCl_3$ in a monolayer with fully extended alkyl chains that are perpendicular to the surface. The length of the alkyl chains has been shortened for clarity. The $C_1–C_4$ axis of the phenyl ring has a 35° tilt (β) from the substrate surface normal (*z* axis).**

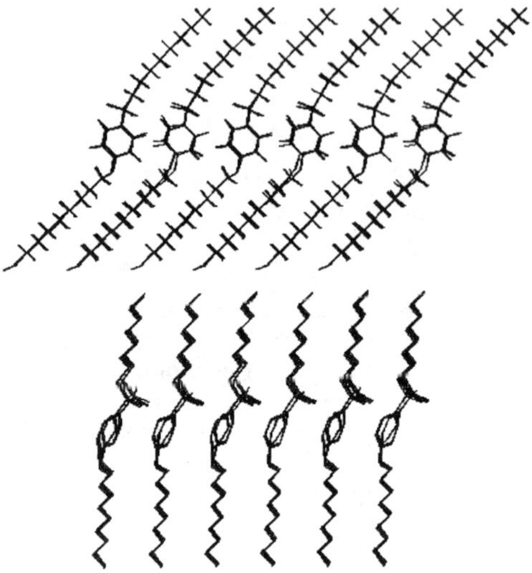

Figure 2.8 **Two side views of 6 × 6 assembly of octyl 4-(8-mercaptooctyl-oxy)phenyl sulfone molecules (three simulation cells by three simulation cells) after constant stress minimization under monolayer constraints.**

further information on the structure of the alkyl chain is available in this case because no modeling was carried out.

Self-assembled monolayers of $CH_3(CH_2)_7–SO_2–C_6H_4–O–(CH_2)_8–SH$ on gold were studied both experimentally[38] and using molecular dynamics (MD) simulations.[39] Energy minimizations show that within the minimized assembly the chains are nearly all trans with an overall *lean* of ~30° from the surface normal (Figure 2.8) in a plane about parallel to that of the phenyl ring. A very small *tilt* in a plane that is about perpendicular to that of the phenyl ring is observed, mainly in the upper chain. This is in agreement with the rectangular shape of the phenyl projection, with its different dimensions. Thus, because the phenyl–phenyl interactions determine the short chain spacing, a small rotation of the alkyl chain with respect to the aromatic moiety will bring the spacing to that which is required for an effective tilt. On the other hand, the larger spacing in the perpendicular direction explains the lean angle of ~30°. Noticeable chain distortions can be seen near the sulfone groups. Similar distortions have been observed in MD simulations of alkanethiolates on Au(100).[13] Hence, alkyl chains use a similar mechanism to overcome mismatches within the assembly, whether it results from a stratum of different symmetry or one with units of a different size.

2.6 Conclusions

We can expect nanodevices to become the industry's standard because they are based on the ability to manipulate the structure and properties of matter at the molecular level. Such devices will be essential not only for information processing, memories, and communication systems, but also for environmental sensors and biosensors that can detect changes in very small cell clusters.

The beauty of SAMs is in the spontaneous association of molecules *under equilibrium conditions* that gives stable, structurally well-defined two-dimensional aggregates. From the practical point of view, SAMs have a clear advantage because they are inherently manufacturable. Therefore, molecular engineering and self-organization may become key to molecular-based nanodevices. However, the fundamental understanding of self-organization phenomena—a prerequisite for any technological advance in this area—still is very limited. Other important issues to be explored are the development of new surface reactions[40, 41] and the formation of patterned SAMs.[42] Even so, building materials from the molecule "up" requires control of the structure and properties of matter. The work presented in this chapter touches a number of aspects of this important subject; much more needs to be done, both experimentally and by using molecular simulations.

References

1 A. Ulman. *Introduction to Thin Organic Films: From Langmuir–Blodgett to Self-Assembly.* Academic Press, Boston, 1991.

2 Y.-T. Tao. *J. Am. Chem. Soc.* **115**, 4350, 1993.

3 Y.-T. Tao, M.-T. Lee, and S.-C. Chang. *J. Am. Chem. Soc.* **115**, 9547, 1993.

4 S. R. Wasserman, G. M. Whitesides, I. M. Tidswell, B. M. Ocko, P. S. Pershan, and J. D. Axe. *J. Am. Chem. Soc.* **111**, 5852, 1989.

5 S. R. Wasserman, Y.-T. Tao, and G. M. Whitesides. *Langmuir.* **5**, 1074, 1989.

6 N. Tillman, A. Ulman, and T. L. Penner. *Langmuir.* **5**, 101, 1989.

7 R. G. Nuzzo and D. L. Allara. *J. Am. Chem. Soc.* **105**, 4481, 1983.

8 C. W. Sheen, J. X. Shi, J. Martensson, A. N. Parikh, and D. L. Allara. *J. Am. Chem. Soc.* **114**, 1514, 1992.

9 C. D. Bain, E. B. Troughton, Y.-T. Tao, J. Evall, G. M. Whitesides, and R. G. Nuzzo. *J. Am. Chem. Soc.* **111**, 321, 1989.

10 M. D. Porter, T. B. Bright, D. L. Allara, and C. E. D. Chidsey. *J. Am. Chem. Soc.* **109**, 3559, 1987.

11 Buck, M. Grunze, Eisert, Fischer, and Träger. *J. Vac. Sci. Technol. A.* **10**, 926, 1992.

12 Hähner, Wöll, Buck, and M. Grunze. *Langmuir.* **9**, 1955, 1993.

13 R. G. Nuzzo, F. A. Fusco, and D. L. Allara. *J. Am. Chem. Soc.* **109**, 2358, 1987.

14 P. E. Laibinis, G. M. Whitesides, D. L. Allara, Y.-T. Tao, A. N. Parikh, and R. G. Nuzzo. *J. Am. Chem. Soc.* **113**, 7152, 1991.

15 C. D. Bain, H. A. Biebuyck, and G. M. Whitesides. *Langmuir.* **5**, 723, 1989.

16 R. G. Nuzzo, B. R. Zegarski, and L. H. Dubois. *J. Am. Chem. Soc.* **109**, 733, 1987.

17 R. G. Nuzzo, L. H. Dubois, and D. L. Allara. *J. Am. Chem. Soc.* **112**, 558, 1990.

18 Y. Li, J. Huang, R. T. McIver, Jr., and J. C. Hemminger. *J. Am. Chem. Soc.* **114**, 2428, 1992.

19 C. A. Widrig, C. Chung, and M. D. Porter. *J. Electroanal. Chem.* **310**, 335, 1991.

20 M. A. Bryant and J. E. Pemberton. *J. Am. Chem. Soc.* **113**, 3630, 1991.

21 M. A. Bryant and J. E. Pemberton. *J. Am. Chem. Soc.* **113**, 8284, 1991.

22 L. H. Dubois and R. G. Nuzzo. *Annu. Rev. Phys. Chem.* **43**, 437, 1992.

23 L. Strong and G. M. Whitesides. *Langmuir.* **4**, 546, 1988.

24 C. E. D. Chidsey and D. N. Loiacono. *Langmuir.* **6**, 709, 1990.

25 L. H. Dubois, B. R. Zegarski, and R. G. Nuzzo. *J. Chem. Phys.* **98**, 678, 1993.

26 C. E. D. Chidsey, G. Liu, Y. P. Rowntree, and G. Scoles. *J. Chem. Phys.* **91**, 4421, 1989.

27 C. A. Alves, E. L. Smith, and M. D. Porter. *J. Am. Chem. Soc.* **114**, 1222, 1992.

28 N. Camillone, C. E. D. Chidsey, G.-Y. Liu, and G. Scoles. *J. Chem. Phys.* **98**, 3503, 1993.

29 G. E. Poirier and M. J. Tarlov. *Langmuir.* **10**, 1994.

30 H. Sellers, A. Ulman, Y. Shnidman, and J. E. Eilers. *J. Am. Chem. Soc.* **115**, 9389, 1993.

31 M. Majda. Private communication.

32 H. Sellers. *Surf. Sci.* **294**, 99, 1993.

33 M. Grunze. *Physica Scripta.* In press.

34 C. Schönenberger, J. A. M. Sondag-Huethrost, and L. G. J. Fokkink. To be published.

35 E. Sabatani, J. Cohen-Boulakia, M. Bruening, and I. Rubinstein. *Langmuir.* **9**, 1993.

36 K. S. Liang, P. Fenter, and P. Eisenberger. *Phys. Rev. Lett.* **70**, 2447, 1993.

37 N. Tillman, A. Ulman, J. S. Schildkraut, and T. L. Penner. *J. Am. Chem. Soc.* **110**, 6136, 1988.

38 S. D. Evans, E. Urankar, A. Ulman, and N. Ferris. *J. Am. Chem. Soc.* **113**, 4121, 1991.

39 Y. Shnidman, A. Ulman, and J. E. Eilers. *Langmuir.* **9**, 1071, 1993.

40 S. D. Evans, A. Ulman, K. E. Goppert-Berrarducci, and L. J. Gerenser. *J. Am. Chem. Soc.* **113**, 5866, 1991.

41 N. Balachander and C. N. Sukenik. *Langmuir.* **6**, 1621, 1990.

42 C. S. Dulcey, J. H. Georger, Jr., V. Krauthamer, D. A. Stenger, T. L. Fare, and J. M. Calvert. *Science.* **252**, 551, 1991.

Part II

Analysis of Film and Surface Properties

3

Spectroscopic Ellipsometry

ROBERT W. COLLINS, DAVID L. ALLARA,
YEON-TAIK KIM, YIWEI LU, and JIANOU SHI

Contents

3.1 Introduction and Overview

Ellipsometry is a technique used to determine the optical properties and physical structure of thin film and bulk materials.[1] In this measurement, one determines the polarization change that an incident polarized, monochromatic beam of light incurs when it reflects from the specular surface of a material at a non-normal angle of incidence, θ. The power of ellipsometry derives from its ability to provide two parameters (ψ,Δ), called the ellipsometric angles, in a single measurement. For an ideal isotropic substrate that is opaque and creates a single perfect interface to the ambient medium, both real (ε_{1s}) and imaginary (ε_{2s}) parts of its complex dielectric function, $\varepsilon_s = \varepsilon_{1s} + i\varepsilon_{2s}$, can be calculated directly from (ψ,Δ) as long as θ and ε_a, the dielectric function of the ambient medium, are known. The complex dielectric function is related to the complex index of refraction, N, by $\varepsilon = N^2$, where the real and imaginary parts of N are the real index n and the extinction coefficient k, respectively.

Determining accurate dielectric functions of real substrates requires preparing their surfaces so that they approximate as closely as possible a single optical interface. In fact, because of the surface sensitivity of Δ and the high angular precision with which it can be determined, neglecting monolayer thicknesses of oxide, contaminants, or surface roughness can lead to measured dielectric functions that are significantly different than the correct values. This disadvantage, however, becomes a source of information when thin films are deposited under controlled conditions, and ellipsometry is used most extensively in such applications.

In the case of isotropic thin film growth on a substrate of known dielectric function, three unknowns are required to solve the optical problem, namely, the real (ε_{1f}) and imaginary (ε_{2f}) parts of the complex dielectric function of the film, $\varepsilon_f = \varepsilon_{1f} + i\varepsilon_{2f}$, and the film thickness, d. Thus, a single measurement of the two ellipsometric parameters cannot provide sufficient information, and multiple measurements under different conditions are required.[1, 2] Various techniques have been developed in this regard, including spectroscopic ellipsometry (variable wavelength, λ),[3] multiple angle of incidence ellipsometry (variable θ),[2, 4] real time ellipsometry (variable d),[5, 6] and variable ambient ellipsometry (variable ε_a).[2]

Among the techniques in which a single parameter is varied, spectroscopic ellipsometry (SE) has emerged as the most powerful method.[7] Not only does it allow one to solve the optical problem in many cases, but it also provides information on the electronic structure of the film from the inherent λ-dependence of ε_f. Techniques that involve varying two-measurement parameters, such as variable angle spectroscopic ellipsometry (VASE),[4] and real time spectroscopic ellipsometry (RTSE)[5] are even more powerful and are currently used at the forefront of research. VASE is especially useful for characterizing the dielectric function tensor in static anisotropic films. The basic principles of this technique, and applications, are described in the lead volume of this Series, *Encyclopedia of Materials Characterization.* RTSE is useful in characterizing the dynamics of film growth, including the evolution of morphology and electronic properties with film thickness. In this chapter, an overview of the modern methods of SE and RTSE is presented, and examples are drawn from in situ studies of organic thin films.

3.2 Theory of Ellipsometry

The polarization states of fully polarized light beams incident on (i) and reflected from (r) an assumed isotropic surface are defined by the ratios of the complex electric field components: $\chi_i = E_{pi}/E_{si}$ and $\chi_r = E_{pr}/E_{sr}$. Here, the subscripts p and s denote the axes parallel and perpendicular, respectively, to the plane of incidence, which together form a right-hand coordinate system (looking into the beam). An ellipsometer measures the polarization state change as the complex amplitude reflection ratio ρ between reflected and incident states. The ellipsometry angles ψ and Δ are defined in terms of ρ by

$$\tan\psi\exp(i\Delta) \equiv \rho \equiv \chi_r/\chi_i \tag{3.1a}$$

$$= (E_{pr}/E_{sr})/(E_{pi}/E_{si}) \tag{3.1b}$$

$$= r_p/r_s = (|r_p|/|r_s|)\exp[i(\delta_p - \delta_s)] \tag{3.1c}$$

where r_p and r_s in Equation 3.1*c* are the complex amplitude reflection coefficients given by $r_p = |r_p|\exp(i\delta_p) = E_{pr}/E_{pi}$ and $r_s = |r_s|\exp(i\delta_s) = E_{sr}/E_{si}$. In these expressions, $|r_\upsilon|$ and δ_υ ($\upsilon = p,s$) are the magnitudes and phases of the reflection coefficients, respectively.

For a single interface between an ambient medium and an isotropic substrate, with dielectric functions ε_a and ε_s, respectively, r_p and r_s can be expressed in terms of θ, ε_a, and ε_s using the Fresnel equations. These equations can be inverted to find that

$$\varepsilon_s = N_s^2 = \varepsilon_a \sin^2\theta\{1 + \tan^2\theta[(1-\rho)/(1+\rho)]^2\} \tag{3.2}$$

Thus, in theory, if one knows θ and $\varepsilon_a(\lambda)$, then an SE measurement of (ψ,Δ) at M wavelengths provides $\{\varepsilon_s(\lambda_i);\ i = 1,\ldots,M\}$ through Equations 3.1*a* and 3.2. Application of Equation 3.2 involves a leap of faith since the measured surface may not necessarily be film free, atomically smooth, and abrupt. To indicate such uncertainty, ε_s deduced from Equation 3.2 is often called the pseudo-dielectric function and denoted $\langle\varepsilon_s\rangle$. Sometimes Equation 3.2 is applied to (ψ,Δ) regardless of the known multilayered nature of the sample; the term pseudo-dielectric function is used in this situation as well.

For a single substrate-supported isotropic film of dielectric function ε_f, the situation is more complex since there are two interfaces in the problem. In this case, the complex amplitude reflection coefficients of Equation 3.1*c* are given by

$$r_\upsilon = [Zr_{\upsilon,fs} + r_{\upsilon,af}]/[Zr_{\upsilon,af}r_{\upsilon,fs} + 1]; \qquad \upsilon = p,s \tag{3.3a}$$

where

$$Z = \exp[4\pi i(\varepsilon_f - \varepsilon_a\sin^2\theta)^{1/2}\delta/\lambda] \tag{3.3b}$$

Here, $r_{\upsilon,af}$ and $r_{\upsilon,fs}$ ($\upsilon = p,s$) are the reflection coefficients for the separate ambient–film (*af*) and film–substrate (*fs*) interfaces from the Fresnel equations. Equations 3.3 cannot be solved uniquely for the three unknowns ε_{1f}, ε_{2f}, and d in a single measurement, since ψ and Δ provide only two conditions. Typically, if one of the three parameters is known (e.g., a trial choice for d), then the equations can be inverted numerically using an algorithm such as Newton's method.[8] Solving the problem with SE data—that is, ψ and Δ at M wavelengths—is not trivial either, since $2M + 1$ unknowns must be determined, namely d, $\varepsilon_{1f}(\lambda_i)$, and $\varepsilon_{2f}(\lambda_i)$; $(i = 1,\ldots,M)$. In this case, as well, if d is known then point-by-point numerical inversion can provide $\varepsilon_{1f}(\lambda_i)$ and $\varepsilon_{2f}(\lambda_i)$; $(i = 1,\ldots,M)$. Values for d obtained by an independent measurement, however, are not usually accurate enough to solve the problem. In Section 3.4, techniques will be discussed for extracting the $2M + 1$ unknowns from the $2M$ measurements.

For surfaces with more than two interfaces, the matrix methods of optical multilayer analysis are most convenient for determining r_p and r_s in Equation 3.1*c* from λ, θ, the layer thicknesses, and all dielectric functions.[9] Thus, $\psi(\lambda)$ and $\Delta(\lambda)$ can be simulated by computer for any multilayer system. If all information except for the jth dielectric function ε_j is available, then a numerical inversion program can be applied to the SE data point-by-point to determine $\varepsilon_{1j}(\lambda_i)$ and $\varepsilon_{2j}(\lambda_i)$; $(i = 1,\ldots,M)$.

When linearly polarized light with the electric field along either the *p* or *s* direction is incident on a specular, stratified surface composed of any number of isotropic media, then the same linear *p* or *s* polarization state emerges. When any one of the media is anisotropic, however, a special problem is presented. In the general case, the incident linear *p* or *s* states result in both *p* and *s* electric field components in the reflected wave. This leads to three complex amplitude reflection coefficients, r_{pp}, r_{ss}, and r_{ps}. Thus, two complex amplitude reflection ratios, r_{pp}/r_{ss} and r_{ps}/r_{ss}, must be determined in the experiment to extract meaningful information about the surface. In this situation, measurements at different angles of incidence or orientations of the sample are required. A detailed discussion is beyond the scope of this chapter, but the reader is referred to a recent review with an extensive list of references.[10]

3.3 Instrumentation

Ellipsometers generally consist of the following sequence of optical components: light source, collimator, polarizer, compensator, sample, analyzer, and detector. For SE measurements, a spectrometer is placed either between the source and collimator or between the analyzer and detector, depending on the instrument design. In manual null ellipsometers, the angle of the polarizer is adjusted so that the sample converts incident elliptical polarization, generated by the polarizer–compensator combination, to linear polarization.[1] Then the irradiance reaching the detector can be extinguished by a proper adjustment of the analyzer angle. In practice, the two angles P and A [measured in a counterclockwise sense with respect to the plane of incidence (*p*-direction)] are adjusted alternately to reach the null settings P_0 and A_0 in a process that takes several minutes. When the compensator angle is set at 45°, one of two possible nulls provides ψ and Δ according to $\Delta = 270° - 2P_0$ and $\psi = A_0$, where $0 \leq P_0, A_0 \leq 90°$. Because of the measurement time limitation, it is not rational to use the manual-null ellipsometer for continuous SE measurements.

Thus, SE became viable only with the development of automatic ellipsometers. The most common are the rotating-element, polarization-modulation, and automatic-null ellipsometers.[11] Because a rotating-element principle forms the basis for the latest commercial instruments that perform RTSE and VASE, we will discuss this design in detail. In this ellipsometer, either the polarizer or analyzer element is mounted within the hollow shaft of a motor and is rotated continuously at frequency $\omega/2\pi$ (typically from 10 to 100 Hz) during measurement.[12, 13] One encoder output provides 1 pulse per rotation (Z_0), and another typically provides 60–180 pulses per rotation, both for triggering data collection. For the rotating polarizer and analyzer versions, the spectrometer is placed at the detector and source sides, respectively, because of the polarization sensitivity of the spectrometer throughput.

In addition, the compensator is optional in both versions and is not included unless the measurement conditions demand it (e.g., transparent films on transparent substrates for which $\Delta \approx 0{,}180°$).[14]

For the rotating-polarizer instrument, the output irradiance at the detector for any given wavelength band is given by:

$$I(t) = I_0\{1 + \alpha \cos 2(\omega t - P_s) + \beta \sin 2(\omega t - P_s)\} \tag{3.4}$$

where α and β are normalized Fourier coefficients that depend on λ. In most instruments for SE, a photomultiplier tube detector is used, and its output is sampled typically every 2–6° of polarizer rotation, triggered by the encoder output. Then α and β are calculated by Fourier analysis. In Equation 3.4, $-P_s$ is the polarizer angle at $t = 0$ defined by the Z_0 encoder output and determined in calibration. In the absence of a compensator, ψ and Δ can be obtained from α and β:

$$\tan\psi = [(1+\alpha)/(1-\alpha)]^{1/2} \tan A \tag{3.5a}$$

$$\cos\Delta = \beta/(1-\alpha^2)^{1/2} \tag{3.5b}$$

where A is the fixed-analyzer angle. Note that the average signal level I_0 in Equation 3.4 is not needed; this is a characteristic of polarization state analysis. For the rotating-analyzer configuration, P_s in Equation 3.4 is replaced by the rotating-analyzer calibration angle A_s, and the fixed-analyzer angle A in Equation 3.5*a* is replaced by the polarizer angle P.

With such systems, a single measurement of (ψ,Δ) at a given λ can be obtained in 5–50 ms from the 30–90 points collected over an optical cycle (one-half mechanical cycle). Typically, results from an even number of successive optical cycles are averaged to obtain higher precision in a measurement time of 0.1–10 s. The time for the collection of a full spectrum then depends on the number of desired spectral points and the scanning speed of the spectrometer, but it is typically several minutes or more.

For RTSE, a spectrograph, in combination with a photodiode-array detector, has been incorporated into the rotating-polarizer configuration for the parallel acquisition of ~100-wavelength positions from 275 to 825 nm or 4.5 to 1.5 eV (see schematic in Figure 3.1).[15, 16] Because the array integrates the incident irradiance between two successive readouts, α, β, and a consistency check can be computed from four detector readouts acquired over the four quadrants of an optical cycle. Even though this detection system acquires data in parallel for all wavelengths, the readout is not simultaneous, and the time-zero defined for each successive λ value along the array is shifted. Thus, P_s in Equation 3.4 is a linear function of pixel number along the array. For this ellipsometer, a spectrum in (ψ,Δ) has been obtained in a single optical period of 16 ms ($\omega/2\pi = 31$ Hz); however, accumulating results obtained over an even number of optical cycles improves the signal-to-noise ratio. Spectra collected in 3.2 s have shown a precision in (ψ,Δ) of (0.003°, 0.007°) at the optimum photon energy of 2.5 eV.

3.4 Determination of Optical Properties

In the first part of this section, the discussion centers on approaches designed to solve the optical problem for a single semitransparent layer using one experimental (ψ,Δ) spectrum. This involves deducing $\varepsilon_{1f}(\lambda)$, $\varepsilon_{2f}(\lambda)$, and d from $\{\psi(\lambda),\Delta(\lambda)\}$. In the second part, the basic goal is the same, but the solution requires a series of spectra collected in real time at different film thicknesses during growth. The latter approach is valid only if the dielectric function of the film is either thickness independent or can be described by one or more thickness-dependent free parameters that are independent of λ. The examples that demonstrate these approaches have been drawn from recent in situ studies of the electrochemical deposition of conducting polymers.

Analysis of Single Ellipsometric Spectra: Direct Inversion Methods

Direct inversion methods involve choosing a series of closely spaced trial values for the thickness of the film, substituting the experimental $\{\psi(\lambda),\Delta(\lambda)\}$ spectra into Equations 3.3, and numerically inverting the equations in order to determine a series of trial $\{\varepsilon_{1f}(\lambda), \varepsilon_{2f}(\lambda)\}$ solutions. Only one such solution, obtained with the correct thickness, is itself correct.

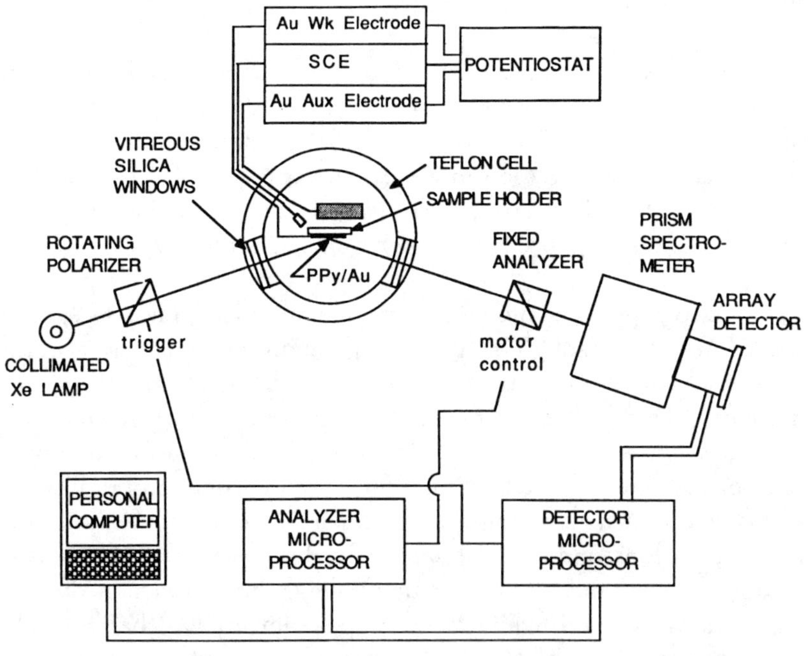

Figure 3.1 **Rotating-polarizer spectroscopic ellipsometer designed for real-time applications. In this schematic, a cell is mounted at the axis of the ellipsometer for in situ observations of the electrochemical deposition of conducting polymer films.**

There are three possible criteria that can be applied to determine which of the thickness choices is correct, based on an inspection of the form of the trial $\{\varepsilon_{1f}(\lambda), \varepsilon_{2f}(\lambda)\}$ solutions: (1) It may be known that $\varepsilon_{2f} = 0$ over some range of wavelength (e.g., below the band gap of a material), and only the correct thickness may provide an $\varepsilon_{2f}(\lambda)$ that satisfies this constraint. (2) If the substrate exhibits a sharp feature in ε_s over a narrow range of λ where ε_f is featureless, then the correct value of d can be identified based on the elimination of artifacts in the inverted $\{\varepsilon_{1f}(\lambda), \varepsilon_{2f}(\lambda)\}$ that arise from ε_s.[17] Substrate-related artifacts only appear when d is chosen incorrectly. (3) If the film is thick and transparent enough so that Z in Equation 3.3*b* passes through a number of periods over a detectable range of λ without complete damping by absorption, then interference oscillations will appear in $\{\psi(\lambda), \Delta(\lambda)\}$. As in criterion 2, the correct thickness can be identified based on the elimination of interference-related artifacts in $\{\varepsilon_{1f}(\lambda), \varepsilon_{2f}(\lambda)\}$.[18]

In these procedures, some experience and knowledge about the thin film is helpful to narrow the trial thickness range. In addition, starting dielectric functions for the film are required in order to initiate each numerical inversion sequence, which is iterative by nature. Since multiple solutions to Equations 3.3 may exist, the starting dielectric functions must be sufficiently close to the desired branch. Furthermore the single-film model must be valid, and if additional unaccounted layers are present, the artifacts that appear in the $\{\varepsilon_{1f}(\lambda), \varepsilon_{2f}(\lambda)\}$ spectra with approaches 2 and 3 cannot be completely eliminated for any choice of d. Usually, limitations on the accuracy of all these methods stem from unaccounted interfacial and surface layers of different optical properties, such as oxides and roughness at the ambient–film and film–substrate interfaces. In some cases—for example, film growth on a fully characterized oxide-covered substrate surface—the same direct inversion approach can be used to solve the thin film problem. Instead of applying Equations 3.3, however, a two-film matrix computation is used as the core of the numerical inversion.

As an example of approach 2, results are presented for the optical analysis of a polypyrrole film in the oxidized state (0.62 V versus SCE) electrochemically deposited onto a Au-coated Si wafer in an aqueous solution of 0.1 M KNO_3.[19] Because the Au was opaque and prepared under conditions optimized for surface smoothness, a one-film optical model could be applied first. The single pair of (ψ, Δ) spectra employed in the analysis was collected in real time, 600 s after the onset of film growth, using a multichannel ellipsometer with a 3.2 s acquisition time. In this time, an average thickness of 2.5 Å accumulates.

The second derivatives of the trial ε_{1f} spectra, calculated with different trial thicknesses from 450 to 550 Å are shown in Figure 3.2. The derivatives are taken in order to enhance the substrate-related artifact, which in this case arises from the onset of d-band to Fermi level transitions in Au. The second derivative of ε_{1s} for Au is also shown in order to identify the position of the artifact at 2.5 eV. It is clear that a thickness of 490 Å leads to the smoothest value of ε_{1f} throughout the 2.4–2.6 eV range, and is thus deemed to be correct within an uncertainty of ±5–10 Å. The trial

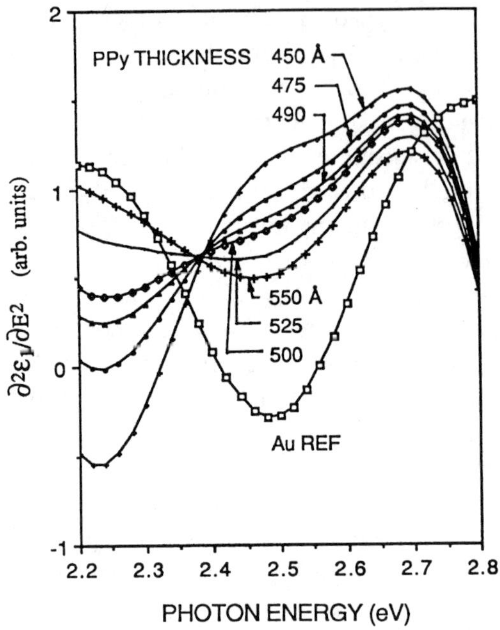

Figure 3.2 **Second derivatives with respect to photon energy of the real parts ε_1 of trial dielectric functions for thin film polypyrrole. Spectra in ε_1 were obtained by mathematical inversion from the (ψ,Δ) spectra collected 600 s after the onset of growth using a one-layer model with the thicknesses shown. Second derivative spectra for the Au substrate are also shown for reference. (After Reference 19.)**

dielectric function generated with this thickness (shown in Figure 3.3) is then also correct. Further support for the validity of this result is to be presented shortly.

Analysis of Single Ellipsometric Spectra: Least-Squares Regression Analysis Method

The least-squares regression analysis method is important because it can be applied to determine simultaneously the thickness of the single film and an interpretation of its dielectric function in terms of electronic transitions.[19–21] In this case, one assumes that $\{\varepsilon_{1f}(\lambda), \varepsilon_{2f}(\lambda)\}$ can be expressed as an analytical function of a number of wavelength-independent parameters. For example, if N_o electronic transitions are considered, each described as a damped harmonic oscillator (Lorentzian lineshape), then ε_f can be written[22]

$$\varepsilon_f = \varepsilon_\infty + \sum_{j=1}^{N_o} [F_j/(E_j^2 - E^2 + iE\Gamma_j)] \tag{3.6}$$

where E is the photon energy (hc/λ) and E_j, F_j, and Γ_j denote the resonance energy, strength, and damping factor, respectively, for the jth transition. In addition, ε_∞ is

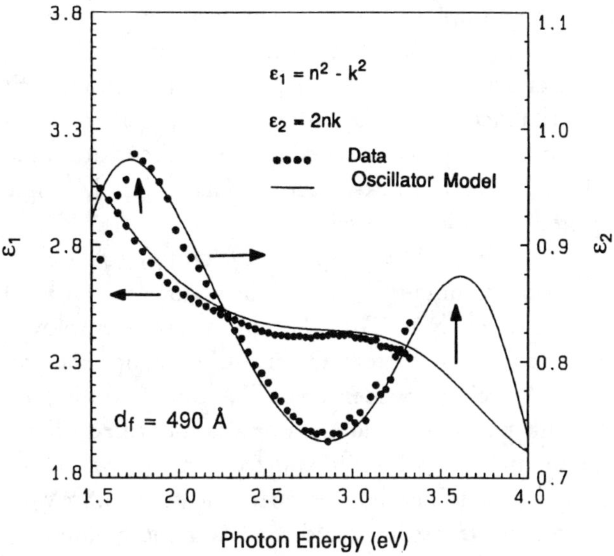

Figure 3.3 **Comparison of the real and imaginary parts of the dielectric function of the polypyrrole obtained by the direct inversion method (points) with a thickness of 490 Å, and by regression analysis assuming three electronic transitions described by damped harmonic oscillators. In the latter analysis, the thickness is determined to be 475 ± 20 Å. (After Reference 19.)**

a real quantity that includes contributions from transitions well above the upper energy limit of the measurement. Other lineshapes can also be employed to simulate the optical features.

Thus, in order to solve the optical problem with Equation 3.6, one must determine $3N_o + 2$ wavelength-independent parameters including the film thickness, d. This is an ideal problem for least-squares regression analysis.[23] Generally one starts with an expression for the thin film dielectric function with a minimum number of free parameters, and initial values for all parameters are assigned to start the iterative process. The regression analysis successively adjusts the parameters in an attempt to minimize σ, the unbiased estimator of the mean-square deviation:

$$\sigma = (M - p - 1)^{1/2} \left\{ \sum_{j=1}^{M} \left[|\tan\psi_e(\lambda_j) - \tan\psi_c(\lambda_j)|^2 + |\cos\Delta_e(\lambda_j) - \cos\Delta_c(\lambda_j)|^2 \right] \right\}^{1/2} \tag{3.7}$$

Here the subscripts e and c designate the experimental and best-fit calculated spectra, respectively, and p is the number of wavelength-independent free parameters. Once a best fit is found, the cross-correlation coefficients and 90% confidence limits on the free parameters are calculated, and the quality of the fit is considered in view of experimental errors. The complexity of the expression for ε_f may be increased at this time using additional free parameters if warranted, and the regression analysis is repeated in further efforts to improve σ and obtain an acceptable fit while maintaining reasonable confidence limits and correlation coefficients. If an adequate fit cannot be obtained without highly correlated variables, then additional measurements (i.e., VASE or RTSE—see the next section, "Analysis of Multiple Ellipsometric Spectra") are required to improve certainty.

For comparison purposes, the same pair of (ψ,Δ) spectra for polypyrrole on Au that were analyzed by direct inversion to obtain the results in Figures 3.2 and 3.3 (points) were also studied by regression analysis. Three Lorentzians were employed in the latter, resulting in a total of 11 free parameters. In this fit, resonance energies of 1.65 ± 0.15, 2.3 ± 0.9, and 3.8 ± 0.2 eV were identified. The 2.3 eV transition, which is somewhat uncertain in this particular analysis, was also identified unambiguously in additional measurements of less highly oxidized films in which the stronger 1.65 eV transition is diminished. The highest energy feature (3.8 eV) is attributed to the onset of band-to-band transitions, and the lower energy ones are attributed to transitions from the valence band to defect states in the energy gap associated with NO_3 doping.[24]

In addition to the information on the electronic transitions, the best-fit film thickness was determined to be 475 ± 20 Å, in agreement with the direct inversion method within the 90% confidence limits. Figure 3.3 includes the polypyrrole dielectric function calculated from the best-fit parameters of the regression analysis (solid line). The difference here is a good indication of the errors that arise from the 15 Å difference in thickness. Overall, however, the agreement lends support to both methods for solving the optical problem, in this case from a single pair of $\{\psi(\lambda),\Delta(\lambda)\}$ spectra.

Analysis of Multiple Ellipsometric Spectra

The above direct-inversion method may not be possible if (1) the film is absorbing throughout the photon energy range ($\varepsilon_{2f} \neq 0$ for all λ), (2) ε_s for the substrate exhibits no sharp features (or such features occur near the wavelengths of the features in ε_f), and (3) the film is too thin to exhibit interference oscillations. Furthermore, the regression analysis may not be possible if the analytical form of the dielectric function of the film is too complex to be described by a few oscillators. In such cases, multiple (ψ,Δ) spectra are typically needed to solve such problems. In the absence of real-time measurements, the methods of VASE can be applied.[4] Here one approach will be described that has been developed using RTSE data.[25]

To begin this analysis, one selects a single pair of (ψ,Δ) spectra collected at t_0 from among those obtained in real time during growth. Next, it may be possible to assume that in some time interval including t_0 defined by $[t_1,t_2]$, ε_f is independent of thickness (or time). As in the direct inversion approaches, a series of trial thicknesses is chosen for the film at time t_0, and Equations 3.3 are inverted upon substitution of the selected (ψ,Δ) spectra. The result of these inversions is a series of trial $\{\varepsilon_{1f}(\lambda), \varepsilon_{2f}(\lambda)\}$. Selecting one trial $\{\varepsilon_{1f}(\lambda), \varepsilon_{2f}(\lambda)\}$, regression analysis can be applied to each of the (ψ,Δ) spectra in turn over the interval $[t_1,t_2]$, since the only parameter remaining to be determined is the thickness, which depends on time. The unbiased estimators of the mean-square deviation (see Equation 3.7) are then summed for each of the best fits. This entire regression analysis procedure is repeated for each trial film dielectric function, and the correct one is chosen based on minimizing the sum of the unbiased estimators. As a result, one obtains the global best fit to the real-time spectra over the interval $[t_1,t_2]$ using a fixed thin film dielectric function and a thickness that varies with time.

A further complication can be introduced into this procedure based on the theory presented later in "Microstructural Evolution in Thick Film Growth" in Section 3.5. The dielectric function of the film may not necessarily be independent of thickness over the $[t_1,t_2]$ interval. This problem may be revealed by poor fitting of the real-time (ψ,Δ) spectra in the individual regression analyses, resulting in increasing σ values for the best fits to spectra with increasing $|t - t_0|$, even for the best global fit. In this case, it may be possible to account for the problem by introducing another free parameter in the regression analyses (in addition to d), namely, the void volume fraction in the film measured relative to that at t_0. The relative void volume fraction, f_v, can be used as a free parameter in the analysis, since it defines the dielectric function of the film through an effective medium theory (see "Microstructural Evolution in Thick Film Growth" in Section 3.5). Thus, the global best fit in this case can provide $\{\varepsilon_{1f}(\lambda), \varepsilon_{2f}(\lambda)\}$ at the chosen time t_0, and both d and f_v as a function of time over the $[t_1,t_2]$ interval.

This more complicated analysis procedure was applied to characterize thin film polyaniline electrochemically deposited onto a Au-covered Si wafer in 0.1 M Na_2SO_4.[26] The Au was opaque over the full photon energy range, so the one-film optical model could be employed. Trial dielectric functions for the film at $t_0 = 500$ s were obtained from trial thicknesses and a single $\{\psi(\lambda),\Delta(\lambda)\}$ pair collected in 3.2 s at t_0. Figure 3.4 (inset) shows the sum of the unbiased estimators $(\Sigma\sigma)$ obtained by fitting 28 spectra from 230 to 500 s by regression analysis using a dielectric function for the film obtained by assuming the thickness plotted along the abscissa. Although the minimum in $\Sigma\sigma$ is broad, a thickness of 460 Å at $t_0 = 500$ s minimizes $\Sigma\sigma$ and provides the best global fit.

Figure 3.4 also includes the dielectric function of the 460 Å thick polyaniline film deduced in this analysis. A complex optical structure is revealed, particularly in the 2.0–3.0 eV range, and this has made interpretation by the substrate artifact minimization technique (discussed in the "Direct Inversion Methods" part of this section) more difficult. In the analysis of Figure 3.4, the void volume fraction and thickness of the film are also obtained over the specified time interval. Such features of film growth will be discussed in greater detail, along with the theoretical background, in "Microstructural Evolution in Thick Film Growth" in Section 3.5. Finally, it is important to apply complementary analysis techniques to solve the

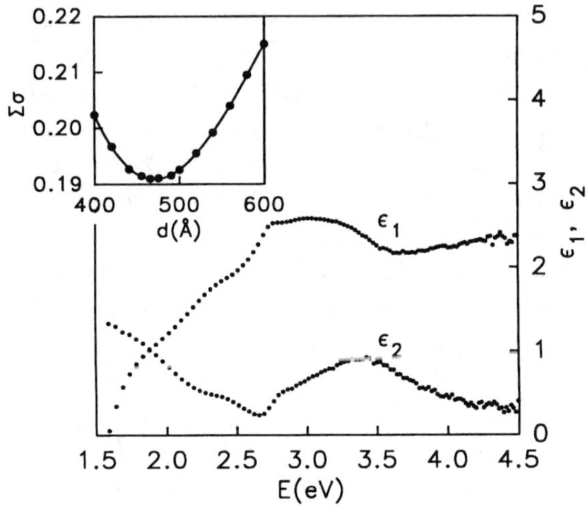

Figure 3.4 **Sum of the unbiased estimators of the mean-square deviation ($\Sigma\sigma$) obtained in regression analysis fits to 28 pairs of (ψ,Δ) spectra collected in real time during the electrochemical deposition of polyaniline, plotted versus trial thickness for the pair of spectra collected at t_0 = 500 s (inset). A global best fit is obtained with a thickness of 460 Å. The dielectric function of polyaniline is calculated by inversion from (ψ,Δ) at t = 500 s with this thickness value. (After Reference 26.)**

optical problem (such as in Figure 3.3) in order to assess potential errors due to deviations in the real film structure from the idealized optical model.

3.5 Determination of Thin Film Structure

For many organic films, the constituent molecular groups exhibit no interesting electronic features, that is, there are no chromophores. For these cases, one may not be interested in the dielectric function of the film, but rather in information on structural characteristics such as the film thickness and density distribution. For an adsorbed polymer, for example, one may be interested in a density gradient away from the substrate surface or, in a deposition process, the thickness at which initial growth nuclei coalesce to form a continuous film. In addition, for organized molecular films, one may be interested in the orientation of a dominant molecular axis with respect to the surface plane, which in principle can be inferred from the thickness for asymmetrically shaped molecules packed at maximum density. Although such information can often be deduced from ellipsometry at a single wavelength, systematic errors can occur. Application of the spectroscopic approach can substantially reduce these errors and thus can increase confidence in the deduced structure.

Thickness Determination for Monolayers

Single-wavelength ellipsometry has been applied extensively over the years to determine the thickness of organic and biological films[27] and, in particular, organized monolayers.[28, 29] As in earlier discussions, the usual procedure involves assuming that the pseudo-dielectric function for the opaque, "bare" substrate, calculated from the right-hand side of Equation 3.2, corresponds to the true dielectric function ε_s. This result is used in a one-layer analysis of (ψ,Δ) measured after film formation (see, for example, Reference 29). In fact, if the two measurements are performed in situ before and after monolayer adsorption, without changing sample alignment, errors due to alignment irreproducibility are avoided.

For isotropic films that are sufficiently thin, one can express the change in the amplitude reflection coefficients, r_p and r_s, in Equations 3.3 as two power series to the first order in d, and then calculate the change in the ellipsometric parameters.[3, 30] The results are

$$\delta\Delta = (4\pi d n_a \cos\theta/\lambda)\ \mathrm{Re}(X) \tag{3.8a}$$

$$(2/\sin 2\psi_0)\delta\psi = (4\pi d n_a \cos\theta/\lambda)\ \mathrm{Im}(X) \tag{3.8b}$$

where

$$X = (\varepsilon_s/\varepsilon_f)[(\varepsilon_f - \varepsilon_a)(\varepsilon_s - \varepsilon_f)]/[(\varepsilon_s - \varepsilon_a)(\varepsilon_s \cot^2\theta - \varepsilon_a)] \tag{3.8c}$$

Here, ψ_0 is the ellipsometric spectrum for the bare substrate and $n_a = \varepsilon_a^{1/2}$. For an organic film on a Si wafer, for example, this expansion is valid for $d < 15$ Å.

If the substrate is a dielectric (e.g., Si in the visible) and the monolayer is also nonabsorbing, then X is real, $\delta\psi \approx 0$, and only Δ will change upon film formation. A similar result occurs if $|\varepsilon_s| \gg \varepsilon_a$, and $\varepsilon_f \approx \varepsilon_a$, which is typical of organic monolayer films on metal substrates. In such cases, it is impossible to determine ε_{1f} and d uniquely from Equation 3.8*a*, and even spectroscopic measurements cannot remove the correlation. Often d can be determined, however, if ε_{1f} is estimated from independent measurements, for example, by using the properties of the molecules in their bulk state. Most often one assumes that the molecular packing density is the same on the surface and in the bulk—a favorable case for calculating good thickness values. If this assumption is incorrect, then the deduced d is an effective thickness which will be different than the physical thickness of the monolayer. For larger molecules corresponding to thicknesses greater than 20 Å, the first order approximation of Equations 3.8 breaks down, higher-order terms become measurable, and consequently, more information can be extracted from the observed changes in both ψ and Δ.

The simple advantage of SE over the single-wavelength null-ellipsometry approach is the redundancy of multiple measurements that provides better experimental statistics, and in turn a better assessment of the confidence limits. If the sample deviates from the ideal assumptions employed in the optical analysis, however, the improved statistics will not increase confidence in the resulting physical parameters. By analysis of the underlying origin of the deviations, one may be able to arrive at an improved physical description of the film. Such deviations can result from rough substrate surfaces, or from a modification of the initial surface electronic states or atomic-scale structure of the substrate. In a single-wavelength measurement, small deviations go undetected because their λ-dependence is not evident; thus, an essential physical feature is missed.

As an example, an in situ SE investigation of the spontaneous organization of alkane thiols [$CH_3(CH_2)_{n-1}SH$] on evaporated Au(111) will be discussed. This self-assembled monolayer system has been studied extensively since its initial discovery,[31, 32] and single-wavelength ellipsometry has been applied in attempts to determine accurate film-thickness values.[33, 34] The sulfur groups are known to bond to the Au surface with the chains extended as near-rigid rods, the chain axes being an average angle of 26–28° from the surface normal.[35] This system is interesting because the structure is believed to be essentially unaffected by the number of carbon atoms in the chain (n). Thus, the physical thickness of the film should be precisely linear with n. Several groups have made such correlations using ex situ ellipsometry performed at a single λ, but the measurement errors tend to be on the order of the thickness increment corresponding to a unit increment in n.

In a recent study,[36] SE spectra were collected in situ before and after the adsorption of alkane thiols, with $n = 12$, 16, 18, and 22, from absolute methanol. In order to interpret the spectra, ε_f for the film phase was assumed to be independent of n and was initially calculated from the monolayer film corresponding to $n = 22$. The determination of d for $n = 22$ by regression analysis required independent information about the film dielectric function to reduce the number of free parameters. First, since the $CH_3(CH_2)_{n-1}$ chains exhibit no electronic transitions between 1.5 and 3.5 eV, it was assumed that $\varepsilon_{2f} = 0$, thus ignoring the effects of the sulfur group. Second, it was assumed that ε_{1f} is dependent on λ with $\varepsilon_{1f}^{1/2}(\lambda_{\text{He-Ne}} = 632.8 \text{ nm}) = 1.47$, as estimated from measurements on bulk thiols or disulfides. The wavelength dispersion used is that of Sellmeier, which is obtained using one term in Equation 3.6 with $\Gamma = 0$. From the resulting dielectric function, the thicknesses for the other alkane thiols were determined by regression analysis. Such an analysis led to relatively poor fits to the data for $n = 12$ and 16 (as indicated by σ—see Equation 3.7 and Figure 3.5) that could not be improved by relaxing the condition that $\varepsilon_{2f}(632.8 \text{ nm}) = 1.47$.

To better understand the origin of the problem, the optical structure for $n = 12$ was assumed to consist of an overlying layer of thiol having the best-fit thickness obtained in the earlier analysis (17 Å), and an underlying monolayer-thick (3 Å) interface with unknown optical properties. With this assumption for the structure, the dielectric function for the interface layer could be extracted. The result showed reasonable behavior: weak absorption in ε_2, increasing to higher energies, and an associated anomalous dispersion characteristic in ε_1. The regression analyses of the other three data sets ($n = 16$, 18, 22) were repeated using variable thiol and interface-layer thicknesses and the fixed dielectric properties for the layers. For $n = 16$, a similar thickness interface improved the fit considerably, whereas the two larger n values required a weaker interface effect (~1 Å thickness).

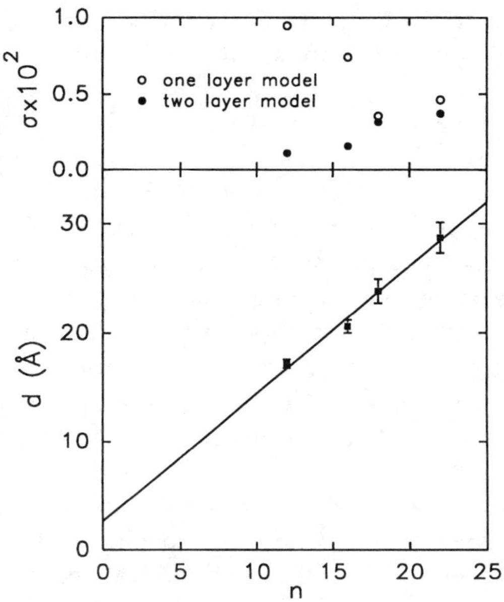

Figure 3.5 **Thicknesses of adsorbed layers of alkane thiol molecules on Au in a methanol ambient as a function of the number of C atoms in the chain for four separate experiments with n = 12, 16, 18, 22. The results were obtained by in situ SE. In order to determine the thicknesses, a two-layer optical model was employed including thiol and modified interface layers. The unbiased estimators of the mean-square devation for the best fits are included at the top for the one-layer (no interface) and two-layer models. (After Reference 36.)**

Finally, the entire analysis was repeated, choosing different constraints on the thiol refractive index at 632.8 nm in a narrow range about $n_f = 1.47$ to account for any possible deviations from the bulk value. A small improvement was obtained with $n_f = 1.50$. The final results as a function of n are given in Figure 3.5. Linear behavior in the thiol thickness is observed as a function of n with a slope and intercept of 1.2 Å/CH_2 and 2.6 Å, respectively. With the previously described structural model for the thiols, a slope of 1.1 Å/CH_2 is expected from a simple bond length accounting for the molecule. The expected intercept is less clear, but should be approximately the size of the S atom. The agreement is quite good and provides support for the validity of the optical model. However, it must be realized that the intercept inherently contains an uncertainty because of the possible partitioning of the thin interface layer between the thiol layer and the Au.

The unbiased estimator, σ, plotted in the upper part of Figure 3.5 for best-fit models with and without the interface layer shows the significant improvement that the latter provides. This also leads to greater confidence in the deduced thickness values over measurements at single wavelength. However, the 90% confidence limits on the thicknesses in Figure 3.5 are determined from the ~50 independent measurements versus λ, rather than from repetitive measurements at a fixed λ.[33, 34] Further experimental and theoretical studies are needed to exploit the additional information spectroscopic ellipsometry provides. In principle, insights into charge transfer at the interface may be extracted from the spectral dependence of the interface layer optical properties. In practice, however, changes in the atomic-scale structure of the Au surface upon adsorption may also give rise to the observed interface effects. Separating such effects poses a challenge for future research (see Section 3.6).

Further challenges arise from the expectation, confirmed by observation in some cases,[10] that oriented molecular monolayers are optically anisotropic. In fact, single nonabsorbing films that are uniaxial, with the optic axis perpendicular to the film plane, can be fully characterized using measurements at different thicknesses (assuming the dielectric tensor does not depend on d).[37] Neglecting anisotropy has been shown to lead to an erroneous nonzero ε_{2f} value for nonabsorbing films.[38] Such does not appear to be the case for the alkane thiols, since the role of the absorbing interface layer is to improve σ for the shorter chain molecules If the problem were anisotropy, the thicker films would be affected most strongly. In cases where the situation is not clear, possible anisotropic effects can be identified by modeling spectra collected for a range of incidence angles.

Microstructural Evolution in Thick Film Growth

If one knows the dielectric functions of the substrate and ambient, as well as that of a thick film in its bulk density form, then the RTSE data set obtained during the growth process can be analyzed in terms of the microstructural evolution. In the simplest case of a single-film optical model, one can characterize the film by its thickness and average density relative to the bulk film. In such a model, the dielectric function of the film in its density deficient form ε_f' is calculated from the bulk film dielectric function ε_f and the density deficit f_v using an effective medium theory (EMT), given generally by[39, 40]

$$\frac{\varepsilon_f' - \varepsilon_h}{\varepsilon_f' + 2\varepsilon_h} = (1 - f_v)\frac{\varepsilon_f - \varepsilon_h}{\varepsilon_f + 2\varepsilon_h} + f_v\frac{\varepsilon_a - \varepsilon_h}{\varepsilon_a + 2\varepsilon_h} \tag{3.9}$$

Here the film is interpreted as a two-phase composite, consisting of a volume fraction f_v of void space having a dielectric function equal to that of the ambient, and a volume fraction $1 - f_v$ of material having a dielectric function equal to that of the bulk-film phase. Equation 3.9 can be solved for the complex-valued ε_f' as long as the other information is known. In Equation 3.9, ε_h is the dielectric function of the phase to be considered as the host. EMTs differ in the choice of the host material from among the three possibilities of ε_f', ε_a, and ε_f. The Bruggeman EMT, corresponding to $\varepsilon_h = \varepsilon_f'$, is a reasonable choice for modeling the aggregate microstructures of organic films.

There are two assumptions involved in using an EMT. The first is that the solid-phase component of the composite exhibits a dielectric response identical to that of bulk density material. This assumption tends to be valid for the interband optical transitions in disordered materials, since the dielectric function is determined by the local molecular bonding, the electron lifetime is relatively short, and the phase boundaries have little effect. In single crystalline materials with long electron lifetimes, however, size effects occur whereby the lifetime is controlled by scattering at the phase boundaries. This gives rise to a broadening of electronic transitions and the corresponding optical features, and this is not taken into account in the EMT. The second assumption is that the size scale of the inhomogeneous structure must be small in comparison to the wavelength of light in the film. For very thin films (<1000 Å) this is probably a valid approximation.

The first application of the EMT approach to the characterization of organic thin films was in an analysis of RTSE data collected on polypyrrole.[19] In this study, it was first assumed that the dielectric function deduced at the end of the growth process ($d = 490$ Å; see Figure 3.3) was also valid at all stages, meaning that the film grew uniformly without any changes in the density with thickness. From this assumption, regression analysis procedures can be applied to determine d versus time. To assess the validity of this approach, σ is plotted versus time in Figure 3.6 (curve 1). The fit at the beginning and end of film growth is perfect because these points correspond to the spectra that were used to calculate the Au and polypyrrole

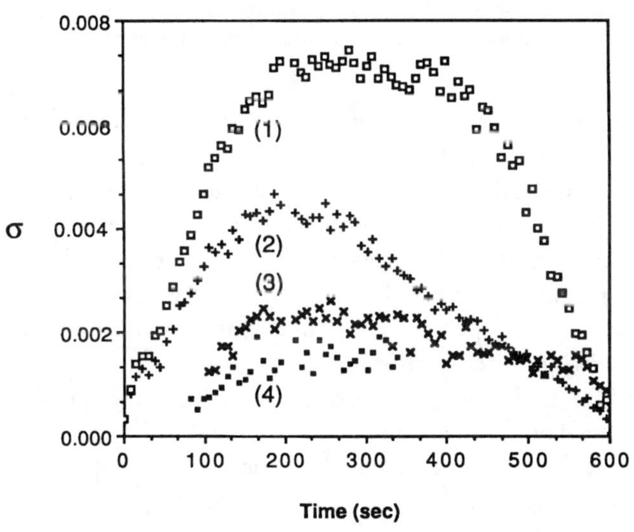

Figure 3.6 **Comparison of the unbiased estimators of the mean-square deviation as a function of time deduced from best-fit optical and microstructural models of polypyrrole film growth by electrochemical deposition. For curves 1 and 2, a one-layer model is assumed, and for 3 and 4, a two-layer model is assumed [(bulk film)/(substrate interface)]. For model 1, the dielectric function of the film is assumed to be independent of thickness and the only free parameter is film thickness. For model 2, the average void volume fraction in the film is assumed to evolve with time during growth, and two free parameters, thickness and void volume fraction, are deduced (see Figure 3.7). (After Reference 19.)**

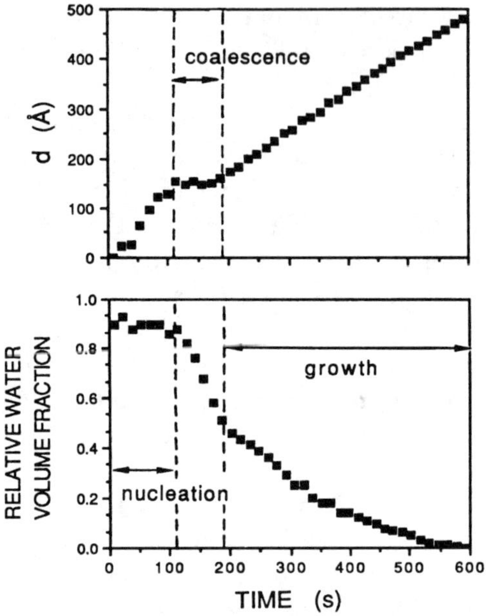

Figure 3.7 **Relative void volume fraction (i.e., water content) and thickness as a function of time deduced from a regression analysis of real time SE data collected during the electrochemical deposition of polypyrrole. The unbiased estimator of the mean-square deviation for this analysis is included in Figure 3.6 (curve 2).**

dielectric functions. In the intermediate stages of growth (150–450 s), the fit is relatively poor considering the precision and accuracy of the ellipsometer.

The fit can be improved considerably by assuming that the average volume fraction of void in the film varies with the thickness, as would be expected for nucleation, gradual coalescence, and bulk-film growth processes. In this case, two wavelength-independent free parameters are used in the regression analysis, d and f_v. As usual, in fitting each one of the spectra versus time, initial values must be assigned to these parameters to begin the iteration procedure. Once f_v is set, then the dielectric function of the film, ε_f', is calculated from the bulk dielectric function through Equation 3.9. With the dielectric function of the film and d, (ψ,Δ) can be calculated for comparison with the experiment. On the basis of this comparison, adjustments are applied to d and f_v in order to further minimize σ. Figure 3.6 shows the minimum $\sigma(t)$ (curve 2) obtained by this method, and Figure 3.7 shows the evolution of thickness and void volume fraction (water content, in this case).

The time dependence of σ in Figure 3.6 (curve 2) shows significant improvement over the uniform growth model (curve 1), except in the earliest stage of the growth. In more extensive studies, the problem has been attributed to changes in roughness at the Au–polypyrrole interface, that lead to a two-film model (curves 3 and 4 in Figure 3.6). The addition of this complexity does not have a significant effect on the interpretation of the microstructural evolution, as shown in Figure 3.7 for the one-film model (curve 2). Figure 3.7 shows that initial nucleation of the film is characterized by a very high void fraction (~0.9), and this low density phase persists for the first 100–150 Å of film growth. At 150 Å, the film densifies as the void space between initial structures is filled with polymer. Above 150 Å, there is a gradual densification of the material to the point where deposition is terminated. Since further densification seems possible, the void fractions quoted are those measured relative to the final material. This example reveals the ability of real time SE to provide information on the expected microstructural integrity of the deposit.

3.6 Future Prospects

With the recently developed methods described here, it is anticipated that SE will become more widely used to characterize the optical, electronic, and microstructural properties of organic thin films. Because of the spectroscopy aspect, it is expected to be particularly useful in studies of electronically active thin films. For such films, the effects of the preparation process on the density of electronic states and the microstructure are of great interest. Because RTSE is an in situ probe, the potential problems associated with post-process modification due to interaction with the ambient atmosphere are avoided. Ultimately, optical techniques such as ellipsometry may find an important niche in controlling the structure (e.g., thickness) and electronic properties (e.g., doping level and electron concentration) of the films. As the applications for these films become more demanding, such techniques can be applied to ensure that certain desired specifications are met. Finally, for the electronically active films, accurate dielectric functions free of artifacts, as provided by RTSE, are a critical component for understanding the electronic processes in such films.

For films that are inherently electronically inactive, the characterization of the physical structure will be the major application of optical SE. SE is recommended over single-wavelength ellipsometry as a means to increase the level of information and the statistical confidence obtainable. The optical nature of the substrate–film interface for chemically attached monolayers will also be an area of fruitful future research as a result of the interesting alterations in the electronic states of metallic or semiconducting substrates upon film formation. However, these perturbations to the optical model are typically small, making it difficult to solve interface-related problems. In addition, changes in interface optical properties arising from a physical composite generated by the nonplanarity of the interface may be difficult to distinguish from changes arising from bond formation or charge transfer.

There is a general need to correlate SE with other techniques having greater chemical specificity, such as infrared reflection and photoelectron spectroscopies. However, it should be noted that the other techniques are severely limited for in situ studies. Photoelectron spectroscopies are strictly vacuum techniques; infrared spectra can be performed through liquids, but serious constraints are placed on solvent choices—only nonabsorbing solvents are useful.

Acknowledgment

This preparation of this article was made possible by funding from the National Science Foundation under Grant Nos. DMR-8957159 (RWC) and DMR-9001270 (DLA).

References

1 R. M. A. Azzam and N. M. Bashara. *Ellipsometry and Polarized Light.* North-Holland, Amsterdam, 1977. A popular text on ellipsometry.

2 F. L. McCrackin and J. P. Colson. In *Ellipsometry in the Measurement of Thin Films.* (E. Passaglia, R. R. Stromberg, and J. Kruger, Eds.) National Bureau of Standards Misc. Publ. 256, Washington, DC, 1964, p. 61.

3 D. E. Aspnes. In *Optical Properties of Solids: New Developments.* (B. O. Seraphin, Ed.) North-Holland, Amsterdam, 1976, chapt. 15.

4 J. A. Woollam and P. G. Snyder. *Mater. Sci. Eng. B.* **5**, 279, 1990.

5 R. W. Collins. *Rev. Sci. Instrum.* **61**, 2029, 1990.

6 S. Gottesfeld. In *Electroanalytical Chemistry: A Series of Advances,* Vol. 15. (A. J. Bard, Ed.) Marcel Dekker, New York, 1989, p. 143.

7 "Proceedings of the First International Conference on Spectroscopic Ellipsometry." (A. C. Boccara, C. Pickering, and J. Rivory, Eds.) *Thin Solid Films.* **233**, 1993.

8 W. G. Oldham. *Surf. Sci.* **16**, 97, 1969.

9 M. Born and E. Wolf. *Principles of Optics,* 3rd ed. Pergamon, New York, 1965, p. 51.

10 R. H. Muller. In *Techniques for Characterization of Electrodes and Electrochemical Processes.* (R. Varma and J. R. Selman, Eds.) Wiley, New York, 1991, chapt. 2.

11 P. S. Hauge. *Surf. Sci.* **96**, 108, 1980.

12 B. D. Cahan and R. F. Spanier. *Surf. Sci.* **16**, 166, 1969.

13 D. E. Aspnes and A. A. Studna. *Appl. Opt.* **14**, 220, 1975.

14 P. Chindaudom and K. Vedam. To be published in *Thin Solid Films.*

15 I. An and R. W. Collins. *Rev. Sci. Instrum.* **62**, 1904, 1991.

16 I. An, Y. M. Li, H. V. Nguyen, and R. W. Collins. *Rev. Sci. Instrum.* **63**, 3842, 1992.

17 H. Arwin and D. E. Aspnes. *Thin Solid Films.* **113**, 101, 1984.

18 D. E. Aspnes, A. A. Studna, and E. Kinsbron. *Phys. Rev. B.* **29**, 768, 1984.

19 Y.-T. Kim, R.W. Collins, K. Vedam, and D. L. Allara. *J. Electrochem. Soc.* **138**, 3266, 1991.

20 S. Orzeszko, J. A. Woollam, D. C. Ingram, and A. McCormack. *J. Appl. Phys.* **64**, 2611, 1988.

21 S. Y. Kim and K. Vedam. *Thin Solid Films.* **166**, 325, 1988.

22 F. Wooten. *Optical Properties of Solids.* Academic, New York, 1972, chapt. 3.

23 D. E. Aspnes. *Proc. Soc. Photo-Opt. Instrum. Eng.* **276**, 188, 1981.

24 G. B. Street. In *Handbook of Conducting Polymers*, Vol. 1. (T. A. Skotheim, Ed.) Marcel Dekker, New York, 1986, chapt. 8.

25 Y. Cong, I. An, K. Vedam, and R. W. Collins. *Appl. Opt.* **30**, 2692, 1991.

26 Y.-T. Kim, Y. Lu, R. W. Collins, and D. L. Allara. Unpublished data.

27 A. Rothen. In *Ellipsometry in the Measurement of Thin Films.* (E. Passaglia, R. R. Stromberg, and J. Kruger, Eds.) National Bureau of Standards Misc. Publ. 256, Washington, DC, 1964, p. 7.

28 A. Ulman. *An Introduction to Ultrathin Organic Films From Langmuir–Blodgett to Self-Assembly.* Academic, New York, 1991.

29 D. L. Allara and R. G. Nuzzo. *Langmuir.* **1**, 52, 1985.

30 Y. Bouizem, F. Chao, M. Costa, A. Tadjeddine, and G. Lecayon. *J. Electroanal. Chem.* **172**, 101, 1984.

31 R. G. Nuzzo and D. L. Allara. *J. Am. Chem. Soc.* **105**, 4481, 1983.

32 R. G. Nuzzo, F. Fusco, and D. L. Allara. *J. Am. Chem. Soc.* **109**, 2358, 1987.

33 M. D. Porter, T. B. Bright, D. L. Allara, and C. E. D. Chidsey. *J. Am. Chem. Soc.* **109**, 3559, 1987.

34 C. D. Bain, E. B. Troughton, Y. T. Tao, J. Evall, G. M. Whitesides, and R. G. Nuzzo. *J. Am. Chem. Soc.* **111**, 321, 1989.

35 P. E. Laibinis, G. M. Whitesides, D. L. Allara, A. N. Parikh, Y. T. Tao, and R. G. Nuzzo. *J. Am. Chem. Soc.* **113**, 7152, 1991.

36 J. Shi, D. L. Allara, and R. W. Collins. Unpublished data.

37 D. den Engelsen. *J. Opt. Soc. Am.* **61**, 1460, 1971.

38 M. J. Dignam, M. Moskovits, and R. W. Stobie. *Trans. Faraday. Soc.* **67**, 3306, 1971.

39 G. A. Niklasson, C. G. Granqvist, and O. Hunderi. *Appl. Opt.* **20**, 26, 1981.

40 D. E. Aspnes. *Thin Solid Films.* **89**, 249, 1982.

4

Infrared Spectroscopy in the Characterization of Organic Thin Films

DAVID L. ALLARA

Contents

4.1 Introduction

Traditionally, infrared spectroscopy (IRS) has been a highly successful method for characterizing chemical bonding and structure in organic and organometallic compounds.[1–3] With the growing importance of thin organic films there has been a growing need to extend the capabilities of IRS analysis to these types of samples. However, this task has not been straightforward because of the frequent needs for quantitative analyses with monolayer sensitivity involving a variety of types of film materials and substrates as well as multiple-film structures. Fortunately, within the past few years advances in both instrumentation and theory have made such analyses possible, and a variety of extensions of IRS have been made to surface and thin film analysis.[4–7] This chapter presents the fundamental principles that govern the IRS thin film experiment, and the capabilities and limitations of the method in terms of these principles with respect to the specific task of characterizing organic films. The emphasis is on the application of the technique to provide quantitative analysis. In the first section, some general characterization needs for organic thin films are presented. With this background, a qualitative description of the principles of the IRS thin film experiment is presented, and the capabilities of the technique are discussed within this context. Following this, a recently developed generalized method for the quantitative treatment of IRS thin film data is presented. In the next section, a description of experimental considerations is given. Finally, a few recent examples of applications of thin film IRS are given with an emphasis on quantitative work.

Specific Needs for Characterizing Organic Thin Films

In the analysis of an organic thin film there are eight major characteristics of interest: coverage, thickness, density, atomic composition, chemical group composition, chemical structure, ordering, and defect content. Each of these has specific variations of interest for specific applications. In general, the above characteristics are also of interest in inorganic films, but the ranking of importance is different for these two classes of films. In particular, for organic films the concentrations and arrangements of functional groups is of interest. In contrast, for inorganic thin films the periodic arrangements of the constituent atoms, the nature of atomic-scale defects, and the presence of impurity atoms are often of interest, but these characteristics are usually not important in typical organic films. The additional complexity in the chemical bonding variations introduced by the presence of entire functional groups, for example, $CONH_2$ or aromatic rings, usually makes questions of atomic structure a tangential issue in organic materials. In order to understand the analysis requirements in organic films, the eight characteristics are discussed briefly. This background will be useful as a means to evaluate the capabilities of IRS to solve characterization problems in organic films.

Coverage, thickness, and density are three closely related features which describe the amount of film present. For films of thickness below a molecular diameter, the amount of film often is given in terms of coverage, either as the fraction of a monolayer or as a surface density of chemical species per unit surface area; for thicker films, thickness is usually used to describe the amount. The ability of a technique to detect monolayer to submonolayer quantities is termed surface sensitivity. Even if the amount of material can be analyzed, since coverage, thickness, and density are average quantities, one can never accurately describe film structure without additional knowledge of density distributions, including such features as roughness and interior voids. For organic films, the distribution of functional groups or chemical bonds is the important measure of density. In addition, other subtleties in structure may be very important, such as conformational sequences and hydrogen-bonding interactions. With regard to such molecular fine-structure features, and all others as well, it is critical that the analysis be nondestructive. In some cases, particularly for Langmuir–Blodgett or self-assembled monolayer assemblies of alkyl-chain molecules, the film structure may exhibit some degree of ordering with respect to both uniform orientations and uniform spacings of chemical bonds or groups. In these cases, the film can be characterized in terms of orientational and translational ordering, which in turn are given properly in terms of the dimensions and symmetry of the simplest structural unit and a correlation length for the persistence of this structure. In the usual case of a film attached to a substrate, the film–substrate interface is characterized by the nature of the bonding (physisorption or chemical bonding), the orientations of the bonds, and the relationship between the location of substrate surface sites and film bonding sites (epitaxial, commensurate, etc.). In all types of films, the presence of defects and impurities usually presents a severe challenge in quantitative analysis. Major types of defects include voids and boundary regions between ordered domains, and in some cases these defects may be correlated with substrate defects. A significant impurity associated with many organic films on hydrophilic substrates, such as SiO_2, is water adsorbed in the film–substrate interfacial region.

For films of thickness greater than several chemical groups, the distribution of the above characteristics as a function of distance from the surface can be of interest. Even if the analysis technique is monolayer sensitive, there is still the issue of whether the technique can selectively analyze one region of the film, such as the film–substrate interface or the outer-surface layer. The ability of a technique to selectively analyze a specific depth region is termed surface or interface selectivity. Ultimately, it is desirable to compile analyses for a series of depths in the film(s), which is usually called depth profiling. Monitoring film characteristics as a function of lateral location on the surface is referred to as imaging. A goal in surface analysis is to obtain three-dimensional characterization, including both lateral and depth distribution of important features.

The most tractable analyses are those involving vacuum-isolated samples and unlimited sampling time. However, the most significant practical problems are usually those involving a sample surrounded by an ambient fluid (gas or liquid) environment in which sample characteristics are highly time dependent. Challenging examples for characterization are lubricant films on a metal sheet during a rolling operation and a growing polymer film on an electrode during electropolymerization. In this respect, a goal of surface/thin film characterization is to develop real-time techniques, that is, those with capabilities for following time-dependent changes in a sample directly (in situ) in a processing or service environment.

General Principles and Capabilities of Infrared Spectroscopy for Surface and Thin Film Analysis

In order to understand the ability of IRS to satisfy the different analysis requirements, it is first necessary to understand the general principles of vibrational spectroscopy and thin film optics. Accordingly, the simple theoretical principles underlying thin film IRS are presented qualitatively in the following discussion and then used to explain the basis for the capabilities and limitations of the method. Common terms in surface and thin film analyses which are used throughout the text are defined in this section. In a later section, more detailed theoretical relationships will be given in order to establish the basis for quantitative analysis.

Thin film IRS can be understood on the basis of two types of theory: (1) the quantum theory of radiative excitation and decay between energy states[8] and (2) the classical theory of the interaction of electromagnetic waves with dielectric media.[9] The first theory involves transitions between vibrational energy levels with the accompanying emission or absorption of a quantum of energy in the form of infrared wavelength photons. This spectroscopic information reveals structural features involving the types and symmetries of chemical bonds between groups of atoms,[10] typically defined as functional or molecular groups. It is this aspect which makes IRS so powerful for determining the structures of organic compounds. The second theory takes into account the wavelength (momentum) character of the light and treats the phase interference effects that occur upon the reflection of a propagating light beam at the interfaces between different media in thin film samples.[11] For the long wavelengths of infrared light, uniform density materials appear continuous, with an average electron density, so diffraction from the atomic lattice does not occur and the intrinsic optical response of the bulk of a material can be described solely in terms of an average polarization response. For materials with density fluctuations (or rough surfaces) on a size scale of the wavelength of the light or greater, diffuse scattering of the radiation can occur. For the infrared radiation typical of IRS experiments, this scale is ~2 to 20 μm. The treatment of such scattering effects[9] is beyond the scope of this chapter; only uniform materials with smooth surfaces will be considered. At the levels of radiation power used in IRS, the polarization response of a material is linear with the electric-field amplitude, and can be described quantitatively by means of a frequency-dependent refractive index value—or equivalently, a dielectric function—assigned to the medium of interest. As a general case, the refractive index and equivalent dielectric function are defined as frequency-dependent complex quantities in Equations 4.1 and 4.2, respectively; the relationship between these two quantities is given by Equation 4.3.

$$\tilde{n}(\omega) = n + ik \tag{4.1}$$

$$\varepsilon(\omega) = \varepsilon_1 + i\varepsilon_2 \tag{4.2}$$

$$[\tilde{n}(\omega)]^2 = \varepsilon(\omega) \tag{4.3}$$

The use of the complex quantity $\tilde{n}(\omega)$ allows for the inclusion of both refraction effects using the real part, and absorption (or spectroscopic) effects using the imaginary part. The physical basis of the dielectric function is analogous. With this background material in mind, the factors governing the capabilities of an IRS thin film characterization can be discussed.

The surface selectivity (discussed in the preceding section) of an IRS experiment is determined by the ability to confine the overall distribution of the electromagnetic fields of the interacting light in the surface regions of the target solid. In general, for a medium with a smooth, planar surface, a useful rule of thumb is that the electromagnetic fields cannot be localized at the surface much better than they can be within a depth equal to a major fraction of the wavelength of the probing light. In the simplest experiment, the propagation of light through a medium, the light will continue on infinitely into the medium unless it is absorbed, so there is no surface selectivity, that is, the entire bulk of the sample will contribute to the spectrum. If light is reflected from a surface, a portion of the wave is also transmitted into the medium, and since the reflection and transmission are coupled—that is, they are one event together—the behavior of the transmitted beam deep in the medium (e.g., due to another reflection or absorption by some species) will affect the reflection.

This can be described in terms of Figure 4.1, where a beam of intensity I_0 impinges on the boundary between materials 1 and 2 at an angle of incidence of ϕ and generates a reflected beam of intensity I_{1R} at the exit angle ϕ and also a transmitted beam with intensity I_{1T}. The diagram shown implies that the refractive index (real) ratio $n_2/n_1 > 1$, since the transmitted beam in medium 2 refracts to a steeper angle to the surface normal than the incoming angle ϕ (Snell's law). Thus there is no special surface selectivity in simple reflectivity, and as a consequence a simple infrared reflection measurement on a bulk absorbing ($k > 0$) solid with a thin absorbing overlayer film can be useful for the analysis of the surface film only by first subtracting the signal that arises from the bulk. This problem can be significant, since in some cases the bulk signal could be orders of magnitude more intense than that of the overlayer and also could exhibit spectral features which overlap the surface spectrum.

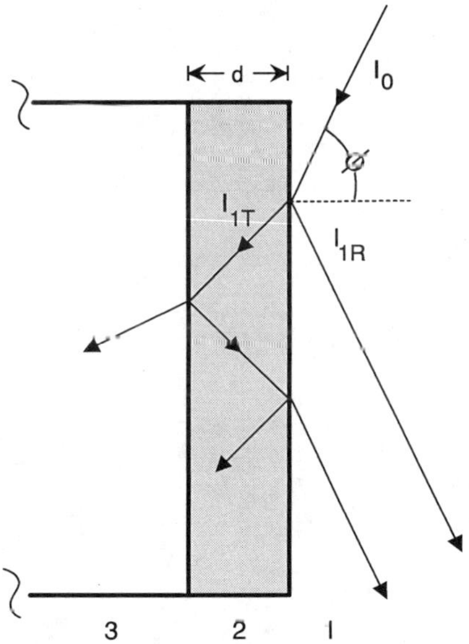

Figure 4.1 A schematic representation of the paths which a beam of light takes when impinging on a thin film sample. The incoming light of intensity I_0 is propagating in medium 1 and approaches the surface at an angle of incidence ϕ. A fraction of the beam of intensity I_{1R} is reflected at the outer boundary, and a fraction I_{1T} is transmitted into medium 2. Part of the beam is shown also propagating into medium 3. Multiple reflections occur in medium 2, which sets up interference between the beams.

However, there are classes of experiments in which the light can be guided along the boundary between two adjacent, planar media with the result that the bounded electromagnetic fields, termed evanescent waves, can be constrained to penetrate into the lower refractive index medium to a depth slightly less than the wavelength of the light.[12] This is shown schematically in Figure 4.2 where the light reflects off the boundary between media 1 and 2. In this case $n_2/n_1 > 1$ as in Figure 4.1 but the beam is propagating through medium 2 now. When the angle of incidence ϕ is greater than a critical value ϕ_c, the transmitted intensity $I_{1T} = 0$, and the electric field of the light decays exponentially away from the phase boundary, as shown schematically in the figure by the shading. In Figure 4.2, a plot of the mean-square value of the evanescent electric field versus distance also is shown. A general equation for this distance dependence is given in Equation 4.4. The term $\langle E_z^2 \rangle / \langle E_{z0}^2 \rangle$ is the ratio of the mean-square value of the electric field at a distance z away from the phase-boundary interface to the value at the interface, c is a collection of constants and d_p is the penetration depth, which is defined in Equation 4.5.

$$\langle E_z^2 \rangle / \langle E_{z0}^2 \rangle = e^{-d_p z c} \tag{4.4}$$

$$d_p = (\lambda / 2\pi n_1)[\sin^2\phi - (n_1/n_2)^2]^{-1/2} \tag{4.5}$$

In the latter equation, λ is the wavelength of the incoming light, n_1 and n_2 are the refractive indices (usually real values) of the adjoining media, and ϕ is the angle of

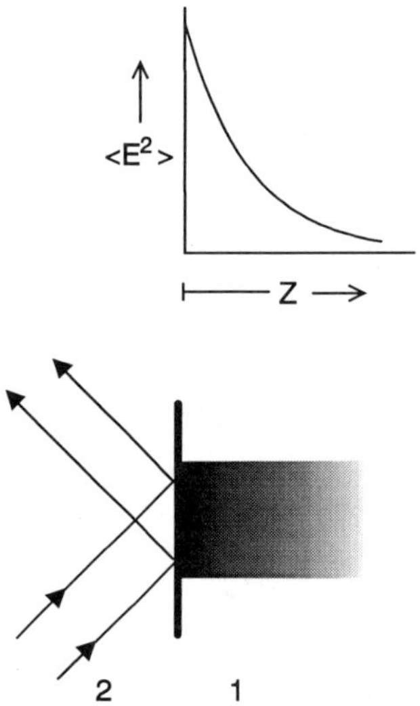

Figure 4.2 A schematic representation of the reflection of a beam of light at the boundary between two media for which the refractive index of medium 2 is largest. When the angle of incidence exceeds the critical angle, the beam is totally reflected and the oscillating electric field set up in medium 2 decays with distance away from the 1–2 boundary, as shown in the plot of mean-square field strength with distance (*top*).

incidence. For such configurations, one can arrange to have the lower index medium be the one of interest; thus some depth region into the medium can be probed selectively.[12]

Although, as the preceding discussion points out, there is no specific surface selectivity because of the tendency of light to be distributed throughout a sample, it is possible to take advantage of the extreme sensitivity of this distribution to the structure of the sample, and make quantitative determinations of the structural features by measurements of the reflectivity and/or transmissivity of the light beam with the sample.[13] The principle is simple, essentially depending upon two fundamental properties of waves propagating through parallel-layer media. The first property is the inherent tendency of the beam to partition some of its intensity between reflection and transmission at every interface or material phase boundary in the sample, as was schematically illustrated in Figure 4.1. In any optical experiment, the exact degree of partitioning, I_R/I_0 and I_T/I_0, depends upon the refractive index difference between the two adjoining phases, the wavelength of the light, the angle of incidence, and the polarization direction of the input beam. The second property is that of interference. This phenomenon is simply described by referring to Figure 4.1, which depicts the beam of light entering the thin film (phase 2) supported on an infinite substrate. The complete interaction of the beam with the sample is understood in terms of a standing wave permeating the entire structure. Within the film layer, it is possible to picture the interaction in terms of the beam bouncing within the phase boundaries to create constructive and destructive interference effects which regulate the intensity of the total output light beam. The exact outcome of these interference effects depends upon the refractive indices of the media and the thickness of the film. Thus if the experimental variables are fixed namely, the angle of incidence and wavelength—the total reflectivity can be related directly to the layer thickness and refractive indices. A similar picture holds for multiple layers. These types of measurements and relationships are the basis of several types of IRS experiments, and the general principles underlie the quantitative interpretation of any experiment in which light interacts with condensed media. Note that the basic equations used to interpret data apply to all wavelengths, from X rays to radio frequencies.

The surface sensitivity of an experiment, as defined in the preceding section, depends completely upon the signal/noise of the measurement with regard to the fraction of the total signal contributed by the surface region of interest. The signal/noise in an IRS experiment with good instrumentation is sufficient in most cases to detect the signal from a monolayer. In many cases, fractions of a monolayer can be detected easily. As expected, the problem often is simply that with no corresponding surface selectivity the surface signal contrast is low, that is, the bulk signal obscures the surface contribution.

One general advantage of IRS, compared to many popular ion- and electron-beam surface-analysis techniques, is the nondestructive nature of the analysis, which is due to the fact that photons possess much different momentum properties than matter and cannot remove material by billiard-ball type collisions, as occurs in sputtering by atoms or ions. Further, most of the energy absorbed from the infrared radiation is dissipated rapidly away from the local area in the form of heat, and the heating from typical beam intensities is insufficient to cause damage.

One specific disadvantage of IRS, however, is the fact that lateral imaging capabilities are limited by the diffraction properties of the light. Thus, a beam of infrared light, using normal far-field optics (the focusing optics located many wavelengths away from the sample), cannot be focused to a point much smaller that the wavelength of the light, typically of the order of several micrometers, without the appearance of undesirable diffraction beams at diverging angles. In comparison, the best imaging capabilities of an ion beam (liquid metal) for an analysis by secondary ion mass spectrometry are about 0.05 µm.[14]

As a final general point, the in situ capabilities of IRS can offer a distinct advantage for cases in which the analysis of a surface is desired under conditions of direct exposure to liquid or gaseous environments, for example, electrochemical, biological, or catalytic systems. However, many liquids and gases, such as water, have strong absorption bands in the infrared region, so in situ IRS is limited to solvents or gases with minimal absorption, to very limited frequency ranges between absorption features, or to very short path lengths which minimize absorption.

4.2 Quantitative Aspects

One of the problems in the quantitation of spectra from thin film samples is that for structures such as an overlayer organic film on a semiconductor wafer or a polymer sheet, reflection spectra can be highly complex, with both positive and negative absorption features associated with the overlayer spectral contribution. In the particular case of a planar parallel-layer sample, these spectra can be understood quantitatively in terms of film thickness[15] and structural characteristics, such as the orientational ordering of organic groups,[13] provided that all necessary optical properties of the constituent layers are known and that these properties can be connected to the molecular structure of the film(s). In this section, the basis of spectroscopic intensities will be discussed, followed by the explicit electromagnetic equations which control the observed reflectivities and transmissivities of the experimental spectrum of a thin film sample. These relationships have been used to construct practical approaches to the quantitative interpretation of thin film IRS spectra.[13]

Spectroscopic Intensities

The physical mechanism of IRS is well understood for the elementary case of a photon–electron dipole transition of an isolated molecule in free space.[8] In the dipole approximation, the power absorbed when an electromagnetic field oscillating at a given frequency interacts with an electric dipole is given by the proportionality

$$I \propto \left|\mathbf{E}_{\text{loc}} \cdot \mathbf{p}\right|^2 \tag{4.6}$$

where I is the power absorbed or the rate of the excitation, $\mathbf{E}_{loc}$ is the electric field vector in the local vicinity of the molecule, and **p**, in quantum mechanical terms, is the electric dipole moment transition matrix element for the given excitation at the given frequency. The unit vector $\mathbf{p}/|\mathbf{p}|$ defines the direction of the transition dipole and $|\mathbf{p}|$ is the magnitude of a single isolated oscillator. For a condensed-phase medium of parallel aligned oscillators, the value of $k = \text{Im}(\mathbf{n})$ defines the intensity of the transition; so it is convenient to use k rather than $|\mathbf{p}|$ as the basis for intensity, since k often can be derived experimentally from spectroscopic measurements on the pure material, whereas $|\mathbf{p}|$ is usually thought of as a property of an isolated species. Therefore, it is convenient to form a vector **k** which carries both the direction of **p** and the intensity of k. Methods for actually making such calculations will be described later. The relationship between **p** and **k** establishes the fundamental connections between the molecular structure of the film medium and the optical response properties. This information is essential for the correct interpretation of the thin film spectra in terms of molecular structure, particularly the orientation of molecular groups in organized organic films.

Electromagnetic Fields in Thin Film Structures

Definition of the fundamental electromagnetic problem The general reflection/transmission experiment of interest here is schematically shown in Figure 4.3. An incident beam of collimated light propagating in infinite medium 1 impinges onto the surface of a stack of semi-infinite, parallel, planar slabs. The angle between

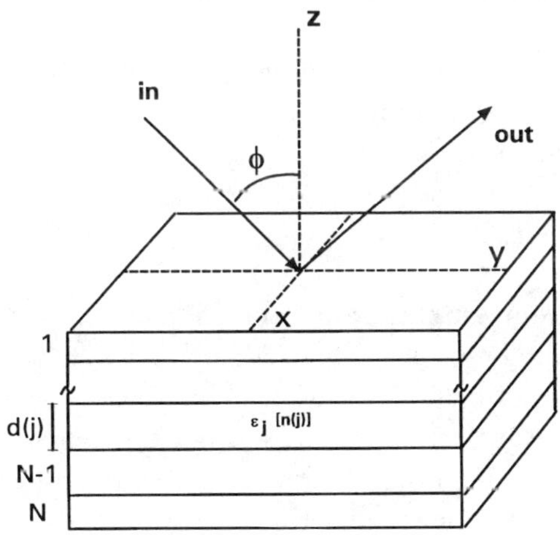

Figure 4.3 **A schematic representation of a beam of light reflecting (also transmitting into the sample, but not shown here) from a stack of $N-2$ parallel-layer thin films. An x–y–z coordinate system is defined in which the incidence plane is y–z and the surface plane is x–y. Each film layer is of thickness d_j and possesses an associated complex dielectric function and a complex refractive index, ε_j and n_j.**

the incoming beam and the surface normal defines the angle of incidence ϕ, which is contained in the plane of incidence. The coordinate system is chosen for convenience to have the slab surface as the x–y plane and the plane of incidence as the x–z plane. The state of polarization of the electric field of the incident beam is defined in terms of the direction of the electric field vector relative to the incidence plane. It is convenient to define two special linear polarization states, s and p, for which the electric field vector is aligned perpendicular or parallel, respectively, to the plane of incidence. The most general case of a polarization state is elliptical, in which the electric field vector can be decomposed into component s and p states with both an amplitude and a phase difference between them. When the phase difference is zero, the light is linearly polarized. As described in Figure 4.1, part of the beam is transmitted into the sample, and if the sample is sufficiently transparent to the radiation, some fraction of the initial beam exits into infinite phase N. The other portion of the incident beam is reflected from the surface. The simplest IRS experiment consists of measuring the reflectivity or transmissivity of a pure s- or p-polarized beam of light as a function of frequency. The data is interpreted in terms of formal relationships between the changes in these reflectivities and material properties by using the electromagnetic theory of the interaction of light with matter. These relationships, which are the basis for the quantitation of spectra, are presented below.

Before proceeding with the presentation of equations, it is necessary to briefly discuss representations. A real scalar quantity s is represented in italics (s) and with a tilde ($\tilde{s}$) when complex. A vector quantity v is expressed in bold ($\mathbf{v}$) and with an arrow ($\vec{\mathbf{v}}$) when complex. A complex tensor t is expressed as $\hat{\hat{t}}$, its complex elements as $\hat{t}_{jk}$, and real elements as t_{jk}.

For a linearly polarized infinite, homogeneous, plane wave of vacuum wavelength λ propagating through a nonabsorbing medium with associated refractive index n_1, the time-dependent electric field vector is

$$\vec{\mathbf{E}} = \mathbf{A}^{\circ}\exp[-i(\omega t - \mathbf{K}\cdot\mathbf{r} + \varphi)] = \vec{\mathbf{E}}^{\circ}\exp[-i(\omega t - \mathbf{K}\cdot\mathbf{r})] \tag{4.7}$$

where $\mathbf{A}^{\circ}$ is the vector defining the maximum amplitude and direction of the electric field, ω is the angular frequency of the light, $\mathbf{K}$ is the wave vector defined as having magnitude $|\mathbf{K}| = 2\pi n_1/\lambda$ and a direction along the wave propagation axis, $\mathbf{r}$ is the location vector of a point along the wave, and φ is a phase angle to relate the phase of the oscillating electric field to some reference. It is also convenient, in order to include the description of ellipsometry experiments, to define the complex vector $\vec{\mathbf{E}}^{\circ}$, which carries the phase factor $e^{-i\varphi}$. In a simple reflection experiment, s- or p-polarized light reflects off the surface (Figure 4.1) of an isotropic material with conservation of the polarization direction. The complex electric fields of the incoming and reflected outgoing linearly polarized waves at the surface are each represented by Equation 4.7, but with different amplitude and phase factors $|\vec{\mathbf{E}}^{\circ}|$ and φ. The complex reflectivity for this special case then is defined as the time-independent ratio of the incoming and outgoing electric field vectors at the specific point of reflection:

$$\frac{|\vec{\mathbf{E}}_{\mathbf{out}}|}{|\vec{\mathbf{E}}_{\mathbf{in}}|} = \frac{|\vec{\mathbf{E}}^{\circ}_{\mathbf{out}}|}{|\vec{\mathbf{E}}^{\circ}_{\mathbf{in}}|} = \frac{|\vec{\mathbf{E}}^{\circ}_{\mathrm{out}}|}{|\vec{\mathbf{E}}^{\circ}_{\mathrm{in}}|} e^{(\varphi_{\mathrm{out}} - \varphi_{\mathrm{in}})} = \tilde{r} \tag{4.8}$$

The complex scalar $\tilde{r}$, a convenient abbreviated notation for complex electric field ratios, is known as the Fresnel coefficient and is designated as $\tilde{r}_s$ or $\tilde{r}_p$, depending on the polarization state. The power reflectivity, the actual quantity measured in a reflection experiment, is defined in terms of the ratio of the absolute squares of the complex electric fields, which here can be either *s* or *p* polarized depending upon the polarization chosen for the experiment:

$$R = \frac{|\vec{\mathbf{E}}_{\mathrm{out}}|^2}{|\vec{\mathbf{E}}_{\mathrm{in}}|^2} = \frac{(\vec{\mathbf{E}}^*_{\mathrm{out}} \cdot \vec{\mathbf{E}}_{\mathrm{out}})}{(\vec{\mathbf{E}}^*_{\mathrm{in}} \cdot \vec{\mathbf{E}}_{\mathrm{in}})} = \tilde{r}^* \cdot \tilde{r} = \frac{I_{\mathrm{out}}}{I_{\mathrm{in}}} \tag{4.9}$$

where $\vec{\mathbf{E}}^*$ denotes a complex conjugate and the *I*'s represent the physically measurable intensities (power) of the electric field. Similarly, one can define related quantities to describe the transmissivity:

$$T = \frac{\tilde{n}_N}{\tilde{n}_1} \cdot \frac{|\vec{\mathbf{E}}_N|^2}{|\vec{\mathbf{E}}_{\mathrm{in}}|^2} = \frac{\tilde{n}_N}{\tilde{n}_1}(t^* \cdot t) = \frac{I_N}{I_1} \tag{4.10}$$

where the transmission Fresnel coefficients are defined as $t = |\mathbf{E}_N|/|\mathbf{E}_{\mathrm{in}}|$ and the subscript *N* indicates the last material phase (exit phase) in the sample (Figure 4.3).

Once the reflectivities or transmissivities have been measured, the next step is to relate these quantities to the material properties and the structure of the sample. This is done by constructing mathematical models that represent the interaction of the light with the sample for the specific experimental conditions, fully specified by the frequency of the light, the angle of incidence ϕ, and the sample alignment in the beam, and also specified by an azimuthal angle β (not shown in Figure 4.3) between the plane of incidence of the light and some important axis associated with the sample structure—such as a crystal axis—for biaxial symmetry samples. The standard models are based on the classical electromagnetic theory of light interacting with an ideal, stratified layer structure of isotropic, homogeneous media. In this approach, the sample is completely described, as shown schematically in Figure 4.3, by specifying for each layer *j* either the associated thickness d_j and a complex refractive index $\tilde{n}_j$, or alternatively, a complex dielectric function $\tilde{\varepsilon}_j$; either of these (which were defined in Equations 4.1–4.3) equivalently describes the intrinsic optical response.

In the general case of anisotropic media, such as an oriented Langmuir–Blodgett film or a biaxial polymer film, these quantities are second-rank tensors ($\hat{\tilde{\varepsilon}}$ or $\hat{\tilde{n}}$), and for isotropic media they reduce to scalars. The essential electromagnetic theory is based on the boundary value relationships of Maxwell's equations, which serve to relate the complex Fresnel coefficients (defined above) to the layer thicknesses and the frequency-dependent optical response functions. The material properties of each layer are accessed by additional relationships between the optical functions and the material structure and composition. A central problem which arises in the above strategy is that although the relevant electromagnetic theory is explicitly defined in detail, in practice, for the most general experimental configurations and sample types, the theory quickly leads to such cumbersome mathematical expressions that analytical solutions are often intractable. This problem is generally circumvented by replacing analytical solutions with numerical computation algorithms based directly on the analytically intractable core electromagnetic equations. This approach involves iterative comparisons of simulated data based on trial solutions, with the experimental data and convergence based on some best-fit criterion.

The most rigorous and comprehensive approaches to spectral computation appear to be the 4×4 transfer matrix methods in which the **E** and **B** (magnetic) field vectors at any depth location in the sample are determined relative to the values at the outer boundaries of the sample by a sequential application of matrix operators across each intervening interface and through each intervening phase. The two most useful approach for the present purposes appears to be that by Yeh.[11] This treatment begins by applying the continuity of the tangential components of the **E** and **B** fields across the interfaces. Further, the same functional form of the matrix transforms apply to the two fields. This is an advantage, since the tangential **B**-field component can be used to calculate the perpendicular **E**-field component at any point, and thereby the need to consider perpendicular field components is eliminated. This simplification directly reduces the electromagnetic variables by half and leads to the 4×4 matrix formalisms of the boundary value relationships of the field vectors. An important point is that this matrix method is sufficiently general that it describes any experiment regardless of the wavelength of the incident light.

Implementation of the 4 × 4 transfer matrix method The basic electromagnetic principles applying to these equations can be found elsewhere.[11] FORTRAN codes based on these principles have been developed[13] for use in the quantitative IRS analysis of oriented organic films. They are not presented here but can be obtained from the author upon request.

The solution of the electromagnetic boundary value equations leads to the following relationship between the electric fields in the final medium (N) and the first medium (1), which contains the light source:

$$\begin{pmatrix} A_s(1) \\ B_s(1) \\ A_p(1) \\ B_p(1) \end{pmatrix} = \tilde{\mathbf{M}} \begin{pmatrix} C_s(N) \\ D_s(N) \\ C_p(N) \\ D_p(N) \end{pmatrix} \tag{4.11}$$

In this equation, A, B, C, and D are the electric field intensities characterized by either s- or p-polarization directions. The A and C fields are associated with waves moving in the direction of phase 1 to phase N, and the B and D fields involve waves moving in the reverse direction. The detailed nature of the 4 × 4 matrix $\tilde{\mathbf{M}}$ is explicitly defined in terms of the experimental geometry, the sample structure, the sample optical functions, and the frequency of the light. Details of the calculations of the matrix elements in terms of these quantities are given elsewhere.[13]

The solution of the boundary value equations results only in setting relative values of the fields with respect to their location along the z axis in the sample (refer to Figure 4.3). These values are referenced to the fields in the Nth phase at the last boundary $[N/(N-1)]$, where it is required that the modal field amplitudes $D_s(N)$ and $D_p(N)$ vanish since there can be no back reflection from this infinite phase, and where the value of the outgoing fields $C_s(N)$ and $C_p(N)$ can be set to unity for convenience. The complex reflectivities then can be computed easily by taking the ratio of the reflected amplitudes to the incident amplitudes, respectively, for a desired input polarization of the beam. For instance, the fractional complex amplitude of an s-polarized input beam which appears as a reflected s-polarized output beam can be computed as follows:

$$\tilde{r}_s = \left[\frac{C_s(1)}{A_s(1)}\right]_{(A_p(1)=0)} = \frac{m_{21}m_{33} - m_{23}m_{31}}{m_{11}m_{33} - m_{13}m_{31}} \tag{4.12}$$

where m_{jk} represents the corresponding element of the matrix $\tilde{\mathbf{M}}$. Similarly, the p-polarization coefficient r_p can be calculated. The power reflectivities and transmissivities then can be calculated easily from Equations 4.9 and 4.10. The calculations are repeated for the entire range of frequencies of interest to provide a simulated spectrum.

Generation of optical function tensors This section considers the relationship between the optical function tensors of a medium and the properties of the oscillators which comprise the medium.

For a condensed-phase medium of parallel aligned oscillators, the value of $k = \mathrm{Im}(\mathbf{n})$ defines the intensity of the transition, and so it is convenient to use k rather than $|\mathbf{p}|$ as the basis for intensity since k often can be derived experimentally from spectroscopic measurements on the pure material, whereas $|\mathbf{p}|$ is usually thought of as a property of an isolated species. Therefore, for the purposes at hand it is convenient to form a vector $\mathbf{k}$ which carries both the direction of $\mathbf{p}$ and the intensity of k. In a real physical material, of course, it would need to be shown that the direction of $\mathbf{p}$ along some molecular (internal) coordinate system is not perturbed by matrix effects relative to the direction for an isolated oscillator. Since the intensity of a transition is usually derived from measurements on an isotropic sample, the following relationship is convenient:

$$|\mathbf{k}| = 3.0k_{\text{iso}} \tag{4.13}$$

where **k** represents the effective response of the material when all the oscillators are aligned in the direction of the exciting electric field, and k_{iso} is the value of **k** for a random orientation of the oscillators in an isotropic sample. To form a tensor, the vector quantity **k** is decomposed into components along the external coordinate axes. The imaginary parts of the diagonal elements of the optical function tensor of the medium composed of these oscillators is generated by the equations

$$\begin{aligned} k_{xx} &= |\mathbf{k}| \sin^2\alpha \cos^2\beta \\ k_{yy} &= |\mathbf{k}| \sin^2\alpha \sin^2\beta \\ k_{zz} &= |\mathbf{k}| \cos^2\alpha \end{aligned} \tag{4.14}$$

where α is the tilt angle of the transition dipole from the surface normal (z) and β is the azimuthal angle in the surface plane (x–y), as shown in Figure 4.4 in terms of a standard Eulerian coordinate system. If more than one type of independent oscillator were present in the medium, the total value of k_{ee} at a given frequency, where $e = x$, y, or z, would be given by the sum of k^j_{ee} terms over all oscillators, where j is the index for the jth type of oscillator.

Each complex diagonal element $\hat{n}_{ee}$ of the tensor $\hat{\hat{n}}$ is given by $\text{Re}(\hat{n}_{ee})$, $\text{Im}(\hat{n}_{ee})$ or n_{ee}, and k_{ee}. A useful approximation is that $\text{Re}(\hat{n})$ consists of an oscillator contribution, $n_{\text{osc}}(\nu)$, which is direction and frequency dependent, and an isotropic background response, $n_\infty(\nu)$, which varies slowly or is constant with frequency. The latter quantity is essentially the constant specified in the Kramers–Kronig transform,[13] and the former is the quantity given by the integral term in the transform. Once the diagonalized tensor elements are known, the nondiagonal tensor for any

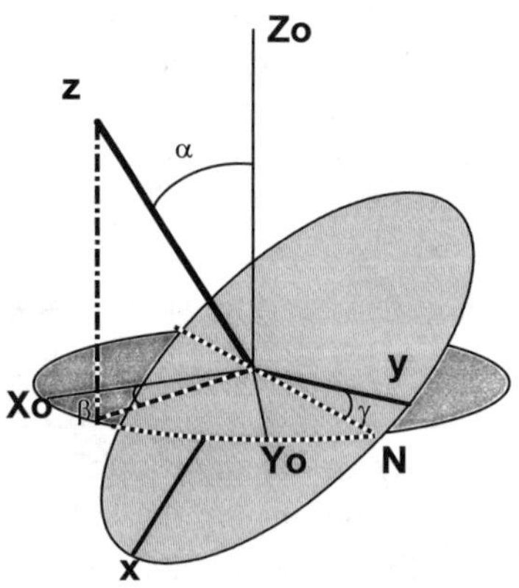

Figure 4.4 **A diagram of a Eulerian coordinate system which serves to describe the coordinates of transition dipole moments associated with an oriented molecule on a surface for different orientations with respect to the surface coordinates X_0–Y_0–Z_0. As the molecule reorients from its original alignment on the surface, its coordinate system X–Y–Z changes from the surface coordinates by a tilt angle α, an azimuthal rotation angle β in the X_0–Y_0 plane, and a twist angle γ around the Z axis.**

other orientation of the oscillator, as given by **k**, can be obtained by methods of simple matrix rotation

$$\hat{\hat{\varepsilon}}_{3\times 3} = \mathcal{R}^{-1}\hat{\hat{\varepsilon}}\mathcal{R} \tag{4.15}$$

where $\mathcal{R}$ is the rotation matrix

$$\mathcal{R} = \begin{pmatrix} \cos\alpha\cos\beta\cos\gamma - \sin\beta\sin\gamma & \cos\alpha\sin\beta\cos\gamma + \cos\beta\sin\gamma & -\sin\alpha\cos\gamma \\ \cos\alpha\cos\beta\sin\gamma - \sin\beta\cos\gamma & -\cos\alpha\sin\beta\sin\gamma + \cos\beta\cos\gamma & \sin\alpha\sin\gamma \\ \sin\alpha\cos\beta & \sin\alpha\sin\beta & \cos\alpha \end{pmatrix} \tag{4.16}$$

for which α, β, and γ are tilt, azimuth, and twist, respectively, in terms of the Eulerian coordinates as defined in Figure 4.4.

4.3 The Infrared Spectroscopy Experiment

General Instrumentation

To carry out the simplest IRS experiment one requires several key components. For an absorption experiment, these include a source of IR radiation of variable frequencies, a radiation detector, and a mechanism for separating the response of the sample for each frequency interval of light (monochromatic response). To set up this experiment there are a large number of choices for these components; for example, a synchrotron can provide an excellent source of broad-band, collimated far-IR radiation. By far, the most common instrument for the majority of surface/thin film analyses which involve mid-IR frequencies is a Fourier transform infrared (FTIR) spectrometer.[16] Typically the instrument configuration obtained commercially consists of a blackbody-type emission source, an interferometer, and a semiconductor photon detector. Usually the specific items are a ceramic glo-bar (peaked at ~10 mm energy), a Michelson interferometer, and a HgCdTe, liquid nitrogen–cooled detector. However, one should note that other specific combinations of components in some limited number of cases may solve critical analysis problems better than typical commercially available instruments.

Experimental Modes

There are a variety of experimental methods for which infrared radiation can be directed to and/or away from a sample in order to perform a spectroscopic measurement. For an absorption experiment, these methods can be classified into three standard types or modes: (1) external reflection, (2) internal reflection, and (3) transmission. Each of these modes can be used to characterize surfaces and thin films, and each displays strengths optimized for different types of sample structures. The strategies for obtaining spectra from the above thin films vary with the exact nature of the sample. Consider the case of a thin film supported on a smooth,

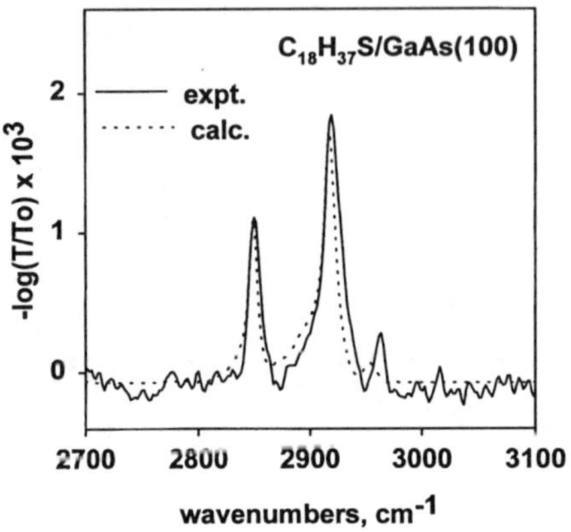

Figure 4.5 **A normal incidence transmission infrared spectrum of a self-assembled monolayer film of octadecanethiolate on a GaAs(100) substrate. The film was grown on both sides of the substrate. The dashed line indicates the simulated spectrum for an average all-trans chain tilted at an angle of ~57° from the normal to the GaAs surface.**

planar substrate. If the substrate were a highly infrared-absorbing (opaque) material, such as the reflective metals Au and Al, external reflection would be the only *simple* method appropriate. On the other hand, when the substrate is transparent to infrared radiation, such as with Si, then transmission or external reflection can be used, and when the substrate is a high refractive index material with a prism-like shape, internal reflection also can be used. It is also possible to obtain emission spectra. Emission is discussed below, but in general it provides only very weak spectra for room-temperature samples and is only used for thick film spectra or for heated samples. For a given combination of mode and sample type, the ability to quantitate structural features depends upon the ability to apply electromagnetic theory accurately, as outlined in the preceding section. A general point to be made with regard to instrumentation is that good FTIR systems operating in the mid-IR under medium-resolution conditions typically can provide ≥0.1% precision in intensity per scan. The signal-to-noise ratio can be increased by up to nearly two orders of magnitude by signal averaging. This level is more than sufficient to detect monolayer quantities of a film for all the modes mentioned above except emission.

Transmission This experiment is by far the most well known of all modes. It has a long history of application to the analysis of gases, liquids, and solid powders. Since detection depends upon transmitted light, the sample must be sufficiently IR transparent. For thin film analysis, the sample typically consists of either a very thin free-standing film (e.g., a 2 μm polymer film) or a very thin film supported on a transparent substrate (e.g., a monolayer on a Si wafer). In the past, transmission was used very little because the spectra are quite weak for monolayer coverage films of typical organic materials. However, recent advances in FTIR instrumentation now

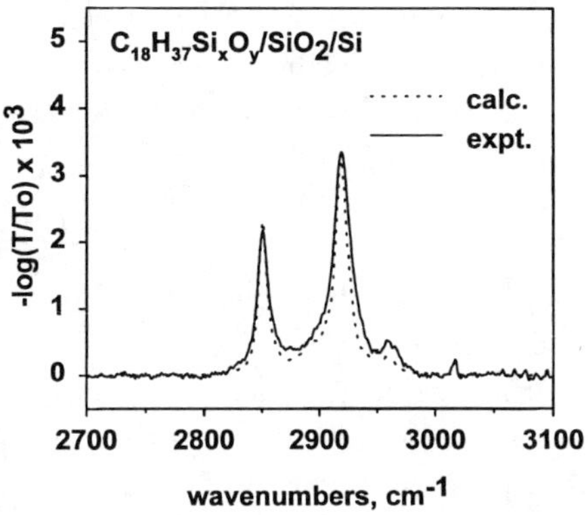

Figure 4.6 **A normal incidence transmission infrared spectrum of a self-assembled monolayer octadecylsiloxane film on an oxidized Si surface. The film was grown on both sides of the silicon substrate. The dashed line indicates the simulated spectrum for an average all-trans chain tilted at an angle of 11° from the normal to the silicon surface.**

have made such analyses feasible[17, 18] because of improved signal-to-noise ratios. The detection and quantitative analysis of moderately absorbing monolayers in the latter sample configuration can be considered routine for good quality commercial instrumentation in the case of very smooth, planar substrates. Examples of transmission spectra of monolayers on highly smooth transparent substrates (both sides covered with the monolayer) are given in Figures 4.5 and 4.6 for samples of self assembled monolayers of $C_{18}H_{37}S$– on GaAs[17] and $C_{18}H_{37}SiO_x$ on oxidized Si.[19] The spectra appear quite similar in spite of the very different substrates and show that transmission spectra are reasonably uncomplicated with respect to changes in materials. However, the intensities of the spectra differ, which is due to different coverages and alkyl chain orientations. The quantitative aspects of these spectra are discussed in the next section.

Rough surfaces scatter light and reduce the signal-to-noise ratio as well as the ability to do quantitative analysis. Curved or nonplanar surfaces distort the beam and also reduce quantitative capabilities. In all absorption spectroscopy experiments, the analyses critically depend upon reliable reference spectra, for example, a "clean" silicon wafer as a reference for a monolayer-covered wafer. Any slight difference (parts in 10^4 or so) in optical characteristics between sample and reference substrates can introduce damaging spectral artifacts, so sample/reference strategies are extremely important in transmission analyses. The same can also be said, in general, for all absorption modes discussed below. The removal of these artifacts is a pervasive issue, but corrective strategies are too case specific to be discussed in detail here.

If good quality spectra are obtained, then quantitative analyses can be carried out using either Beer's law expressions as approximations or classical electromagnetic theory, as outlined in a previous section, for more exactness. The Beer's law approximation is based on a model in which a light beam passing through a slab containing a low density of noninteracting molecules is attenuated by absorption according to the relationship

$$I/I_0 = e^{-\varepsilon Cl} \tag{4.17}$$

where I and I_0 are the power transmitted with and without the molecules present, respectively, ε is the extinction coefficient (or absorption cross section), C is the volume concentration of the molecules, and l is the path length of the light beam that contains the molecules. The quantities ε and k, the imaginary part of the complex refractive index, are both measures of the absorption properties of the molecules of interest, but the former generally is used for dilute mixtures while the latter is used for pure, condensed-phase media. For low refractive index substrates (e.g., KBr or NaCl) Beer's law approximations are reasonably accurate, whereas for high index materials such as Si, errors of >30% can result for the Beer's Law approach. Significant changes in peak positions and line widths also occur as a function of the substrate refractive index, and a classical electromagnetic theoretical approach is necessary to treat these optical effects properly to distinguish them from chemical effects.

There are two final points to make about transmission experiments. First, depth profiling is not possible, since any sample appearing in the IR beam can absorb light, which allows no way to differentiate the depth location of an absorbing group in the beam. Second, lateral imaging is possible. High quality IR spectroscopic microscopes are commercially available that provide spatial resolution to near the diffraction limit (~10 μm).[20] Imaging is best done when the angle of incidence is near normal to the surface and when the dependence of the spectra on angle of incidence is not strong, a factor which arises since high focusing requires highly conical beam shapes. This is suitable for transmission, which uses normal incidence and for which the angle dependence of the spectra is not strong.

External reflection From a practical point of view, this experiment is much more common than the transmission experiment because most overlayer films of interest for surface analysis are supported on absorbing substrates, for which the transmission experiment is not possible. External reflection from metals is probably the most common example of IRS thin film analysis[21] and has been utilized for a variety of studies including, for example, self-assembled monolayers,[22–25] Langmuir–Blodgett films[26–28] and thin polymer films.[29, 30] These studies attest to the the fact that for cases of reflective, planar metal substrates, good quality IR spectra of organic films of monolayer thicknesses can be generated by reflection off the sample surface. Classical electromagnetic theory reveals that the experiments with metal substrates are done optimally at grazing angles of incidence (≤10° off grazing) with p-polarized light. Under such conditions sufficient signal-to-noise can be obtained to allow detection of submonolayer quantities of material, and most importantly, the spectra can be quantitatively analyzed using the theoretical methods described earlier. An example of an external reflection spectrum from a metal substrate

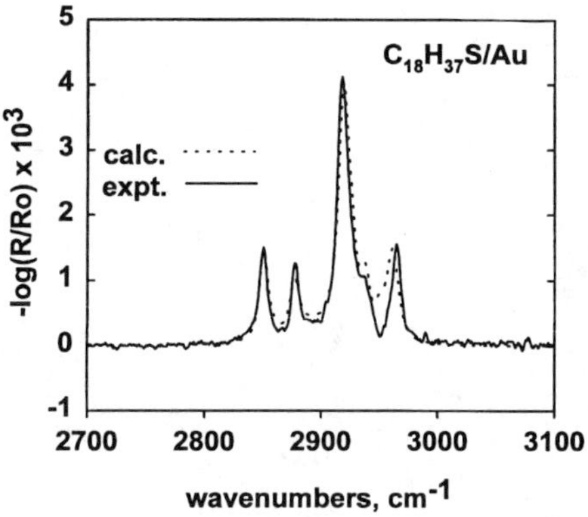

Figure 4.7 **An 86°-angle-of-incidence, *p*-polarized external reflection infrared spectrum of a self-assembled monolayer of octadecanethiolate on a gold substrate. The dashed line indicates the simulated spectrum for a model of two nearly all-trans chains in a quasi-orthorhombic unit subcell, with an average chain tilt of 26° from the surface normal and a chain twist of 55°. In this model structure, the terminal CH_2 group has been allowed to assume a fractional population of gauche conformers.**

is given in Figure 4.7, where the C–H stretching modes of a self-assembled monolayer of an alkanethiolate on a gold surface are shown.[13] This spectrum was taken with *p*-polarized light at an angle of incidence of 86°. Less frequently encountered are cases in which films are supported on IR-absorbing but nonmetallic or weakly metallic substrates, such as silicate glasses and highly doped Si semiconductor wafers. Quantitative monolayer sensitivity analyses are also possible on these substrates using reflection techniques together with commercial instrumentation and the correct application of theory both to optimize the experimental conditions and to interpret the data. In general, the signal-to-noise ratio can be as much as an order of magnitude less than for spectra of equivalent films on highly reflective, highly metallic substrates due to the intrinsic nature of the electromagnetic fields. Again, as for transmission, the planar character of the substrate allows quantitation of molecular and bond orientation relative to the surface plane. These analyses are viable for any external reflection experiment for which the optical properties of the sample are well defined. Examples of external reflection spectra from nonmetallic substrates are shown in Figures 4.8 and 4.9 for samples of Langmuir–Blodgett monolayers of cadmium arachidate.[31] Figure 4.8 is the *p*-polarized spectrum taken at 55° incidence of a five-layer film sample on a glassy carbon surface. Figure 4.9 is the *s*-polarized spectrum of a one-monolayer sample on a poly(methyl methacrylate) substrate taken at 55° incidence. Note the simultaneous presence of positive- and negative-going peaks, an indication of the complexity which can arise in reflection spectra from nonmetals. External reflection IRS spectra have been reported recently for self-assembled monolayers on materials such as semiconductors.[17, 32]

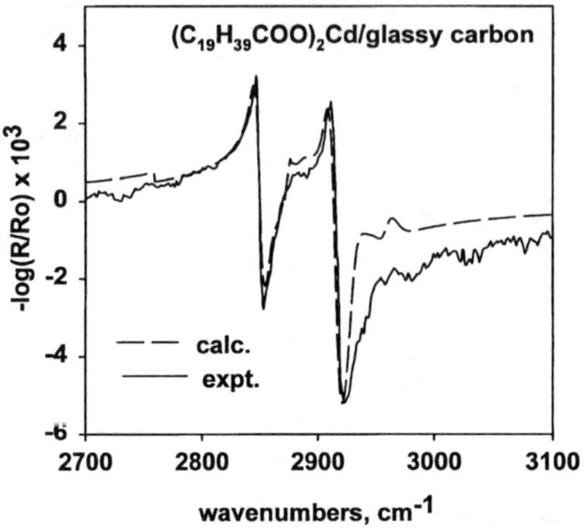

Figure 4.8 **A 55°-angle-of-incidence, *p*-polarized infrared spectrum of a five-monolayer Langmuir–Blodgett film of cadmium arachidate on a glassy carbon substrate. The dashed line indicates the simulated spectrum for a model of an all-trans chain with an average tilt of 15° from the surface normal and a twist of 45° around the chain axis.**

Lateral imaging is also possible on planar substrates using IR microscopy, but for the case of metallic surfaces the requirement of grazing angles of incidence precludes very high spatial resolution. Further, the angle dependence can be quite strong, which leads to quite serious errors in the quantitation of spectra unless the precise spreads of angles and accompanying intensities are known. For nonmetallic

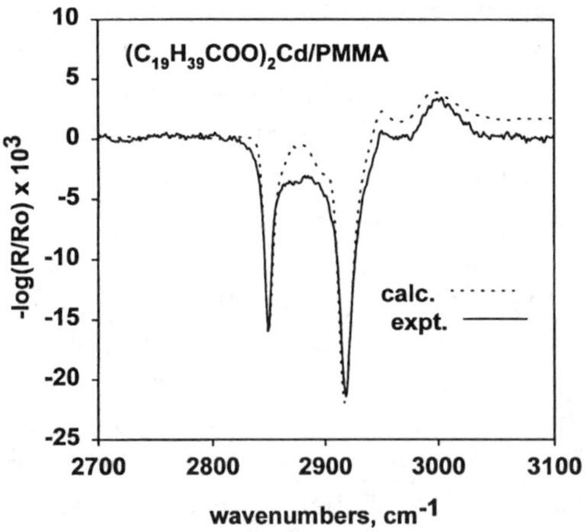

Figure 4.9 **A 55°-angle-of-incidence, *s*-polarized infrared spectrum of a one-monolayer Langmuir–Blodgett film of cadmium arachidate on a poly(methyl methacrylate) substrate. The dashed line indicates the simulated spectrum for a model of an all-trans chain with an average tilt of 8° from the surface normal and a twist of 45° around the chain axis.**

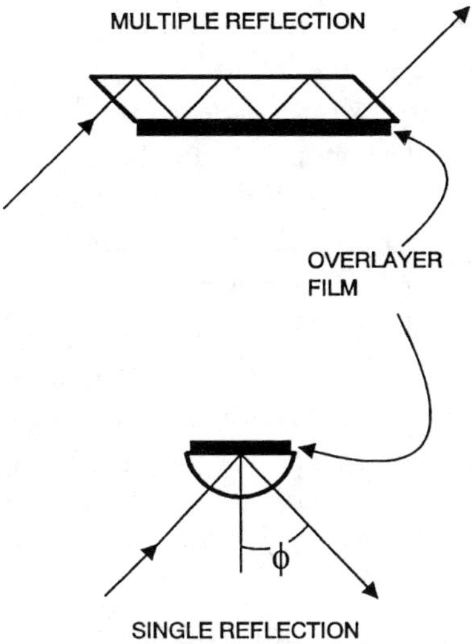

Figure 4.10 **Schematic drawings of internal reflection experiments for single and multiple reflection configurations. An overlayer film is shown on the two reflection elements to indicate the use of the surface evanescent electric field to probe the near surface region of the film.**

or weakly metallic substrates, angles of incidence in the range of normal to ~60° are possible, depending upon polarization, and reasonably good resolution can be expected, again limited by diffraction effects as for the transmission experiment. Commercial microscope instruments with reflection optics are available for such measurements.[20] Finally, the external reflection technique does not allow depth profiling, since the electromagnetic fields generated at the surface extend as far away from the surface as the width of the incoming beam (usually several millimeters) and are too slowly varying as a function of distance—typically, micrometer-scale modulations—to allow interpretation of the spectra in terms of vertical mapping of signal contributions at any useful distance resolution, that is, nanometers. Some recent examples of surface analyses by external reflection IRS have been reviewed elsewhere.[21]

Internal reflection In principle, the size of the signal from a single-reflection internal reflection experiment should be about the same as that of a transmission experiment. Thus the same signal-to-noise problems should exist with internal reflection as with transmission, as mentioned above. However, this problem was overcome a number of years ago[12] by using a long, thin element or prism with parallel sides capable of allowing the beam, when passing through the prism in the long direction, to bounce off the internal surfaces a number of times before exiting from the prism, as shown in Figure 4.10. These elements are fabricated by cutting an infrared-transparent material into a slab (trapezoid or parallelogram) or cylindrical rod (conical ends) which is then polished to give very smooth surfaces. Typical numbers of these multiple internal reflections are 10–50, and the transmission efficiency can be made to be acceptably efficient by adjusting the input angle to the prism such that the reflections occur at values above the critical angle, thus guaranteeing total internal reflection and no transmission leakage out the sides. Since each reflection provides an additional contribution to the total signal, the spectral intensities can be increased over a single reflection by at least an order of magnitude.

Internal reflection prisms work best when the refractive index is high, and optimal materials include Si, Ge, KRS5, and ZnSe. One obvious problem with internal reflection is that many substrates of interest, with respect to their relevance in bonding or adsorbing organic films, are not suitable for utilization as internal reflection elements. However, a major advantage of the internal reflection experiment is that under conditions of total internal reflection, the only light which appears outside the sides of the prism is in the form of a decaying evanescent wave. This advantage is useful in three ways.

First, if a sample of polymer is pressed into firm contact with a side face of the prism, the evanescent wave will penetrate a micrometer or so, according to Equation 4.4, into the surface region, and infrared spectra can be obtained. However, one should note that when the contact of the film with the element surface is incomplete, errors in quantitative analyses arise accordingly. Although this method certainly is not a surface analysis in the typical sense of a few layers of material, it does offer a way of limiting the depth region of the surface studied, a decided advantage over a transmission or external reflection experiment, which both sample the entire bulk of the material. This application is used extensively to characterize the deep surface regions of polymers modified by various treatments.[33] Further, since the direction of the electric field is precisely defined at the surface of the prism when the input beam is polarized, the spectra can be used—when the directions of the vibrational motions of the atoms and their associated dipoles are known—to determine the orientation of molecular groups in the overlayer surface region. This approach has been used to analyze for orientation effects in polymer surfaces.[34–36]

A second advantage is that in situ adsorption studies can be carried out in liquid (or gas) media much easier than with external reflection or transmission. In these experiments, one face of the prism can be exposed to the medium, and the spectral influence will be limited to the effective penetration depth, thus minimizing spectral contributions from the medium. This approach has been increasingly used for polymer adsorption studies.[37, 38] External reflection IRS has also been used under special circumstances for the analysis of electrode surfaces in an electrochemical cell.[39, 40]

The third advantage of the evanescent wave in internal reflection is that depth profiling of attached polymer films is possible by carefully utilizing the functional form of the decaying electric field in the film overlayer. In Equation 4.4, the decay function can be varied by changing element materials and angle of incidence for appropriate types of prisms, which allows the generation of a set of independent spectra each dependent upon concentration gradients in the film interior in a measurably different way.[41] Solution of the equations allows, in good cases, depth resolutions of better than tens of nanometers. In general, however, such applications are rarely feasible for practical problems. The most common practical problem that arises in depth profiling a polymer surface region is the lack of completely reproducible, uniform element–polymer contact from element to element. This problem precludes quantitative profiling analyses, although qualitative analyses should be successful. Many examples exist for the analyses of adsorbed films, and these include Langmuir–Blodgett monolayers,[42, 43] self-assembled monolayers,[44, 45] and thin polymer films.[38, 46, 47]

Emission Emission spectroscopy is a potentially useful, underutilized mode for obtaining infrared spectra of surface films considerably thicker than monolayer coverage.[5, 16, 48, 49] All the preceding modes have referred to experiments in which spectra were obtained by measuring the fraction of light beam power absorbed by various infrared active groups in the irradiated sample. However, every sample always emits a very weak but continuous infrared spectrum from the decay of thermally populated excited vibrational quantum states, and this spectrum contains contributions from the same individual groups as seen in absorption. Thus it is possible, in principle, to obtain a spectrum from a film-covered sample just by admitting the radiation into an infrared spectrometer. However, there are two reasons why this does not work well. First, the radiation is very weak at room temperature and is emitted away in all directions. Second, most of the flux is useless background (blackbody-like) radiation from the substrate and the film, with only a small fractional contribution from the infrared vibrations of interest. Related to the latter point is that all the parts of the spectrometer itself (mirrors, apertures, etc.) emit interfering background radiation. However, strategies have been developed to overcome these problems, primarily by using samples at substantially higher temperatures than the spectrometer and making very accurate corrections for sample background radiation by using film-free reference samples. Commercial instrumentation is available for these experiments for use with heated samples.

Monolayer detection is not possible by such strategies because of the large background corrections. However, it is possible through the use of cryogenically cooled spectrometers to detect monolayers on room-temperature samples,[5, 50, 51] but such instrumentation is not available commercially. For cases in which samples can be heated above room temperature, a major advantage of emission spectroscopy, compared to reflection, is that the sample can be of any geometrical shape and surface roughness. This makes the method extremely attractive for certain applications, such as examining surface films on oddly shaped machine parts that, in fact, may be normally hot. Of course, such sample characteristics do not allow quantitative analyses. In general, without extreme effort emission spectroscopy cannot be performed in a quantitative manner relative to the cases of transmission and reflection.

Spectroscopic ellipsometry Finally, we discuss the method of infrared spectroscopic ellipsometry. This mode is really a special case of the external reflection experiment in which changes in the polarization state of the infrared beam are measured rather than the fraction of beam power reflected, the traditional reflectivity measurement. The major advantage of the ellipsometry measurement is that the polarization state change contains information about both the amplitude change and the phase change of the light, whereas reflectivity measurements only contain intensity information. The appearance of the additional variable of the phase shift allows two unknowns of the sample to be evaluated, for example, sample film thickness and density, whereas reflectivity allows only one—just sample film thickness with the density evaluated independently.

The calculations for ellipsometry are based on standard classical electromagnetic theory. Ellipsometric spectra have the added advantage that the large amount of data accumulated from the total numbers of spectral points can be used to accurately determine wavelength-independent sample characteristics such as thickness, density, surface and interfacial roughness, voids, and even density gradients perpendicular to the surface plane (depth profiling). This information is in addition to the normal spectral information involving chemical structure, bond orientation and other features which affect vibrational modes and the absorption cross sections. By using the depth profiling capability, composition gradients could be measured, in principle.

The best demonstrations of the power of ellipsometry as an incisive analysis tool for thin film structures have been made for spectra obtained in the ultraviolet and visible regions. Here, for example, thicknesses ideally can be measured to within 0.1 Å or better. Recently, interest in extensions into the infrared region has accelerated,[52–55] and commercial instrumentation is now available. However, although it is possible to obtain the spectra of monolayer quantity films, because of signal-to-noise restrictions it is not yet clear whether these films can be analyzed quantitatively with present ellipsometry instrumentation. Studies also have been reported in which limited polarization measurements have been made on polymer and Langmuir–Blodgett films.[7, 56] For various reasons related to the basic physics of the experiment, it would appear that IRS-ellipsometry is most ideal for thin films on nonmetallic or weakly metallic substrates as opposed to highly conductive metals. In addition, since the quantitative value of the experiment derives from extremely high precision measurements, only highly smooth, planar surfaces lead to high precision analyses. However, even for rough surfaces, ellipsometric measurements are of more value than reflectivity measurements simply because of the extra information.

Finally, because of the longer wavelength of the light, the ability of IRS ellipsometry to measure thickness and perform depth profiling is significantly less than in the ultraviolet–visible regions by at least an order of magnitude. In general, considering all the above points, it is always an advantage to use ellipsometric spectra rather than reflectivity spectra if possible, regardless of sample roughness or substrate material type.

Additional Aspects

An important aspect of IRS thin film analysis is that of analysis time. Characterization of film processing, such as plasma modification, or specific application uses, such as gas adsorption, may require analysis in real times which could vary in principle from microseconds to hours. The long times present no problems, but the short times do. While it is possible to operate well below the microsecond time scale by using fast laser-based systems, it is generally impossible to obtain full spectra on these time scales, since wavelength-tuning of the lasers is comparatively slow. Measurements can also be done on the microsecond time scale using time-resolved FTIR, for example, through the use of step-scan instrumentation[57] now commercially available. However, in order to utilize this mode of time resolution, it is necessary that the experiment be capable of being rerun repeatedly in exactly the same manner with the same results. This is fine for repeated pulsed experiments in the laboratory but not for one-shot processing.

A very practical time scale, however, of about one second is obtained with the use of fast, commercial Fourier transform systems. Complete spectra can be obtained in ~10 ms with limited resolution (~16 cm^{-1}) and with signal-to-noise sufficient for thin film detection (~20 nm) but not for monolayers. However, with slower scan speeds and multiple-signal averaging, good resolution spectra (~2 cm^{-1}) of strongly absorbing monolayers in appropriate sample structures (e.g., a reflective, planar sheet metal) can be obtained on the minute time scale. This time scale is quite reasonable for monitoring typical surface processes such as photocuring, sputtering, and plasma deposition. There are ample opportunities for developments of IRS strategies and techniques in this area. The development of fast Fourier transform emission spectroscopy for heated samples would be of interest.

4.4 Examples of Applications

Self-Assembled Monolayers on Gold by External Reflection

One very useful application of quantitative IRS has been the determination of the structure of alkyl chains in highly organized self-assembled monolayers. One example of this can be given by a recent study of alkanethiolates on gold surfaces[23] in which the theoretical methods[13] presented in the earlier section were applied to the analysis of spectra in order to determine the average conformational structure and orientation of the chains. Figure 4.7 shows the experimental external reflection spectrum of a monolayer of $C_{18}H_{37}S/Au$ taken at an angle of incidence of 86° with *p*-polarized light. This is a spectrum of the C–H stretching mode region, and the main peaks were assigned to the symmetric and antisymmetric CH_2 group modes (2850 and 2918 cm^{-1}, respectively) and the symmetric and asymmetric CH_3 group modes (2890 and ~2960 cm^{-1}, respectively).

In order to determine the structure of the chains from the spectrum, according to the methods presented earlier, it was first necessary to obtain the optical function spectrum $\hat{n}(\nu)$ of the alkanethiolate compound. Since pure samples of bulk gold alkanethiolate cannot be obtained, it is necessary to chose a model compound with a structure as close to identical as possible. This was most easily done by using the pure alkanethiol and di-alkyl disulfide compounds as models for a pure alkanethiolate phase and obtaining spectra of the isotropic, polycrystalline solids in the form of quantitative mixtures in KBr pressed into pellet form for transmission spectra. These choices are quite reasonable since the spectra indicate that the packing of the alkyl chains of the pure compounds and the monolayer appear quite similar. The spectra of the former were converted to the isotropic optical function spectra straightforwardly by an iterative procedure which utilized the simulated transmission spectrum of the KBr mixture sample.

It was necessary to assign each feature in both the monolayer and the $\hat{n}(\nu)$ spectra to a specific vibrational mode, curve-resolve the spectra into the peaks from each individual mode, and then assign to each mode a direction for the transition dipole moment in terms of the molecular coordinates. With this information, the scalar optical function can be converted to a tensor for any selected orientation of the chains on the surface of the gold.

A variety of tensors for trial orientations were created. Simulated spectra were generated from these tensors and from the value of the measured thickness of the monolayer, the experimental spectral conditions (polarization and angle of incidence), and the known optical functions of gold. Then the simulated spectra were compared with experiment until the best fit was obtained. Since independent evidence indicated that there were two chains per unit crystal cell in the monolayer, the simulations were made for two independent orientations, and the final simulated spectra was then just the sum of these two.

With regard to the final best fit, it was noted that some loosening of the criteria of all-trans alkyl chains was needed in the form of allowing some gauche defect content in the terminal CH_2 adjacent to the CH_3 group. The dashed line in Figure 4.7 shows the simulated spectrum of the alkanethiolate monolayer in which the simulation was based on a model structure of pairs of all-trans chains, with one chain CCC plane twisted roughly 90° to the other, both chains tilted at 26° from the surface normal, and a small fraction of end gauche defects in a large ensemble of these pairs.[13, 23] Helium scattering data[58] confirm the necessity of the room-temperature disordering of the terminal CH_2 group, and recent X-ray diffraction data[59] set the value of the tilt angle in the range of 28 to 30°. These independent data verify the quantitative nature of the IRS analysis.

Octadecylsiloxane Monolayers on SiO_2 by Transmission

An example of the quantitative analysis of transmission spectra is given by recent studies of monolayer films of octadecylsiloxane ($C_{18}H_{37}SiO_x(OH)_y$ or ODS) on SiO_2 surfaces.[19] Immersion of the clean, oxidized Si substrate in a solution of octadecyltrichlorosilane forms quite dense, hydrophobic structures. The transmission spectrum at normal incidence is given in Figure 4.6. The low value of the CH_2 antisymmetric stretching mode peak at 2917 cm^{-1} indicates that the chains are in highly trans conformations. The monolayer thickness is about 24 Å, according to ellipsometric data, and represents about 90% coverage of what is theoretically possible, with the limit set by closest chain packing in a polyethylene crystal. An analysis of the spectrum can be obtained using optical functions generated from a bulk sample of polymerized starting material that appears to form a well-organized ODS polymer with the expected elemental composition. After curve resolution of the optical function into the component modes and assignments of the transition moment directions, optical function tensors for different orientations of an ODS chain on the SiO_2 surface were generated. The best-fit spectrum, shown as the dashed line in Figure 4.6, was obtained for an all-trans chain tilted at 11° from the surface normal. However, such a tilt is not consistent with the ~90% coverage since this surface density could accommodate a larger tilt (>20°) in order to pack the chains at high density. Since the spectral analysis appears quite reliable, one can conclude that the ODS molecules are packed in islands of near-vertical tilted chains with void regions in between, perhaps of a spacing of only one chain diameter. Further details of this analysis appear elsewhere.[19, 60] This example demonstrates the structural detail which can be achieved by the careful use of IRS thin film spectra.

Langmuir–Blodgett Films on Nonmetallic Substrates by External Reflection

Examples of external reflection spectra of thin films supported on nonmetallic substrates were given in Figures 4.8 and 4.9, which show the C–H stretching mode region of Langmuir–Blodgett films of cadmium eicosonate $[Cd(C_{17}H_{35}CO_2)_2]$ supported on carbon and poly(methyl methacrylate) substrates.[31] Using the procedure described above, isotropic optical functions were obtained for polycrystalline eicosanoic acid, and a series of simulated spectra were generated for different orientations of the eicosonate chains on the carbon surface. Figures 4.8 and 4.9 show the experimental and best-fit simulated spectra. The simulated spectra are based on model structures of single, average all-trans chains tilted at 15° and 8°, respectively, from the surface normal with a twist of the CCC chain axis of 45° and no preferential azimuthal orientation (uniaxial symmetry; see Figure 4.5). Thus, the different substrates play only a minor role in determining the chain orientations and configurations. By glancing at the spectra in Figures 4.8 and 4.9, this conclusion would have not been obvious, thus attesting to the necessity for quantitative interpretation of such spectra in order to obtain accurate molecular structure in organized organic thin films.

Acknowledgment

The author acknowledges the valuable assistance of Dr. A. N. Parikh during the preparation of this manuscript.

References

1 L. J. Bellamy. *The Infrared Spectra of Complex Molecules.* Wiley, New York, 1975.

2 D. Lin-Vien, N. E. Colthup, W. G. Fateley, and J. G. Graselli. *The Handbook of Infrared and Raman Characteristic Frequencies of Organic Molecules.* Academic Press, New York, 1992.

3 K. Nakamoto. *Infrared and Raman Spectra of Inorganic and Coordination Compounds.* 4th ed., Wiley, New York, 1986.

4 *Vibrational Spectroscopy of Molecules on Surfaces—Methods of Surface Characterization.* Vol. 1, (J. T. Yates and T. E. Madey, Eds.) Plenum, New York, 1987.

5 Y. Chabal. *Surf. Sci. Rep.* **8**, 211, 1988.

6 D. L. Allara. *Crit. Revs. Surface Chem.* **2**, 91, 1993.

7 H. Ishida. *Rubber Chem. Technol.* **60**, 497, 1987.

8 I. N. Levine. *Molecular Spectroscopy.* Wiley, New York, 1975, chapt. 3.

9 M. Born and E. Wolf. *Principles of Optics.* 5th ed., Pergamon, New York, 1975.

10 E. Wilson, J. Decius, and P. Cross. *Molecular Vibrations.* Dover, New York, 1955.

11 P. Yeh. *Optical Waves in Layered Media.* Wiley-Interscience, New York, 1988.

12 N. J. Harrick. *Internal Reflection Spectroscopy.* Wiley-Interscience, New York, 1967.

13 A. N. Parikh and D. L. Allara. *J. Chem. Phys.* **96**, 927, 1992.

14 N. Winograd. *Anal. Chem.* **65**, 622A, 1993.

15 M. D. Porter, T. B. Bright, D. L. Allara, and T. Kuwana. *Anal. Chem.* **581**, 2461, 1986.

16 P. R. Griffiths and J. A. DeHaseth. *Fourier Transform Infrared Spectroscopy, Chemical Analysis.* Vol. 83, Wiley, New York, 1986.

17 C. W. Sheen, J. Shi, J. Martensson, and D. L. Allara. *J. Am. Chem. Soc.* **114**, 1514, 1992.

18 S. H. Chen and C. W. Frank. *Langmuir.* **5**, 978, 1989.

19 A. N. Parikh, F. Rondelez, and D. L Allara. Submitted for publication.

20 *Infrared Microscopy.* Practical Spectroscopy Series, Vol. 6. (R. G. Messerschmidt and M. A. Harthcock, Eds.) Marcel Dekker, New York, 1988.

21 M. D. Porter. *Anal. Chem.* **60**, 1143A, 1988.

22 D. L. Allara and R. G. Nuzzo. *Langmuir.* **1**, 52, 1985.

23 P. Laibinis, G. M. Whitesides, A. N. Parikh, Y. T. Tao, D. L. Allara, and R. G. Nuzzo. *J. Am. Chem. Soc.* **113**, 7152, 1991.

24 L. H. Dubois and R. G. Nuzzo. *Annu. Rev. Phys. Chem.* **43**, 437, 1992.

25 M. M. Walczak, C. Chung, S. M. Stole, C. Widrig, and M. D. Porter. *J. Am. Chem. Soc.* **113**, 2370, 1991.

26 D. L. Allara and J. D. Swalen. *J. Phys. Chem.* **86**, 2700, 1982.

27 J. Umemura, T. Kamata, T. Kawai, and T. Takenaka. *J. Phys. Chem.* **94**, 62, 1990.

28 J. F. Rabolt, F. C. Burns, N. E. Schlotter, and J. D. Swalen. *J. Chem. Phys.* **78**, 946, 1983.

29 T. J. Lenk, V. M. Hallmark, J. F. Rabolt, L. Haussling, and H. Ringsdorf. *Macromolecules.* 26, 1230, 1993.

30 K. Kelley, Y. Ishino, and H. Ishida. *Thin Solid Films.* **154**, 271, 1987.

31 A. N. Parikh, C. W. Sheen, G. Hietpas, and D. L. Allara. Submitted for publication.

32 M. R. Krishna, M. Kallury, and M. Thompson. *Langmuir.* **8**, 947, 1992.

33 T. G. Bee and T. J. McCarthy. *Macromolecules.* **25**, 2093, 1992.

34 J. Hobbs, C. S. P. Sung, K. Krishnan, and S. Hill. *Macromolecules.* **16**, 183, 1983.

35 F. M. Mirabella. *Appl. Spectrosc.* **42**, 1258, 1988.

36 D. J. Walls. *Appl. Spectrosc.* **45**, 1193, 1991.

37 P. Frantz and S. Granick. *Phys. Rev. Lett.* **66**, 899, 1991.

38 H. Johnson and S. Granick. *Science.* **255**, 966, 1992.

39 S. M. Stole, D. D. Popenoe, and M. D. Porter. *Electrochemical Interfaces: Modern Techniques for In-Situ Interface Characterization.* 1991, pp. 339–410.

40 C. Korzeniewski and S. Pons. *Prog. Anal. Spectrosc.* **10**, 1, 1987.

41 P. W. Bohn and D. R. Miller. *Crit. Revs. Anal. Chem.* **22**, 1, 1991.

42 D. J. Ahn and E. I. Franses. *J. Phys. Chem.* **96**, 9952, 1992.

43 F. Kimura, T. Umemura, and T. Tanaka. *Langmuir.* **2**, 96, 1986.

44 C. E. McGarvey, D. A. Holden, and M. F. Tchir. *Langmuir.* **7**, 2669, 1991.

45 N. Tillman, A. Ulman, and T. Penner. *Langmuir.* **5**, 101, 1989.

46 S. R. Culler, H. Ishida, and J. L. Koenig. *Ann. Rev. Mater. Sci.* **13**, 303, 1983.

47 H. Ancelein, Z. V. Hauptman, A. J. Banister, and J. Yarwood. *J. Polym. Sci. Polym. Phys. Ed.* **28**, 1611, 1990.

48 *Advances in Applied Fourier Transform Infrared Spectroscopy.* (M. W. Mackenzie, Ed.) Wiley, New York, 1988.

49 S. V. Compton, D. A. C. Compton, and R. G. Messerschmidt. *Spectroscopy.* 35, 1991.

50 D. L. Allara, D. Teicher, and J. F. Durana. *Chem. Phys. Lett.* **84**, 20, 1984.

51 S. Chiang, R. G. Tobin, P. L. Richards, and P. A. Thiel. *Phys. Rev. Lett.* **52**, 648, 1987.

52 A. Roessler. *Nauch. Appar.* **2**, 57, 1987.

53 A. Roessler. *Mikrochim Acta.* **2**, 79, 1987.

54 F. Ferrieu and D. Dutarte. *J. Appl. Phys.* **68**, 5810, 1990.

55 B. Drevillon. *Thin Solid Films.* **163**, 157, 1988.

56 R. T. Graf, F. Eng, J. L. Koenig, and H. Ishida. *Appl. Spectrosc.* **40**, 498, 1986.

57 R. A. Palmer, C. J. Manning, J. A. Rzepiela, J. M. Widder, and J. L. Chao. *Appl. Spectrosc.* **43**, 193, 1989.

58 C. E. D. Chidsey, G. Y. Liu, P. A. Rowntree, and G. A. Scoles. *J. Chem. Phys.* **91**, 4421, 1989.

59 P. Fenter, P. Eisenberger, and K. Liang. *Phys. Rev. Lett.* **70**, 2447, 1993.

60 A. N. Parikh, M. Schively, and D. L. Allara. Unpublished results.

5

Raman Spectroscopic Characterization of Organic Thin Films

JEANNE E. PEMBERTON

Contents

5.1 Introduction

Thin films based on a wide array of organic moieties have become increasingly important in many technological areas. As their importance has increased, so has the need for techniques which can nondestructively provide information about the chemistry of these films. In particular, the vibrational spectroscopies have provided a wealth of molecular information about the structural and orientational characteristics of these films. These probes are uniquely effective due to the extreme sensitivity of molecular vibrations to chemical state and the environment of the probe molecule. Although the most commonly employed vibrational spectroscopic tool is infrared (IR) spectroscopy, Raman spectroscopy has played an increasingly visible role in the characterization of organic thin films.

This chapter introduces Raman spectroscopy as a viable method for organic thin film and monolayer characterization. The chapter starts with a brief review of the fundamentals of Raman spectroscopy followed by a discussion of relevant instrumental considerations. (The basic principles of Raman spectroscopy and applications in materials science and technology can also be found in the lead volume of this Series, *Encyclopedia of Materials Characterization.*) Highlighted in this section is the impact that recent advances in detector technology have had on the analyst's ability to study thin films using Raman spectroscopy. This discussion is followed by a presentation of the various modes of operation in which Raman spectra of thin films can be acquired. Finally, selected examples of Raman spectroscopy of thin organic films are presented to demonstrate the power of this approach for elucidating detailed structural and orientational information about organic thin films.

5.2 Fundamentals of Raman Spectroscopy

Raman spectroscopy is a light-scattering phenomenon based on the inelastic interaction of radiation with molecules. This process can be easily understood in the context of the simplified energy-level diagram shown in Figure 5.1. In this process, incident light of energy $h\nu_0$ impinges on a molecule, originally in its ground vibrational level, causing a momentary excitation to what is termed a virtual state. During this brief interaction, a quantum of vibrational energy is acquired by the molecule from the light. The scattered light, $h(\nu_0 - \nu_n)$, has energy *less* than the original light by an amount equal to that necessary for excitation of a particular vibrational mode, ν_n, of the molecule, and the molecule is left vibrationally excited. This scattering process is called Stokes scattering and is thought to occur on the femtosecond timescale. In the general case, the scattering is not directional and occurs into the 4π steradians of space surrounding the sample. In the case of Raman scattering from surfaces, a preferred direction of scattering can be operative as will be seen below.

The propensity of a vibrational mode to generate Raman scattering depends on the extent to which the vibration generates a significant change in polarizability of the molecule. This parameter is a characteristic of the molecule and is defined in three dimensions. The actual Raman intensity of a particular vibration depends on several parameters, including the intensity of the incident light, the number of scattering molecules, and the Raman scattering cross section, a function of the polarizability. Due to the fact that the Raman scattering cross section for even a moderately strong Raman scattering molecule is on the order of 10^{-29} cm^2/molecule-steradian,

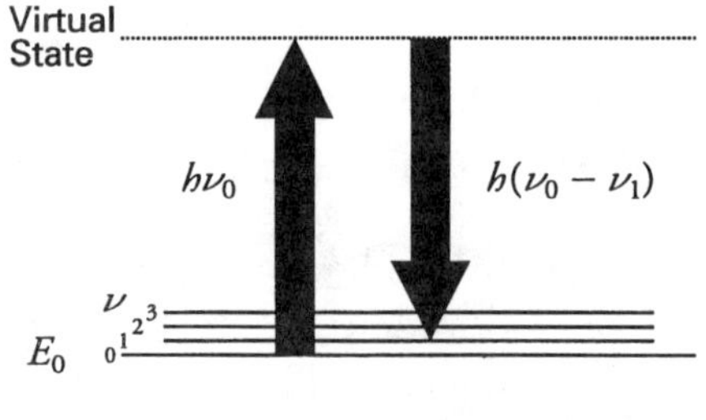

Figure 5.1 Energy level diagram of the Raman scattering process.

the acquisition of Raman spectra from thin films and monolayers on surfaces is not trivial.

Campion has provided a sample calculation for the expected Raman intensity from an organic monolayer on a surface that can serve as a useful baseline for the thin film systems of interest here.[1] This calculation predicts that approximately 250 photons/s of Raman scattered radiation will be generated in a particular vibrational band at the sample surface. After taking into account the approximate 10% transmission efficiency of common spectrometers, an estimated 25 photons/s impinge on the detector. Clearly, this is an extraordinarily low number of signal photons to detect, especially when one considers that numerous sources of background photons are also present. The experimental challenge presented by the sensitivity demands in these studies is obvious.

Raman scattering provides information about molecular vibrations that is complementary to that provided by IR spectroscopy. Although a given molecule has only one set of vibrational modes, not all modes will be observable in both IR and Raman spectroscopy. This fact results from the different fundamental mechanisms of these two vibrational spectroscopies. A change in dipole moment of a molecule is required for a vibrational mode to be IR-active, whereas a change in polarizability is necessary for a mode to be Raman-active. The ramification of this difference is that, although for many molecules several modes are both IR- and Raman-active, certain modes are observed in only one of the two vibrational spectroscopies. Thus, complete vibrational characterization requires that both tools be used to study a given system.

As might be expected, these differences lead to advantages and disadvantages of Raman spectroscopy relative to IR spectroscopy for the study of thin films. One of the most significant advantages of Raman spectroscopy is that it is done in the visible region of the spectrum, making sampling easier from a materials perspective. Glass, Pyrex, or metal oxides can be used as substrates for thin films. These materials are often strongly absorbing in the IR, resulting in a smaller spectral window for analysis of the thin film. In addition, detection is very efficient at all frequencies, making the low frequency region, which can contain direct information about bonding of the thin film to the substrate, readily accessible. The different fundamental mechanisms give rise to strong Raman scattering from organic functionalities which support only weak IR absorption. This difference can be exploited by the analyst to extract structural information from the Raman spectra on these molecules not available from the IR spectra.

Raman spectroscopy also possesses significant disadvantages relative to IR spectroscopy for the study of thin films. Chief among these is the fact that Raman scattering is inherently a very weak effect, making the sensitivity of Raman studies for thin film analysis poor. This feature was responsible for limiting the successful application of Raman spectroscopy to the study of thin films for many years. However, this disadvantage has been somewhat ameliorated recently through the use of charge-coupled device (CCD) detectors for these studies.[2] One additional disadvantage of Raman spectroscopy is the background from fluorescence that can occur. Stokes scattering, on which Raman spectroscopy is based, and fluorescence both occur to the low energy side of the exciting radiation and can be coincident in energy for certain systems. When this happens, the shot noise associated with the much stronger fluorescent signal generally exceeds all but the most intense Raman bands, obliterating the majority of the Raman spectrum. Although this problem can be solved in some cases through the use of near-IR excitation with spectral acquisition in either the normal or Fourier transform (FT) Raman mode, such approaches do not successfully remove fluorescence problems for all systems.

5.3 Instrumental Considerations

In the past decade, Raman spectroscopy has moved out of the realm of the specialist and into the domain of the general analyst. All components of a state-of-the-art Raman spectrometer system can be purchased "off the shelf" from a variety of manufacturers. Furthermore, the development of different operational modes such as FT-Raman spectroscopy have enhanced the user-friendliness of these methods as routine analysis tools.

The basic design of conventional dispersive Raman spectrometers is shown in Figure 5.2. Generally, excitation is provided by laser sources. The Raman scattered

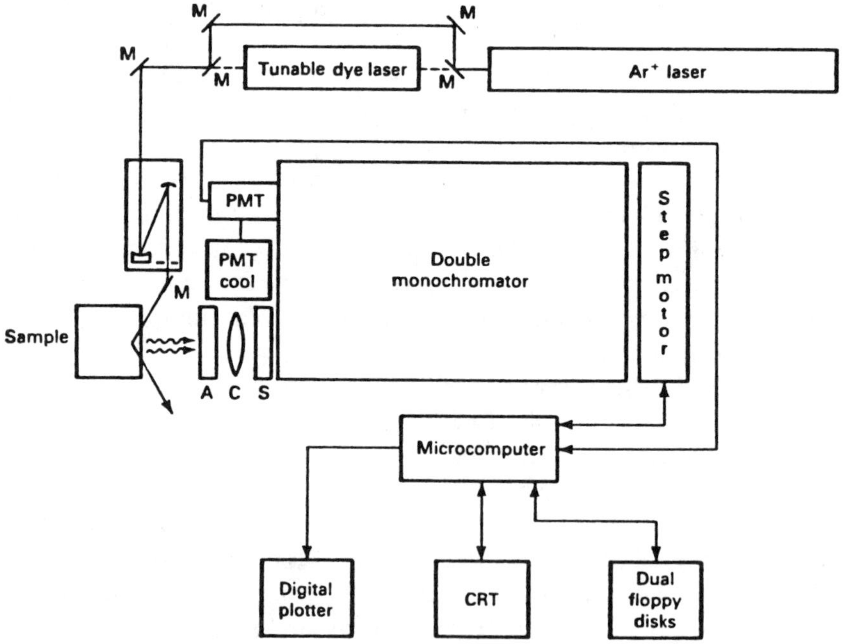

Figure 5.2 Block diagram of a conventional Raman spectrometer system. M = mirror, A = polarization analyzer, C = collection optics, and S = polarization scrambler.

light generated from a sample is collected and focused onto the entrance slits of a spectrometer operating as either a conventional scanning monochromator, as in Figure 5.2, or a spectrograph if being used with a multichannel detector. After dispersion, the light exiting the spectrometer is detected and the signal processed, usually with a computer.

A variety of laser sources are commonly used for excitation. These include the gas ion lasers, Ar^+ for excitation in the green and blue, and Kr^+ for excitation in the blue and red. These gas lasers are also used as optical pumps for tunable dye lasers or Ti:sapphire lasers, both of which generally provide useful radiation in the red and near-IR regions of the spectrum. For FT-Raman spectroscopy, a continuous wave Nd:YAG laser with radiation at 1.064 μm is used.

Spectrometer systems are generally of two types, scanning monochromators and spectrographs, depending on the type of detector employed. Scanning monochromators, devices in which an exit slit is used to isolate a small band of wavelengths that strike a single channel detector, are used with photomultiplier tube (PMT) detectors. Due to the relatively poor sensitivity of these detectors and the detrimental impact of stray radiation from Rayleigh scattering and other sources, these have usually been used with monochromator systems of multiple dispersion stages. Double monochromators (two dispersion stages) have been the norm, although some researchers have used three stages of dispersion for better discrimination against stray radiation.

Spectrographs, devices in which a wide range of wavelengths strike a spatially sensitive detector, are used today with several types of multichannel detectors, including intensified photodiode array (IPDA) and CCD detectors. In order to achieve spectral dispersion over a useful frequency range, spectrographs are typically based on a single stage of dispersion. Stray radiation in these devices, which can be problematic given the weak intensity of the Raman scattered radiation, is usually limited in one of two ways. A multiple-stage filter monochromator preceding the spectrograph stage can be used, as in the Spex Triplemate spectrometer. In this device, the filter stage contains two dispersion devices coupled in subtractive dispersion. This arrangement allows physical discrimination against stray radiation but results in no net dispersion at the end of the second stage. Alternately, a Raman holographic edge filter, which is placed in front of the entrance slit of the spectrograph to prevent the entrance of Rayleigh scattered light, can be employed.

High-sensitivity PMTs, such as the RCA C31034A or the Hamamatsu 943-02, have historically been used with monochromator systems. These devices can be made to have very low dark current noise. However, the rather limited quantum efficiency of these devices has limited the absolute sensitivity that can be achieved for Raman studies of thin films.

Recently, several types of multichannel detectors have been successfully used. Such detectors provide an inherent "multiplex" advantage because they are multichannel in nature. IPDAs, such as those available from Princeton Instruments or EG&G Princeton Applied Research, have been popular. However, these devices have quantum efficiencies comparable to the PMT systems mentioned above. Moreover, they have an additional noise associated with signal processing, the read noise, which limits their absolute sensitivity.

CCD detectors have been recently introduced to Raman spectroscopy and are commercially available as spectroscopic systems from several manufacturers. These devices combine the desirable attributes of high quantum efficiency and very low noise, leading to unparalleled absolute sensitivity. Today, these devices are the detectors of choice for Raman spectroscopic applications demanding high sensitivity, such as the thin film studies that are the subject of this chapter.

FT-Raman spectroscopy is usually performed using a conventional Michelson interferometer–based system fitted with a beamsplitter and detector for the spectral region of interest. Rejection of the Rayleigh scattered light is critical due to the multiplex nature of the spectrometer and is usually achieved with absorption filters. Commonly employed detectors include germanium or InGaAs detectors.

5.4 Raman Spectroscopic Approaches for the Characterization of Organic Thin Films

Table 5.1 is a list of milestones in the development of Raman spectroscopy for the study of organic thin films. As this table attests, a multitude of approaches have been developed over the past two decades for Raman characterization of thin films. These methods range from the total internal reflection methods of the mid-1970s to techniques involving considerable sophistication, such as surface coherent anti-Stokes Raman scattering using integrated optical waveguides, which was developed in the mid-1980s. A brief description of the available techniques introduced in chronological order, their fundamental basis, and major advantages and disadvantages are presented in this section.

Integrated Optical Waveguide Raman Spectroscopy (IOWRS)

In this approach, a light wave is coupled into and reflected back and forth by internal reflection (i.e., guided) through a nonabsorbing thin film. The refractive index of the thin film must be larger than that of the surrounding media. Maxwell's equations govern the physics of the interaction of light with the slab waveguide assembly. These equations are subject to certain boundary conditions which result in trapping and propagation of the light wave only for discrete coupling angles that satisfy the eigenvalue equation. This confined, propagating electromagnetic wave can be used to excite Raman scattering of the thin film waveguide material. The majority of IOWRS experiments are performed in an asymmetric slab waveguide configuration, in which the thin film waveguide is between air and a solid support substrate such as glass. If the thin film of interest is acting as the waveguide, optical confinement requires the film to be of sufficient thickness, typically at least 1 μm given the refractive indices of commonly employed substrate materials.

Year	Chemical System	Method	Ref.
1974	Methylmethacrylate thin films (1 and 6 μm) on glass	Integrated optical waveguide Raman spectroscopy	3
1974	Langmuir–Blodgett multilayers (0.3 μm) of barium stearate on solid substrate	Total internal reflection Raman spectroscopy	4
1974	Pyridine on extensively roughened Ag electrode	Surface enhanced Raman spectroscopy	5
1976	Methyl orange at H_2O/CCl_4 interface (liquid substrate)	Total internal reflection resonance Raman spectroscopy	6
1976	Methylene blue on SnO_2 surface (solid substrate)	Total internal reflection resonance Raman spectroscopy	7
1977	Pyridine on rough Ag (verification of enhancement)	Surface enhanced Raman spectroscopy	8, 9
1977	Langmuir–Blodgett multilayers (80–100) of behenic acid, barium behenate, and barium *cis*-13 erucate on epoxy surface	Normal Raman spectroscopy	10
1977	Methyl orange and crystal violet on rough Ag	Surface enhanced resonance Raman spectroscopy	8
1978	Rose bengal on ZnO	External reflectance resonance Raman spectroscopy	11
	p-Nitroso-dimethylaniline on Ag and Pt	External reflectance resonance Raman spectroscopy	12
1979	Langmuir–Blodgett multilayers of calcium dipalmitoyl phosphatidate on Ag-coated prism in the Otto configuration[13]	Plasmon surface polariton enhanced Raman spectroscopy	14
1980	Langmuir–Blodgett monolayer of cadmium arachidate on Ag grating	Plasmon surface polariton enhanced Raman spectroscopy	15
1980	Langmuir–Blodgett monolayer of cyanine dye on glass waveguide	Integrated optical waveguide Raman spectroscopy	16
1982	Nitrobenzene on Ni(111)	Normal Raman spectroscopy	17
1984	Polystyrene waveguide	Waveguide surface coherent anti-Stokes Raman spectroscopy	18
1986	Langmuir–Blodgett mono-/multilayers of cadmium stearate on glass	Normal Raman spectroscopy with CCD detection	19
1986	Pyridine and phenol monolayers on ZnO	Waveguide surface coherent anti-Stokes Raman spectroscopy	20
1988	Polystyrene on glass 3,3′-Diheptadecyl-2,2′-methylene-bisbenzothiazole perchlorate on glass	FT-Raman spectroscopy	21

Table 5.1 Milestones in the chronological development of Raman spectroscopy for the study of organic thin films.

An evanescent electric field spills out over the physical boundary of the waveguide due to imperfect confinement of the light in real slab waveguide structures. The magnitude of this evanescent field depends on such relevant experimental parameters as the refractive indices of the media and the coupling angle of the light. Extension of the IOWRS approach to the Raman spectroscopic characterization of thin organic films on a waveguide surface (e.g., polymer laminates or monolayer or multilayer films deposited on a polymer waveguide) can be realized through the use of this evanescent tail for excitation of Raman scattering. This was first successfully accomplished using resonant Raman scattering (see below) of a cyanine dye monolayer formed on a polymer waveguide by Langmuir–Blodgett (L–B) deposition.[16]

Several advantages accrue in the use of the integrated optical waveguide Raman method. First, well-defined electric field distributions exist within the waveguide which can be exploited to give spatial information (optical depth profiling) about the Raman scattering species. When polarized radiation is used, the directionality of these fields allows the orientation of molecules within oriented layers and some polarizability tensor components of particular fundamental modes of vibration to be determined. Second, large light intensities exist within the waveguide for exciting the Raman scattering, giving good sensitivity for thin film analyses. Finally, an excitation geometry which is naturally slit-shaped makes the collection of Raman scattered radiation and imaging on a spectrometer slit highly efficient, leading to a further improvement in sensitivity.

Use of this approach also has certain disadvantages. There are fundamental restrictions on the properties of materials that can be used as waveguides. Ideally, these materials must be nonabsorbing and possess a sufficiently high refractive index and must be, at a minimum, approximately 1 μm thick to be able to support the propagating wave. In addition, significant background Raman scattering—from the substrate material for experiments in which the waveguide film is of interest or from the waveguide film in experiments where mono- or multilayers deposited on the waveguide film are of interest—can limit the ultimate sensitivity of these analyses. Fluorescence from either the substrate or the waveguide often represents an additional source of background that limits sensitivity.

Total Internal Reflection Raman Spectroscopy

This method represents the single-reflection limit of the IOWRS approach discussed above. In the total internal reflection mode, Raman scattering is excited in a thin film, with the evanescent wave generated by internal reflection at an interface between media of dissimilar refractive index. In general, these media are nonabsorbing, and the light impinges on the interface from the medium of greater refractive index. This technique represents the Raman spectroscopic analogue of attenuated total reflection (ATR) IR spectroscopy. The total internal reflection method overcomes the minimum film thickness requirement of the IOWRS approach. Recall that for IOWRS, as the critical film thickness is being approached, the Raman scattering contributions of the substrate become very large.

Cipriani and co-workers were the first to employ this method to study a barium stearate L–B thin film 0.3 μm thick.[4] Several years later, the approach was cleverly applied to the resonance Raman characterization of adsorbed dye monolayers at the interface between immiscible solvents by Takenaka and co-workers.[6, 22]

The total internal reflection approach retains the advantage of IOWRS of potential spatial and orientational information about Raman scatterers. In addition, this method can be employed in the study of films at a variety of interfaces, including liquid–liquid and liquid–gas interfaces not readily amenable to study with IOWRS.

Relative to ATR-IR, the total internal reflection approach also possesses the advantages of smaller penetration depth into the sample (i.e., greater surface sensitivity) and a constant thickness of film sampled over the entire spectral region due to the use of visible wavelength radiation. However, Raman scattering from the internal reflection element can be somewhat problematic.

Surface Enhanced Raman Scattering (SERS)

SERS has been used extensively for the acquisition of Raman spectra of organic monolayers on metal surfaces. The genesis of this method dates back to the mid-1970s when Fleischmann and co-workers began a Raman spectroscopic investigation of adsorbed species on highly roughened electrode surfaces.[5] Although the original intent of their roughening procedure was to increase the surface area (and hence, the number of scattering centers), later work by Jeanmaire and Van Duyne[8] and Albrecht and Creighton[9] revealed a connection between the presence of surface roughness and enhanced electric fields at these surfaces. The enhancements in signal intensity that can be realized at roughened metal surfaces of the appropriate metal can reach 10^6. Hence, the technique became known as surface enhanced Raman scattering.

The basis of the phenomenon has received considerable attention by many researchers. It has been reviewed in detail in many other works. Despite the fact that a total quantitative description has not yet been achieved, general agreement about the contributing factors exists. A majority of the enhancement has a purely physical electromagnetic basis, wherein amplification of the incident and scattered electromagnetic fields at the interface between a polarizable solid of some complex dielectric function and some ambient medium of a different dielectric constant occurs. This portion of the enhancement can be calculated from Maxwell's equations, assuming a small particle of known geometry as a model. For a given excitation wavelength, this effect only occurs for certain metals which have the optimum optical properties. In addition to the electromagnetic effect, there is also evidence for contributions from a chemical effect that has a similarity to resonance Raman scattering. This chemical effect is based on photo-assisted charge transfer at chemically unique surface sites, which can exist on these metal surfaces under the right chemical conditions.

As the foregoing discussion implies, the overall enhancement phenomenon is complex. The need for metal surface roughness is undesirable and can be difficult to control experimentally. Moreover, such enhancement is only supported by a relatively limited number of metals. These conditions represent serious disadvantages in using this method. On the other hand, the most compelling advantage is the remarkable sensitivity that this approach provides for the study of organic thin films of thicknesses down to the submonolayer level. Recent work has demonstrated that it is sometimes possible to fabricate surfaces that provide modest electromagnetic enhancement but which remain reasonably well-defined in terms of surface morphology for organic thin film studies. Examples of these substrates include island films or simple mechanically polished, mirrored, polycrystalline surfaces. Such surfaces have allowed the study of L–B and self-assembled monolayer (SAM) thin films.

Surface selection rules operative in SERS dictate that the electric field be largest in a direction along the surface normal and much smaller parallel to the surface. These rules are useful in the determination of molecular orientation if something is known about the polarizability tensor contributions for a given vibrational mode.

Normal Raman Spectroscopy

Normal Raman spectroscopy has been successfully employed in the study of thin films as well. This approach was first used by Lis and co-workers in a study of relatively thick L–B multilayers (80–100 layers) of behenic acid, barium behenate, and barium *cis*-13 erucate on epoxy surfaces.[10] As detector technology improved, however, the normal Raman scattering approach was successfully applied to the study of organic thin films down to the monolayer level. This was first accomplished by Campion and co-workers for a monolayer of nitrobenzene on a Ni(111) surface in 1982 using an IPDA detector.[17] Later work by Murray and Dierker with a CCD detector demonstrated sensitivity adequate for the Raman spectroscopic characterization of a cadmium stearate L–B monolayer on glass.[19]

These studies are extremely difficult to carry out because of the extremely low signal levels that must be detected. In order to achieve adequate sensitivity, all parameters of the experimental system must be appropriately optimized. In addition, even with the best available detectors certain Raman spectral regions are still inaccessible based on signal-to-noise ratio criteria.

Despite these disadvantages, advances in normal Raman spectroscopy of thin films on surfaces have made it possible to acquire spectra on a wider range of thin film materials on surfaces whose morphology can be rigorously controlled. The ability to acquire Raman spectra from films on single crystalline metal surfaces, for example, is important from a fundamental surface science standpoint, since the resolution with which Raman spectra can be acquired greatly exceeds that of high-resolution electron energy-loss spectroscopy (HREELS), one of the alternative approaches. In addition, surface selection rules similar to those for SERS are in effect in normal Raman spectroscopy on metals which can be used to establish orientational information about surface molecules. The progress made to date in this area is significant, suggesting that Raman spectroscopy has the potential to become a routinely applicable tool for the spectroscopic characterization of thin films on a multitude of highly ordered surfaces.

Resonance Raman Spectroscopy

An additional enhancement in the efficiency of the Raman scattering process occurs when the excitation wavelength is coincident with or near an allowed electronic transition of the molecule of interest. The enhancement can reach as high as 10^6 and generally scales with the degree of overlap of the excitation wavelength with the electronic transition. The resulting method is called resonance Raman spectroscopy.

One obvious limitation of this approach is the need for species which absorb the exciting laser radiation. This feature implies the lack of routine applicability of resonance Raman spectroscopic approaches. Despite this drawback, an impressive number of studies have taken advantage of the enhanced sensitivity afforded by this technique for specific chemical systems. Resonance enhancement has been effectively coupled to several of the thin film Raman schemes discussed previously. It has been used effectively for the study of organic thin films on metal and semiconductor surfaces in the normal Raman scattering configuration (i.e., external reflection resonance Raman scattering).[11, 12] Internal reflection resonance Raman spectroscopy has also been used to study adsorbed dye layers at immiscible liquid–liquid interfaces as noted above,[6, 22] as well as on semitransparent semiconductor surfaces.[7] Resonance excitation has also been coupled with surface enhanced Raman scattering to give a technique known as surface enhanced resonance Raman scattering (SERRS). This tool possesses extreme sensitivity for the study of thin films and has been successfully applied to L–B layers.

Plasmon Surface Polariton Enhanced Raman Spectroscopy

Another mechanism for the enhancement of surface electric fields is through the excitation of plasmon surface polaritons (PSPs). Surface plasmons are collective oscillations of conduction electrons in the surface region of a metal. At optical frequencies, these surface plasmons have associated with them intense electromagnetic fields. The combination of the electromagnetic fields and the oscillating surface charge is described in terms of new modes, called plasmon surface polaritons, that are true electromagnetic modes of the interface between a metal and its surroundings. These PSPs can be excited in a Ag optical resonator using the evanescent fields of a prism coupler used in an internal reflection geometry when the resonance condition is met.[13, 14] For excitation in the attenuated total reflection configuration, this resonance condition is satisfied only for well-defined angles of incidence where the momentum components parallel to the surface for light and surface plasmons match. Ag grating couplers can also be used to directly excite PSPs at an interface for Raman scattering without the need for a prism coupler.[15] In general, an enhancement of several orders of magnitude in the electric field intensity at the interface can be realized, conferring greater sensitivity on this method relative to the simple internal reflection (attenuated total reflection) approach. Spatially defined electric fields exist at the Ag surface that are similar to those found in SERS, and these can be used to determine molecular orientation.

This method has advantages and disadvantages. Clearly, the enhanced sensitivity of this approach relative to others is an advantage. However, the need for prism or grating couplers to achieve this enhancement has several consequences, the first being the need for these devices in order to acquire Raman spectra. The second and perhaps less obvious consequence is that it is difficult to use the PSP approach for studies of thin films at single crystalline metal surfaces due to the need for the prism or grating structure. Finally, this approach lends itself to the use of only a relatively narrow group of metals that can support PSPs in the visible wavelength region of the spectrum. It is not amenable to the routine analysis of thin films on substrates other than these metals.

Fourier Transform Raman Spectroscopy

FT-Raman spectroscopy has recently come to the fore as a technique well-suited for the elimination of problems from fluorescence background, inherent to the molecules themselves or from impurities, which often limit Raman spectral analysis of thin films of polymers and certain other organic species. Since the fluorescence yield of these species may be four to eight orders of magnitude higher than the Raman signals, it is often difficult, if not impossible, to acquire spectra from such samples. FT-Raman spectroscopy utilizes excitation in the near-IR region of the spectrum and a Michelson interferometer to spectrally analyze the Raman scattered light. This excitation wavelength is so far removed from the fluorescence-generating absorption characteristics of most organic species that no fluorescence background is excited. Furthermore, the well-known throughput and multiplex advantages of the Michelson interferometer over a standard dispersion instrument contribute to the potential of this approach for the study of thin organic films not amenable to study by other methods. Zimba and co-workers have demonstrated the power of this approach for the study of thin polymer films and Langmuir–Blodgett layers of cyanine dye–containing molecules.[21]

The major drawback of this approach is its sensitivity. This aspect of the FT-Raman experiment is currently limited by several features. First, the available detector technology in the near-IR region is not competitive with the best technology in the visible region. Moreover, the differences between the refractive indices of organic films and commonly employed substrates are smaller in the near-IR region than in the visible, rendering waveguiding less effective. Finally, Michelson interferometers utilize circular apertures which are not as compatible with the linear image provided by the waveguide as are the rectangular slits of monochromators used with visible wavelength excitation. These shortcomings are somewhat offset, however, by the fact that much greater laser powers can be used without sample degradation. In addition, fiber optic sampling can be used to overcome the circular aperture problem. Nonetheless, FT-Raman spectroscopy of thin films is not yet competitive with alternative Raman strategies for the characterization of thin films on the basis of sensitivity.

Waveguide Surface Coherent Anti-Stokes Raman Spectroscopy (WSCARS)

The most commonly used nonlinear Raman method is coherent anti-Stokes Raman spectroscopy (CARS). This method uses two laser beams, a pump beam at ν_1 and a Stokes-shifted probe beam at ν_2, to generate a third laser beam at $\nu_3 = 2\nu_1 - \nu_2$. The probe beam, ν_2, is usually scanned to generate the third frequency using a tunable laser. The power generated at the third frequency is weak until the pump and probe frequencies differ by a vibrational frequency of the molecule, $\nu_n = \nu_1 - \nu_2$. When this resonance condition is achieved, the CARS photon is created, $\nu_{CARS} = \nu_3 = 2\nu_1 - \nu_1 + \nu_n = \nu_1 + \nu_n$.

This method has been developed for use in the analysis of thin films and surface species by coupling it to integrated optical waveguides.[23, 24] The result is a technique called waveguide surface coherent anti-Stokes Raman spectroscopy (WSCARS). Hetherington and co-workers were the first to observe CARS from thin film waveguides of polystyrene.[18] These authors later demonstrated monolayer sensitivity for organics on ZnO waveguide structures.[20]

Coherent Raman scattering benefits from the waveguide geometry more than spontaneous Raman scattering. The geometry intensifies the electric fields by virtue of the small waveguide thickness, resulting in the generation of CARS signals with relatively little laser power. Moreover, the quadratic dependence of CARS on the interaction distance of the two laser beams leads to high conversion efficiencies in the waveguide mode of operation. In addition to these advantages, WSCARS enjoys relative freedom from the background scattering of the substrate that plagues other waveguide Raman measurements. Unfortunately, the technique also suffers from several serious disadvantages. First, the degree of sophistication of the instrumentation required for this method generally renders it a tool of the specialist. Second, waveguide fabrication realities dictate imperfect guided light behavior, diminishing the expected gains in sensitivity for this approach.

5.5 Selected Examples of Thin Film Analyses

The remainder of the chapter is devoted to a discussion of selected examples of Raman spectral characterization of thin organic films. These examples have been chosen to reflect the diversity of the approaches for Raman spectral acquisition and to demonstrate the elucidation of subtle chemical effects in these thin films from the Raman spectral response.

Raman Spectral Characterization of Langmuir–Blodgett Layers of Arachidate and Stearate Salts

Langmuir–Blodgett layers have been studied by means of a variety of Raman sampling strategies. The earliest work involving total reflection Raman spectroscopy was performed on L–B multilayers of alkanoic acid salts of varying chain length.

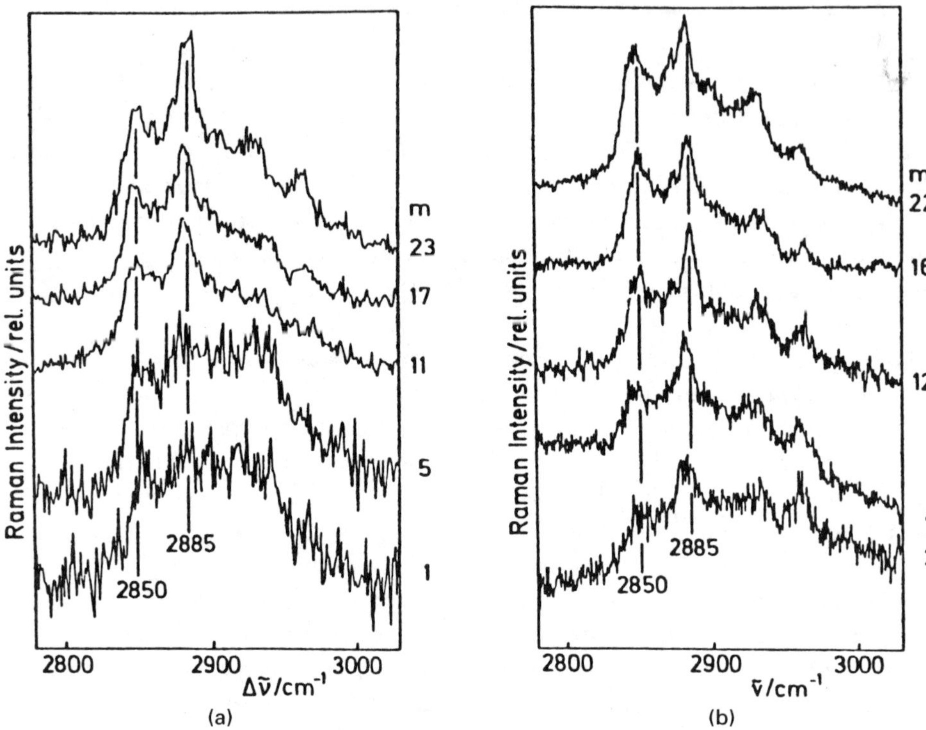

Figure 5.3 **PSP Raman spectra from cadmium arachidate L–B layers of different thicknesses (m = number of layers) on (*a*) SiO_x/Ag and (*b*) bare Ag gratings. (Used with permission.)**

Waveguide, plasmon surface polariton, surface enhanced, and normal Raman spectroscopies have also been used to investigate L–B films of alkanoic acid salts. In many studies, L–B layers of arachidate ($C_{19}H_{39}CO_2^-$) or stearate ($C_{17}H_{35}CO_2^-$) salts have been investigated. Inclusion of the results from several of these studies is instructive.

Two frequency regions are particularly informative in these analyses, because information about the degree of order existing within the L–B layers can be deduced. The region most frequently studied in Raman spectroscopy is that containing the ν(C–H) modes between 2800 and 3000 cm^{-1}. For alkane-containing molecules such as the alkanoic acids, this region is very complex, containing bands due to symmetric and antisymmetric vibrations of the CH_2 and CH_3 groups, a variety of Fermi resonance bands involving both the CH_2 and CH_3 groups, and overtone bands. Despite this complexity, considerable insight into the order of L–B layers can be attained through the analysis of the relative intensities of the CH_2 bands. This is discussed in greater depth below.

The second region of interest is that containing the ν(C–C) vibrations between 1000 and 1200 cm^{-1}. The utility of this region in terms of elucidating the order of

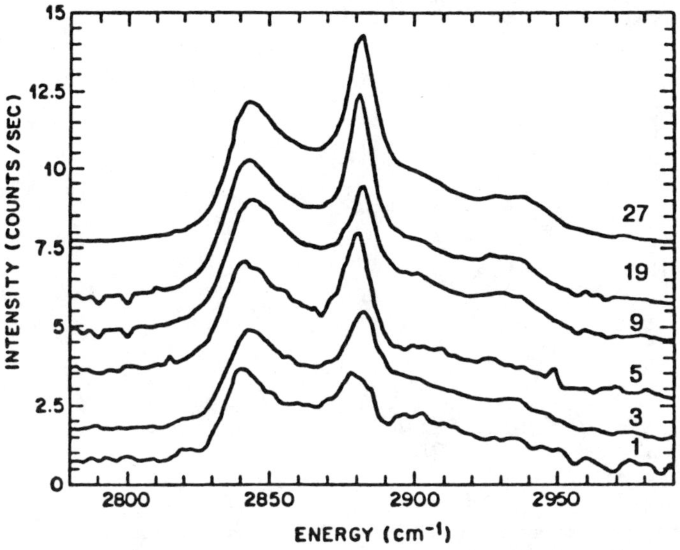

Figure 5.4 **Normal Raman spectra acquired with CCD detection of L–B layers of cadmium stearate of different thicknesses (labeled as number of layers) on fused silica. (Used with permission.)**

these films stems from the fact that discrete bands are observed for vibrations associated with trans and gauche C–C bonds. Thus, ν(C–C) bands are observed for trans conformers at 1060 and 1130 cm^{-1} and for gauche conformers at 1080 cm^{-1}.

Spectra acquired using PSP enhanced Raman spectroscopy from a series of cadmium arachidate L–B films of different thicknesses on a hydrophilic SiO_x coating over a Ag grating are shown in Figure 5.3*a*.[25] Spectra of similar L–B films but on the bare Ag grating are shown in Figure 5.3*b*.[25] Spectra from a similar series of cadmium stearate L–B multilayers on fused silica acquired using normal Raman spectroscopy with CCD detection are shown in Figure 5.4.[19] Several features of these spectra are immediately evident from a cursory examination. First, the signal-to-noise ratio of the spectra increases as the number of L–B layers increases. This behavior is expected and is a consequence of the increased number of scattering centers sampled as the number of L–B layers increases. Second, the spectra acquired using normal Raman spectroscopy with CCD detection are of much better quality than those acquired using PSP enhanced Raman spectroscopy, despite the fact that no enhancement mechanism is operative in the PSP systems. This is a consequence of the truly superior properties of the CCD as a detector. In fact, in later work in which the PSP enhanced Raman approach was coupled to the use of a CCD detector, a signal-to-noise ratio for a cadmium arachidate bilayer comparable to those shown in Figure 5.4 was realized, as demonstrated by the spectrum in Figure 5.5.[26]

Information about the degree of order within these L–B films can be obtained from these spectra.[25] The peak height intensity ratio of the $\nu_a(CH_2)$ band at 2880 cm^{-1}

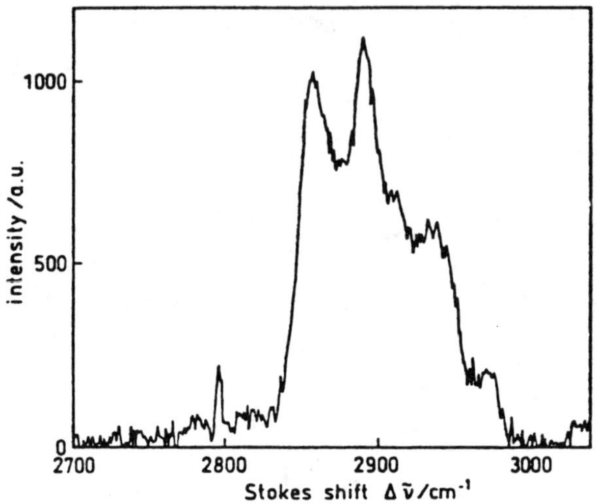

Figure 5.5 **PSP Raman spectrum acquired with CCD detection of two L–B layers of cadmium arachidate on a Ag grating. (Used with permission.)**

to the $\nu_s(CH_2)$ band at 2850 cm^{-1} is a sensitive function of film order. In bulk liquid alkanes, this value is on the order of 0.6–0.7. In the crystalline solid alkanes, this ratio increases dramatically to a value of approximately 2.0. The spectra in Figures 5.3 and 5.4 exhibit an interesting trend in this intensity ratio as a function of film thickness, the ratio increasing noticeably as the number of L–B layers increases. The values range from 1 for the single L–B layer to about 1.5 for the multilayer structures. These observations have been interpreted to indicate that L–B layers composed of one or two monolayers possess a reasonable amount of conformational disorder, probably as a result of underlying substrate surface roughness. This disorder is averaged out of the overall spectral response as the number of L–B layers increases, suggesting that the layers deposited farthest from the surface are ordered to an extent dictated by the van der Waal bonding energy of the system and are not affected by the presence of the surface roughness of the distant substrate.

Temperature has been used as an additional variable for the introduction of disorder into L–B films.[27, 28] Figure 5.6*a* shows spectra acquired using IOWRS for L–B multilayers of cadmium arachidate on glass waveguides at three different temperatures. Clearly, as the temperature increases, the intensity ratio of the $\nu_a(CH_2)$ band to the $\nu_s(CH_2)$ band decreases significantly, indicating a marked decrease in order within the film. In addition, the $\nu_a(CH_2)$ band broadens substantially. These changes have been interpreted in terms of an increase in the gauche conformer population as the temperature approaches the order–disorder phase transition of these materials at 110 °C.

Interestingly, despite the increase in disorder implied by the spectra in the ν(C–H) region, much smaller changes are observed in the spectra in the ν(C–C)

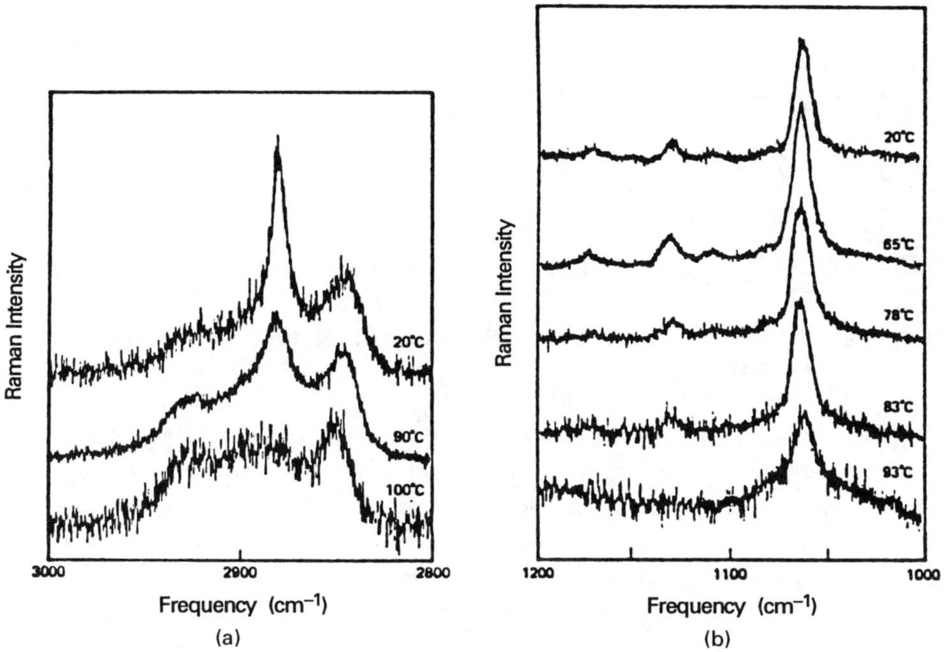

Figure 5.6 **IOWRS spectra in (*a*) ν(C–H) and (*b*) ν(C–C) regions of 13 L–B layers of cadmium arachidate on silica waveguide at different temperatures. (Used with permission.)**

region. Figure 5.6*b* shows the series of IOWRS spectra in the ν(C–C) region that correspond to those shown in Figure 5.6*a* as a function of temperature. The most prominent band in these spectra is the ν_a(C–C) at 1060 cm^{-1}. As the temperature of these films is raised, this band broadens and decreases in intensity, but no discrete band at 1080 cm^{-1} due to gauche conformers is observed. This behavior is interpreted to indicate that, although an increased amount of general conformational disorder is introduced into the film as the temperature increases, gauche conformers do not become a significant portion of the C–C bond population at these temperatures.

Surface enhanced Raman scattering has also been used to study L–B layers of cadmium arachidate on Ag island films.[29] These studies are a further example of the influence of substrate roughness on the spectra in the ν(C–H) region, since these Ag island films are quite rough on a molecular level. The spectrum in Figure 5.7 shows the ν(C–H) region from a monolayer of cadmium arachidate on a Ag

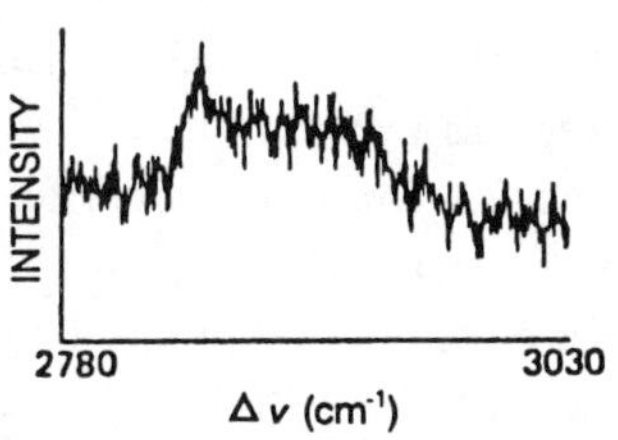

Figure 5.7 **SERS spectrum of L–B monolayer of cadmium arachidate on Ag island film. (Used with permission.)**

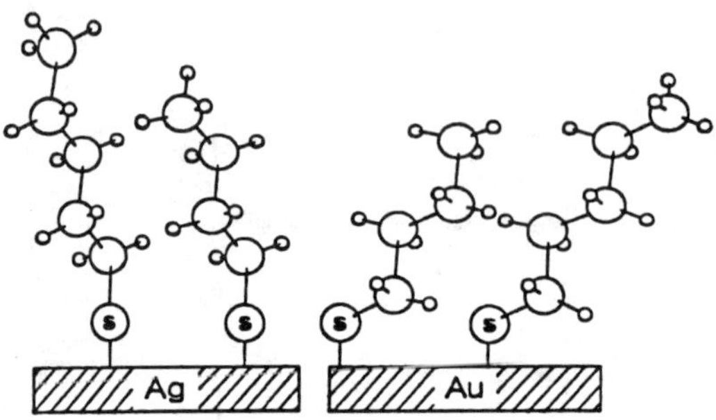

Figure 5.8 **Orientation of alkanethiol SAMs at Ag and Au. (Used with permission.)**

island film. Despite the enhancement afforded by the Ag islands, the signal-to-noise ratio is relatively poor, demonstrating the difficulty in obtaining Raman spectra from single monolayers of these alkyl-containing molecules even with enhancement. Moreover, the intensity ratio of the $\nu_a(CH_2)$ band to the $\nu_s(CH_2)$ band is less than unity, indicating that the monolayer is quite disordered as a result of the morphology of the underlying Ag island film.

Raman Spectral Characterization of Self-Assembled Monolayers of Alkanethiols on Metals

Self-assembled monolayers (SAMs) formed from alkanethiols on noble metals have been the focus of intense study in recent years. Vibrational spectroscopy is a useful tool for analyzing these films, since it provides information about film orientation and order. Raman spectroscopy has been performed on these monolayer films in the author's laboratory in the surface enhanced Raman and normal Raman scattering modes using CCD detection.[30, 31] Information about film order and molecular orientation analogous to that described above for the L–B films can be deduced from these spectra.

One indicator of the degree of order of these films, at least for short alkyl-chain molecules, comes from the existence of an "even–odd" effect for these films at Ag and Au surfaces.[31] This effect results from the tilt angle of the molecules at metal surfaces. A schematic of the orientation of these molecules at Ag and Au surfaces is shown in Figure 5.8. The CH_3 groups at the end of the alkane chains are positioned differently for even- and odd-number carbon-chain lengths. Due to the existence of surface selection rules, the same vibrational mode of the CH_3 group will be observed with different efficiency (i.e., intensity) in these two cases. Furthermore, the even–odd effect expected for these films at Ag is opposite to that expected for the same film at Au, due to the difference in bonding orientation. Figure 5.9 shows Raman spectra in the ν(C–H) region for butanethiol, pentanethiol, and hexanethiol films at Ag and Au surfaces. The CH_3 bands of interest are the $\nu_s(CH_3)$ at 2875 cm^{-1} and the $\nu_a(CH_3)$ at 2965 cm^{-1}. The spectra are plotted normalized to the intensity of the $\nu_s(CH_3)$ band. On Au, the intensity of the $\nu_a(CH_3)$ band

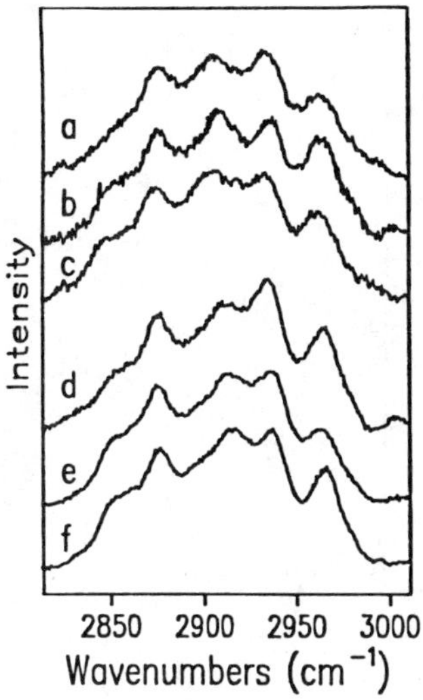

Figure 5.9 **Raman spectra of alkanethiol SAMs at Au (*a, b, and c*) and Ag (*d, e, and f*): (*a and d*) butanethiol, (*b and e*) pentanethiol, (*c and f*) hexanethiol. (Used with permission.)**

relative to the $\nu_s(CH_3)$ band goes from relatively small for butanethiol to relatively large for pentanethiol to relatively small for hexanethiol; hence, an even–odd effect is observed in the relative intensities of these CH_3 bands. The spectra in the bottom half of Figure 5.9 show the opposite even–odd trend for the films at Ag surfaces. This trend can be observed for these films up to chain lengths of approximately ten carbon atoms before the general conformational disorder of the chains and the prominence of the CH_2 vibrations begin to obscure this effect.

Recent work has utilized temperature as a variable in controlling the order of these films.[32] Figures 5.10*a* and 5.10*b* show spectra in the ν(C–C) and ν(C–H) regions, respectively, for monolayers formed on Ag surfaces from C_3H_8SH (propanethiol) at room temperature and after cooling with liquid N_2. The times labeled on the spectra represent time in seconds after emersion from liquid N_2. In the ν(C–C) region at room temperature, bands due to trans C–C bonds are observed at 1025 and 1130 cm^{-1}. A weak band at 1045 cm^{-1} is also observed, which is attributed to a small number of gauche defects within the film. As the film is cooled to liquid-N_2 temperatures, significant changes are noted in the spectrum. The ν_s(C–C) band at 1130 cm^{-1} increases in intensity and the ν_a(C–C) band at 1025 cm^{-1} decreases in intensity as the temperature decreases. This effect is proposed to be the result of a change in orientation of the propanethiol molecules in the film from a greater tilt angle with respect to the surface normal at room temperature to a more vertical position at lower temperatures. This hypothesis is further supported by previous work in which it was shown that the change in polarizability associated with the ν_s(C–C) mode is entirely along the molecular axis, or α_{zz}, and that associated with the ν_a(C–C) mode depends on α_{xx}, α_{yy}, and α_{xy}. Application of surface selection rules for Raman spectroscopy at Ag, which indicate that the ratio of electric field intensity normal to the surface (*z* direction) is 5–10 times the electric field intensity

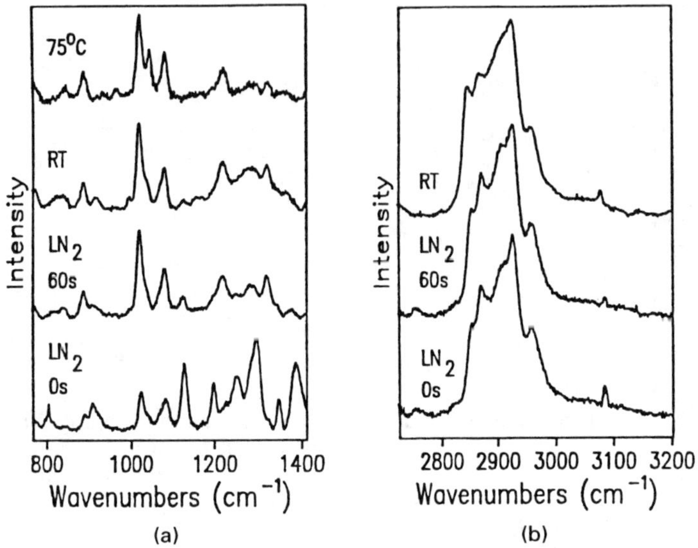

Figure 5.10 **Raman spectra in (*a*) ν(C–C) and (*b*) ν(C–H) regions of SAMs formed from propanethiol at Ag at room and liquid-N_2 temperatures.**

parallel to the surface (*x–y* plane), leads one to the conclusion that the increase in intensity of the ν_s(C–C) mode must be the result of a more vertical orientation of the molecular axis. This reorientation may occur in response to a decrease in the lattice spacing of the Ag surface at liquid-N_2 temperatures. Obviously, excess space within the monolayer on the surface must exist in order for this to occur, implying that the propanethiol films possess a considerable degree of disorder.

When the propanethiol film is heated to 75 °C, a significant band due to gauche conformers is observed in the ν(C–C) region at 1045 cm^{-1}. Obviously, this temperature is high enough to impart sufficient energy to the propanethiol film for C–C bond rotation. The result is a film that possesses a significant degree of gauche disorder.

In the ν(C–H) region, subtle changes are also observed in the spectra from the propanethiol films as the temperature is lowered. The $\nu_a(CH_2)$ band at 2880 cm^{-1} becomes noticeably sharper and more intense relative to the $\nu_s(CH_2)$ band at 2850 cm^{-1} as the temperature is lowered, indicative of a greater degree of conformational order. This behavior is not as clear as that seen for the cadmium arachidate L–B layers discussed above, because the CH_3 vibrations contribute a larger relative intensity for this short alkyl chain molecule, which obscures the response of the CH_2 units somewhat.

Spectra in the ν(C–C) and ν(C–H) regions from octadecanethiol ($C_{18}H_{38}SH$) SAM films on Ag are shown in Figures 5.11*a* and 5.11*b*, respectively. For these films in the ν(C–C) region at room temperature, two prominent bands are observed at 1060 and 1130 cm^{-1} which are assigned to the ν_a(C–C) and ν_s(C–C) for trans C–C

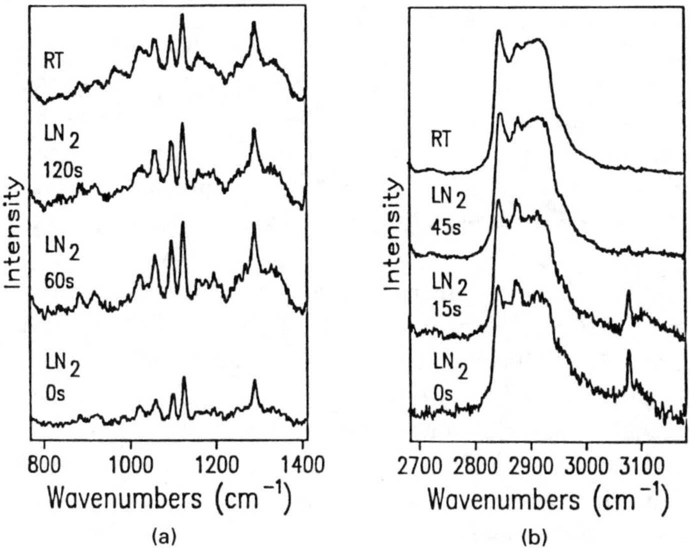

Figure 5.11 **Spectra in (*a*) ν(C–C) and (*b*) ν(C–H) regions of SAMs formed from octadecanethiol on Ag surfaces at room and liquid-N_2 temperatures.**

bonds, respectively. No gauche conformer band is observed at 1080 cm^{-1} for the octadecanethiol film, implying greater order in this film than the propanethiol film at room temperature. In contrast to the propanethiol SAM film behavior, the intensity ratio of the ν_s(C–C) and ν_a(C–C) bands does not change significantly with a decrease in temperature. This outcome is interpreted to indicate that this film is already reasonably close-packed and ordered at room temperature, such that even upon exposure to liquid-N_2 temperatures, no excess room exists to accommodate reorientation of the octadecyl chains to a more vertical position. However, spectra in the ν(C–H) region suggest that subtle changes in the general conformational order of the film do occur as the temperature is lowered. That this is, in fact, occurring is indicated by the small increase in the intensity ratio of the $\nu_a(CH_2)$ band at 2880 cm^{-1} to the $\nu_s(CH_2)$ band at 2850 cm^{-1} from a value of 0.9 at room temperature to a value of about 1 at liquid-N_2 temperature.

Surface Enhanced Resonance Raman Spectral Characterization of Langmuir–Blodgett Layers of Phthalocyanines

Langmuir–Blodgett layers continue to attract wide interest due to their potential utility in areas such as molecular electronics. In particular, L–B layers of phthalocyanines (Pc's) have received considerable attention due to their semiconductor and electrochromic properties. For mixed L–B layers containing submonolayer amounts of the phthalocyanine of interest, sensitive molecularly specific analytical techniques are needed to study the molecular basis of these properties.

Aroca has been particularly active in the development of SERS and SERRS for the analysis of such films. For example, he has demonstrated that SERRS is a unique, nondestructive method for selective vibrational characterization of buried layers in multiple-layer film assemblies. By tuning into molecular resonances and plasmon resonances of the enhancing surfaces, and using the fact that electromagnetic enhancement exists over extended distances away from an enhancing surface, the spectral properties of specific components in multilayers of monolayers can be selectively probed.

A demonstration of this spatial spectroscopic tuning was reported for mixed L–B layers of vanadyl tetra-*tert*-butylphthalocyanine [(*t*-butyl)$_4$VOPc] (abbreviated as VOPc) and *N*-octyl, 3,4:9,10-perylenebis-(dicarboximide) (abbreviated as Oc-PTCDNH) on glass slides covered with Ag-coated Sn spheres.[33] Spectra were acquired with two wavelengths, the 647.1 nm line of a Kr^+ laser, which is in resonance with the VOPc, and the 514.5 nm line of an Ar^+ laser, which is in resonance with the Oc-PTCDNH. In this sampling scenario, excitation with the 647.1 nm line should result in spontaneous Raman scattering from the Oc-PTCDNH and resonant Raman scattering from the VOPc. Alternately, excitation at 514.5 nm should produce spontaneous Raman scattering from the VOPc and resonant Raman scattering from the Oc-PTCDNH. Using the extraordinarily large signals from SERRS (estimated to be on the order of two orders of magnitude larger than SERS signals), selective spectroscopy can be done on either of the two layers, depending on the excitation wavelength chosen. The predicted selectivity is, in fact, observed for a sample consisting of a monolayer of Oc-PTCDNH deposited directly onto the Ag-coated Sn spheres covered by one monolayer of VOPc. SERRS spectra for this system acquired with 647.1 and 541.5 nm radiation are shown in the top and bottom of Figure 5.12, respectively. The 647.1 nm radiation is in resonance with the VOPc; hence, the SERRS spectrum in the top of Figure 5.12 is

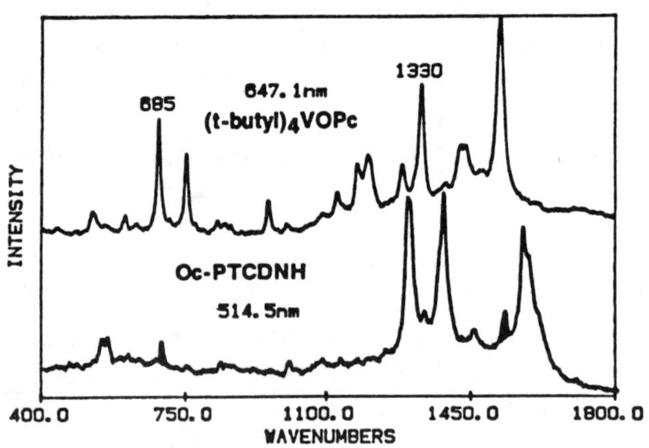

Figure 5.12 **SERRS spectra from Ag/Oc-PTCDNH/VOPc for two excitation wavelengths. (Used with permission.)**

dominated by vibrational bands from this molecule, despite the fact that the Oc-PTCDNH layer is closer to the enhancing substrate. When 514.5 nm radiation is used, which is resonant with the Oc-PTCDNH, bands from this molecule dominate the spectrum shown in the bottom of Figure 5.12. Thus, this approach is demonstrated to be a viable method for the nondestructive vibrational characterization of buried L–B layers.

Later work from this same group demonstrated that changes in the chemical properties of Pc L–B films upon interaction with gas-phase species could also be selectively monitored using SERS and SERRS.[34] Figure 5.13 shows the spectra of a mixed L–B layer containing lutetium bisphthalocyanine (abbreviated $LuPc_2$) and arachidic acid deposited onto glass slides covered with Ag-coated Sn spheres acquired with 488.0, 514.5, and 647.1 nm excitation. Figure 5.14 shows similar spectra after exposure of the film to a gas mixture of NO_2/N_2O_4 (abbreviated as NO_x), which is known to oxidize the $LuPc_2$ film upon chemisorption. In all of these spectra, the vibrational bands observed are from the $LuPc_2$.

The spectra acquired with the 488.0 and 514.5 nm excitation are characteristic of the oxidized form of the initial film. Interestingly, the oxidation process drastically affects the relative intensities of modes associated with the central macrocycle of the $LuPc_2$, but not the frequencies of these modes. The macrocycle breathing vibration at 679 cm^{-1} and the pyrrole stretching vibrations between 1300 and 1500 cm^{-1} are most strongly affected by the change in electronic distribution due to NO_x chemisorption. The SERRS spectrum obtained on the $LuPc_2$ layer with 647.1 nm excitation after exposure to NO_x exhibit relatively minor changes compared to the

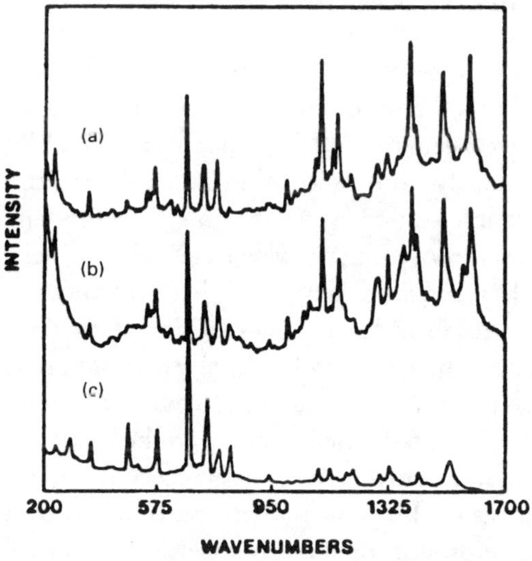

Figure 5.13 Raman spectra of $LuPc_2$ in arachidic acid in a molar ratio of 1:5 on Ag with excitation of (*a*) 488.0, (*b*) 514.5, and (*c*) 647.1 nm. (Used with permission.)

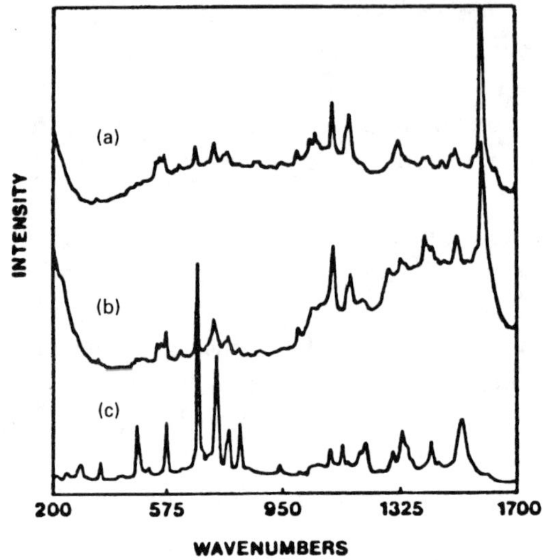

Figure 5.14 Raman spectra of $LuPc_2$ after exposure to NO_x with excitation of (*a*) 488.0, (*b*) 514.5, and (*c*) 647.1 nm. (Used with permission.)

spectrum prior to NO_x exposure. This observation is thought to be the result of the fact that coupling to molecular vibrations occurs through direct electronic excitation of the molecule, which apparently minimizes the change in electronic distribution induced by NO_x chemisorption.

5.6 Prospects for Raman Spectroscopic Characterization of Thin Films

Tremendous advances in the development of Raman spectroscopy as a characterization tool for thin organic films have been realized within the past decade. Although many thin film systems have already been successfully studied with the battery of Raman spectral methods available, much work remains to bring this technique into the realm of a routine analysis tool for these systems. While this field represents a fertile area for future research, it must be recognized that the immense improvements in instrumentation realized recently, such as back-thinned CCD detectors, have almost brought Raman spectroscopy to its fundamental sensitivity limit. Given the throughput limitations imposed by the imperfect reflectivity of real mirrors in the visible-wavelength region and the limited efficiency of even the best gratings, there is no longer another factor-of-two improvement to be achieved in signal-to-noise ratio. Thus, despite the many areas of thin film Raman spectroscopy that remain to be exploited from a chemical standpoint, the field is largely mature from the perspective of instrumentation. Nonetheless, exciting results on new organic thin film systems will undoubtedly be forthcoming as a result of the power of Raman spectroscopy for elucidating the detailed chemistry of these thin film systems.

Acknowledgments

The author is grateful for the partial support of this work by the the National Science Foundation.

References

1 A. Campion. In *Vibrational Spectroscopy of Molecules on Surfaces.* (J. T. Yates and T. E. Madey, Eds.) Plenum, New York, 1987, chapt. 8.

2 J. E. Pemberton, R. L. Sobocinski, M. A. Bryant, and D. A. Carter. *Spectroscopy.* **5**, 26, 1990.

3 Y. Levy, C. Imbert, J. Cipriani, S. Racine, and R. Dupeyrat. *Opt. Commun.* **11**, 66, 1974.

4 J. Cipriani, S. Racine, R. Dupeyrat, H. Hasmonay, M. Dupeyrat, Y. Levy, and C. Imbert. *Opt. Commun.* **11**, 70, 1974.

5 M. Fleischmann, P. J. Hendra, and A. J. McQuillan. *Chem. Phys. Lett.* **26**, 163, 1974.

6 T. Takenaka and T. Nakanaga. *J. Phys. Chem.* **80**, 475, 1976.

7 M. Fujihara and T. Osa. *J. Am. Chem. Soc.* **98**, 7850, 1976.

8 D. L. Jeanmaire and R. P. Van Duyne. *J. Electroanal. Chem.* **84**, 1, 1977.

9 M. G. Albrecht and J. A. Creighton. *J. Am. Chem. Soc.* **99**, 215, 1977.

10 L. J. Lis, S. C. Goheen, and J. W. Kaufman. *Biochem. Biophys. Res. Commun.* **78**, 492, 1977.

11 H. Yamada, T. Amamiya, and H. Tsubomura. *Chem. Phys. Lett.* **56**, 591, 1978.

12 G. Hagen, B. Simic-Glavaski, and E. Yeager. *J. Electroanal. Chem.* **88**, 269, 1978.

13 A. Otto. *Z. Physik.* **216**, 398, 1968.

14 A. Aurengo, M. Masson, R. Dupeyrat, Y. Levy, H. Hasmonay, and J. Barbillat. *Biochem. Biophys. Res. Commun.* **89**, 559, 1979.

15 A. Girlando, J. G. Gordon, D. Heitmann, M. R. Philpott, H. Seki, and J. D. Swalen. *Surface Sci.* **101**, 417, 1980.

16 J. F. Rabolt, R. Santo, and J. D. Swalen. *Appl. Spectrosc.* **34**, 517, 1980.

17 A. Campion, J. K. Brown, and V. M. Grizzle. *Surface Sci.* **115**, L153, 1982.

18 W. M. Hetherington, N. E. Van Wyck, E. W. Koenig, G. I. Stegeman, and R. M. Fortenberry. *Opt. Lett.* **9**, 88, 1984.

19 C. A. Murray and S. B. Dierker. *J. Opt. Soc. Am. A.* **3**, 2151, 1986.

20 W. M. Hetherington, Z. Z. Ho, E. W. Koenig, G. I. Stegeman, and R. M. Fortenberry. *Chem. Phys. Lett.* **128**, 150, 1986.

21 C. G. Zimba, V. M. Hallmark, J. F. Rabolt, and J. D. Swalen. *Thin Solid Films.* **160**, 311, 1988.

22 T. Takenaga and T. Nakanaga. *J. Phys. Chem.* **81**, 645, 1977.

23 Y. J. Chen, W. P. Chen, and E. Burstein. *Phys. Rev. Lett.* **36**, 1207, 1976.

24 G. I. Stegeman, R. M. Fortenberry, C. Karagulcff, R. Moshrefzadeh, W. M. Hetherington, N. E. Van Wyck, and J. E. Sipe. *Opt. Lett.* **8**, 295, 1983.

25 C. Dusch and W. Knoll. *J. Chem. Phys.* **88**, 4062, 1988.

26 H. Knobloch and W. Knoll. *J. Chem. Phys.* **94**, 835, 1991.

27 E. Barbaczy, F. Dodge, and J. F. Rabolt. *Appl. Spectrosc.* **41**, 176, 1987.

28 J. P. Rabe, J. D. Swalen, and J. F. Rabolt. *J. Chem. Phys.* **86**, 1601, 1987.

29 W. Knoll, M. R. Philpott, and W. G. Golden. *J. Chem. Phys.* **77**, 219, 1982.

30 M. A. Bryant and J. E. Pemberton. *J. Am. Chem. Soc.* **113**, 3629, 1991.

31 M. A. Bryant and J. E. Pemberton. *J. Am. Chem. Soc.* **113**, 8284, 1991.

32 C. E. Taylor and J. E. Pemberton. To be published.

33 R. Aroca and U. Guhathakurta-Ghosh. *J. Am. Chem. Soc.* **111**, 7681, 1989.

34 R. E. Clavijo, D. Battisti, R. Aroca, G. J. Kovacs, and C. A. Jennings. *Langmuir.* **8**, 113, 1992.

6

Surface Potential

STEPHEN D. EVANS

Contents

6.1 Introduction

Measurements of surface potential (SP) have been performed on a wide variety of systems over the past half century. They have been used for a range of studies from the determination of work functions of materials to the study of adsorption at fluid–fluid interfaces.

For descriptive purposes it is convenient to split the discussion of organic thin films at interfaces into two parts, the first dealing with films at the air–water interface and the second with those at solid–air interfaces. While there are a large number of ways for determining SP changes at a solid surface using both direct and indirect methods, there are relatively few ways of determining it at the liquid surface. This has led to the prominance of the vibrating-probe technique, which is suitable for both types of systems.

In spite of the large number of studies performed on monolayers at the air–water interface, there have been far fewer carried out on films at the solid–air interface. This seems somewhat surprising in view of the richness of information available from this technique.

This chapter introduces the origin of the SP and discusses how the technique may be used to characterize monolayers at the air–water and the solid–air interfaces.

Work Function (with zero applied bias)

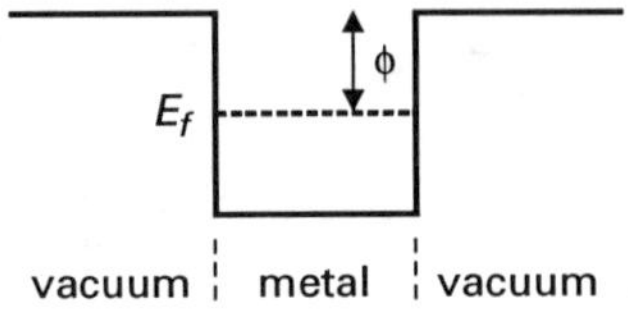

Figure 6.1 **For a conduction electron at the Fermi level to leave the metal it must have a minimum energy, ϕ, known as the work function.**

Emphasis is particularly paid to the study of films at the solid–air interface, where much potential for this technique remains.

6.2 Origins of the Contact Potential Difference and Surface Potential

The Work Function

The conduction electrons in a metal are considered to be "free," that is, they are able to move freely within the confines of the metal and can be treated as a perfect gas obeying Fermi–Dirac statistics. In accordance with quantum theory and the Pauli exclusion principle, such an electron may only have certain discrete values of energy. The probability that an electron occupies any given state of total energy E is given by $F(E)$:

$$F(E) = \frac{1}{e^{[(E-\mu)/kT]} + 1} \tag{6.1}$$

where k is the Boltzmann constant, T is the temperature, and μ is the chemical potential of the electrons. At $T = 0$ K, the chemical potential is equal to the Fermi energy, E_f, where the Fermi energy (level) is defined as the topmost energy level occupied at $T = 0$ K (and μ in Equation 6.1 is usually replaced with E_f). For temperatures greater than 0 K, μ and E_f are not equal, due to the dependence of μ on T. However this variation is usually small, typically 0.01% at room temperature, and for our purposes can be ignored.

If the metal is not subject to any applied fields, then the value of E_f remains constant inside the metal and implies that there must be a barrier at the surface of the metal which prevents electrons from leaving it. The simplest model is to assume that the electrostatic potential energy rises sharply at the edge of the metal to a value sufficiently above the Fermi level that the electrons have insufficient kinetic energy to surmount this barrier, and hence remain trapped (Figure 6.1).

For simplicity the potential is assumed to rise vertically at the metal–vacuum interface to a value which exceeds the Fermi level by an amount ϕ. The quantity ϕ is known as the "work function," and at $T = 0$ K it represents the minimum energy

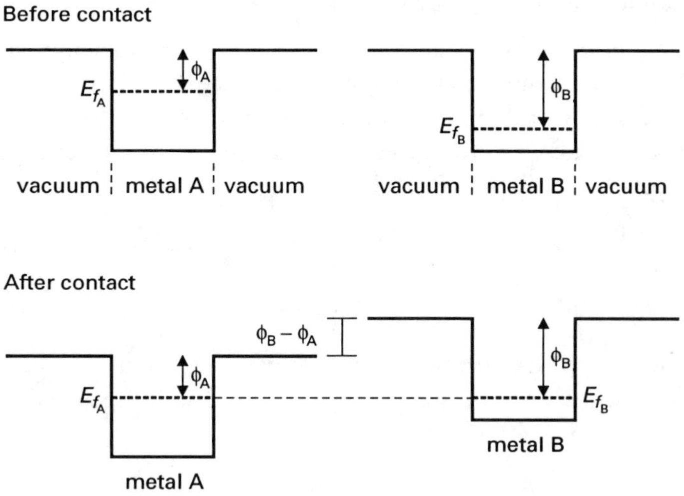

Figure 6.2 **CPD arising from electrical contact being made between two metals: (*a*) before electrical contact is made and (*b*) following electrical contact.**

required to remove an electron from the top of the Fermi level to a point outside the metal, often taken to be at an infinitely large distance from the surface, so as to leave it with zero kinetic energy.† The value of the work function is dependent on the type of metal and the temperature, and is typically of the order of a few electron-volts (eV) for most metals.

Contact Potential Difference and Surface Potential

If two metals, A and B, with work functions ϕ_A and ϕ_B such that $\phi_A < \phi_B$, are connected so that electrons are free to flow between the two metals, then electrons will flow from A to B. This flow of charge sets up a potential difference between the residual positive charge left within A and the negative charge accumulating in B, which increases until the two Fermi levels equalize (Figure 6.2). At coincidence, the electron flow ceases, and the resulting potential difference, Φ, known as the *contact potential difference (CPD)*, is equal to the difference in the work functions:

$$\Phi = \phi_B - \phi_A \qquad (6.2)$$

Such a potential difference will exist whenever two dissimilar materials are brought into contact and allowed to reach equilibrium.

† The work function of a metal is dependent on the surface structure of the metal. To account for this the work function must be redefined to be the minimum energy required to remove an electron from a metal and leave it at rest at a distance outside the metal, so that the image force is negligible but small compared to the distance from another face with a different work function.

Surface Potential Changes Induced by Adsorbates

The adsorption of a molecule onto a clean metal surface produces a change in its work function. For example, if one surface, say that of metal B, is covered by an adsorbed layer which causes its work function to change, say to ϕ'_B, then this change in work function will cause the CPD to change (Equation 6.4), and this change is known as the *surface potential (SP)*, ΔV (Equation 6.7).

$$\phi'_B = \phi_B + \phi_{ads} \tag{6.3}$$

$$\Phi = \phi'_B - \phi_A = \phi_B + \phi_{ads} - \phi_A \tag{6.4}$$

For the situations of most interest to us, that is, in obtaining information about the overlayer, it is necessary to perform measurements before and after the adsorption of the overlayer.

$$\Phi_1 = \phi_B - \phi_A = \phi_B - \phi_A \qquad \text{clean surface} \tag{6.5}$$

$$\Phi_2 = \phi'_B - \phi_A = \phi_B + \phi_{ads} - \phi_A \qquad \text{surface + adsorbate} \tag{6.6}$$

Typically, the change in CPD is referred to as the *change in surface potential*, ΔV, and it gives us information about the electrostatic properties of the adsorbate–adsorbent system.

$$\Delta V = \Phi_2 - \Phi_1 = \phi_{ads} \tag{6.7}$$

Unfortunately, ϕ_{ads} may not be simply a property of the adsorbed layer, but also has components from any modification of the electronic properties of the surface/sub-phase.

The SP arises because of a net spontaneous polarization (dipole/unit volume) in the adsorbate layer.

In the case of two-dimensional molecular assemblies, we can consider each of the adsorbed molecules to have a normal component of its intrinsic dipole moment, μ_n, and for there to be n molecules (or dipoles) per unit area. Following Langmuir's classical treatment we would expect an electrostatic potential to exist at the surface of a metal due to this dipole sheet (Figure 6.3).[1] The change in SP caused by such a double layer may be evaluated from Gauss's Law to be given by

$$\phi_{ads} = n\mu_n/\varepsilon\varepsilon_0 = \mu_n/A\varepsilon\varepsilon_0 \tag{6.8}$$

where ε_0 is the permittivity of free space and A is the area per dipole (usually per molecule). Thus the dipole sheet presents a potential gradient to electrons trying to leave the metal, and depending on the direction of the dipole, such electrons may require a higher or lower amount of energy than if no monolayer were present. ΔV is negative if the negative ends of the dipoles are furthest from the metal–monolayer interface, and it corresponds to an increase in the electrostatic potential barrier for an electron trying to escape from the metal and vice versa.

6.3 Measurement of Surface Potential

Although it is possible to make absolute measurements of work functions and CPDs, these measurements are not commonly made for a number of reasons. First, absolute methods are usually based on one of the following processes: thermionic emission, photo emission, or field emission, which generally require high vacuum facilities and are thus not suitable for studies of monolayer films at liquid–air interfaces or in studying systems at atmospheric pressure (due to the reduction in the electron mean free path).[2, 3] Second, they require the substrates to be either raised to high temperatures, exposed to high energy photons, or subjected to large electric fields. There are additional problems associated with the interpretation of the results obtained using these techniques.[2, 3] There are however a number of indirect techniques which can be used to measure the SP. Of these, the most widely used are the vibrating capacitor (Kelvin/Zisman) technique, which has frequently been used to study films at both the air–water and the solid–air interfaces, and the ionizing probe technique, which has been used mainly to study films at the air–water interface. A number of comparative studies have shown that these (and related) techniques yield the same results to within ±5 mV for monolayers at the air–water interface.[4]

Capacitance Techniques

Vibrating-capacitor methods The majority of capacitance-based techniques for determining the SP are based on modifications of the Kelvin probe technique.[5, 6] Lord Kelvin demonstrated that if any two different materials are arranged as a parallel plate capacitor, the resultant potential difference between the two plates is equal to the CPD (Figure 6.4*a*).

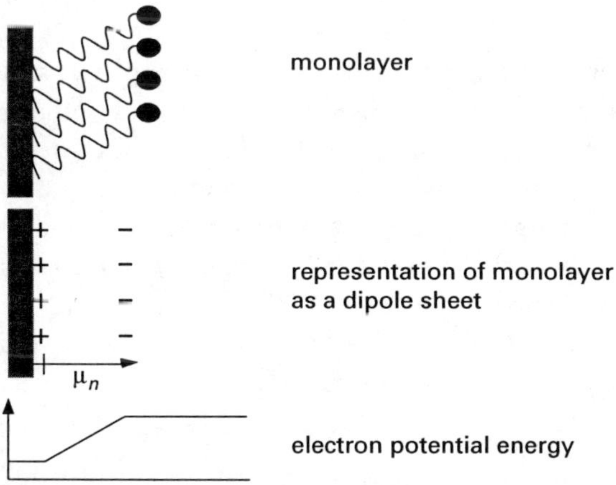

Figure 6.3 Representation of a monolayer as a dipole sheet, and the electron potential energy as seen by an electron moving in the vicinity of the surface.

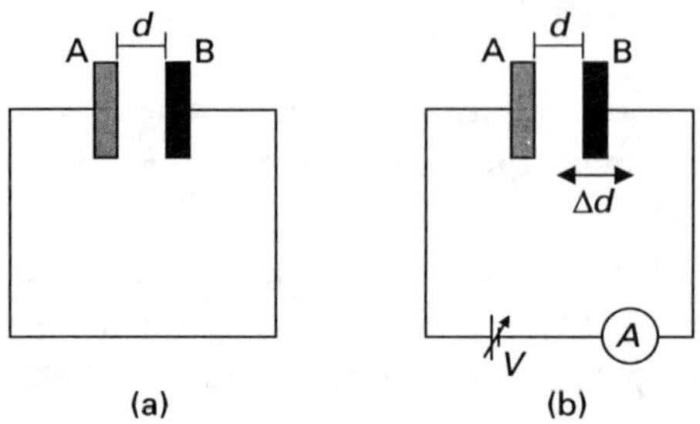

Figure 6.4 (*a*) Equivalent circuit for two materials in electrical contact. The potential difference between the two plates is equal in magnitude to the CPD. (*b*) An equivalent circuit for measuring the CPD and/or the SP of an adsorbed film.

If the plates are separated by a distance d and there is a net potential difference between them Φ, then the amount of charge residing on each of the plates, at equilibrium, is given by Equation 6.9, in which it is assumed that ε is unity and A is the area of the plates.

$$Q = C\Phi = \frac{A\varepsilon_0\Phi}{d} \tag{6.9}$$

Changing the separation between the two plates by Δd requires us to alter the amount of charge on each plate by ΔQ for the potential difference between the two plates to remain constant (Equation 6.10). This means that charge must flow in the circuit; that is, a current is created.

$$\Delta Q = -\frac{A\varepsilon_0 V_c}{d^2}\Delta d \tag{6.10}$$

The current is a transient one and lasts only as long as one of the plates is in motion. If one of the plates is held stationary while the other is varied sinusoidally about some mean separation d, then an alternating current will flow in the circuit ($i = dQ/dt$). Detection and amplification of this alternating current is much easier and leads to a determination of the contact potential. A more direct, widely used approach is to include an external bias in the circuit (see Figure 6.4*b*). The bias is adjusted so that the net potential difference in the circuit is zero, that is, so that no current flows when the plates are moved. When this condition is satisfied, the applied bias, V, is equal (and of opposite sign) to the CPD, Φ.

$$\Delta Q = \frac{-(\Phi + V)A\varepsilon_0\Delta d}{d^2} \tag{6.11}$$

The movement of the plate (reference electrode) can be achieved by mechanical, electromechanical, or electrostatic means.[7–9] The alternating current detected is

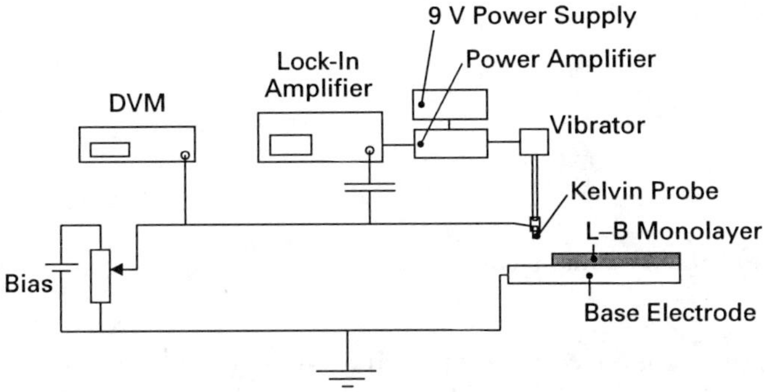

Figure 6.5 **Schematic arrangement for the measurement of SP using the Kelvin probe technique.**

usually phase-locked to the vibrator via a phase-sensitive detector, which makes the determination of the null point much easier. Using suitable circuitry, the ac signal can, after rectification, be fed back as a dc signal to adjust the applied bias, giving a self-balancing (self-nulling) arrangement.

It can be seen from Equation 6.11 that it is desirable to have a small d and a large Δd for a maximum ΔQ. In practice, there is a trade-off between the two.

Since the underlying principles of vibrating Kelvin probes are similar, details of the modifications and improvements used by various workers in the field are not given, but the interested reader should refer to the relevant references. (Also, note that Figure 6.5 shows the sample electrically grounded; however, an alternative arrangement in which the reference electrode is grounded has often been found to lead to more reproducible results).[10–12]

Static-capacitor method This method is more straightforward in many ways than the Kelvin technique and is capable of making measurements on fast time scales for observing dynamic events.[13, 14] The set-up is shown in its equivalent circuit form in Figure 6.6. The upper plate/electrode is placed at a small distance, of the order of a millimeter or so, above the air–water interface and the electrometer is short-circuited, discharging the capacitors and setting V_e to zero. If a monolayer is then spread at the interface, it will introduce a change in the capacitance, C_{pl}, and hence in the electrometer reading. The SP is related to the electrometer reading by Equation 6.12.

$$\Delta V = V_e\left(1 + \frac{C_{par} + C_e}{C_{pl}}\right) \qquad (6.12)$$

Ionizing-Probe Technique

The ionizing-probe technique has been widely used in the study of monolayers at the air–water interface.[15–18] In this method an ionizing electrode is placed above

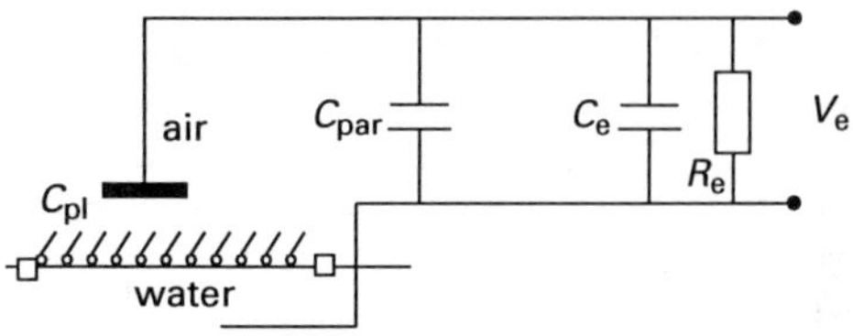

Figure 6.6 Static-capacitor method for determining SP.[14]

the air–water interface and the second electrode is placed in the water subphase (see Figure 6.7). The electrodes are connected via a high impedance electrometer. The ionizing electrode ionizes the air between the subphase and itself, thus completing an electrical circuit in which the air can be treated as if it were a large resistance R of the order of $10^{10}\ \Omega$. In such an arrangement, there will exist a potential difference between the two electrodes due to the CPDs within the circuit.

If a film (monolayer) is spread on the water surface and compressed, it may now be treated as a dipolar sheet, giving a change in the measured potential difference. *This change is defined as the surface potential of the monolayer.*

On spreading the monolayer, taking the SP of the clean water to be the reference value, the potential introduced by the dipolar sheet (assuming that the SP of the reference subphase has remained constant) is given by

$$\Delta V = \frac{\mu_n}{A\varepsilon_r\varepsilon_0} \tag{6.13}$$

where ε_r is the relative permittivity, ε_0 the permittivity of free space, A the area per molecule, and μ_n the normal component of the dipole moment per molecule.

As shown in Figure 6.8, a sealed, gold-coated, 100 μC, americium$_{241}$ alpha foil may be used as the radioactive source (and also as the upper electrode). This is insulated in a PTFE holder and surrounded by a grounded brass surround to reduce stray signals. For the lower electrode, in the subphase, a long bare platinum wire is typically used. The voltage signal can be read either directly using a high input impedance electrometer or can be fed directly into the ADC (anolog-to-digital converter) port of a computer (so long as its input impedance is high; otherwise, an operational amplifier, such as the LF356, should be placed before the ADC,

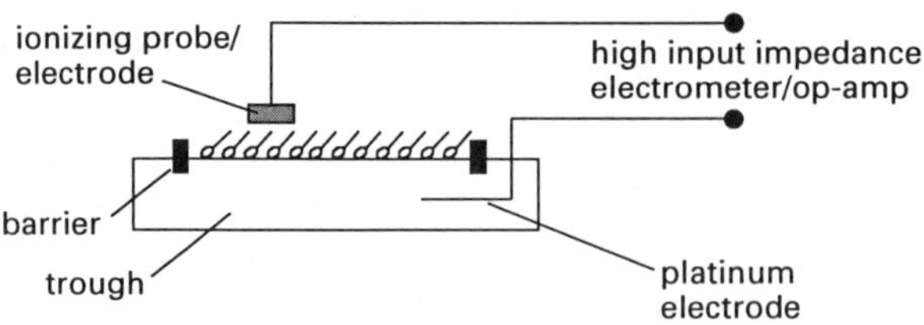

Figure 6.7 Setup for measuring the SP using the ionizing-probe method.

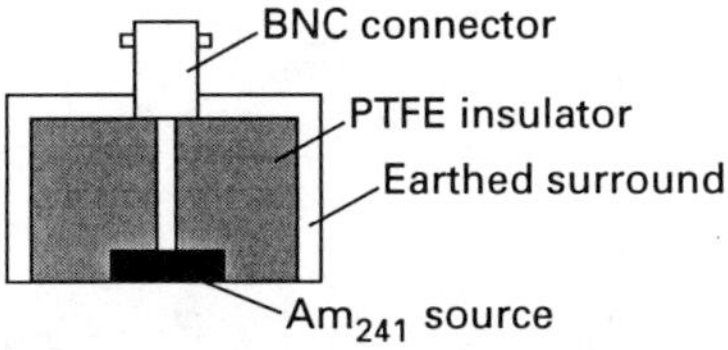

Figure 6.8 Probe head, consisting of the upper electrode and the Am_{241} ionizing source.

since the input impedance must be much greater than R if meaningful measurements are to be made).

Surface potential measurements are susceptible to a wide variety of errors, the main one being the changing of the reference sample with time. Other problems may arise from stray capacitances. The changing reference sample is difficult to avoid, especially when working at atmospheric pressure, but it can be minimized by working under an inert atmosphere. Stray signals are best avoided by enclosing the apparatus entirely within a grounded metal shield (to act as a Faraday cage).

6.4 Surface Potentials of Organic Thin Films

Air–Water Interface: Surface Potential of Langmuir Monolayers

Due to the very large number of studies on the SP of monolayers at the air–water interface, this section cannot hope to give an exhaustive account of the field but will emphasize instead the key developments. Comprehensive lists of work performed in this area up to 1965 are given by Gaines,[16] Davies and Rideal,[19] and more recently by Ulman.[20]

The interpretation of results is best split into two sections: those involving electrically neutral molecules, where one does not have to consider the formation of a double layer as a result of charges in the subphase, and those which are not electrically neutral, which therefore require a treatment of the distribution potential of the ions in the film and the subphase.

Electrically neutral molecules The SP value quoted for a monolayer film at the air–water interface is given as a difference between the clean water surface and that containing a monolayer. However, it is probable that most monolayer-forming molecules will cause a change in the structure/orientation of the water molecules closest to this interface, and this in turn will lead to a change in the contribution of the subphase to the SP. Further it also seems likely that this contribution will vary as the area available per molecule changes and different structural conformations are taken up by the head group. Indirect evidence for this has been shown by Ulman et al., who studied the structuring/orientation of water molecules adsorbed at an OH surface as a function of the OH spacing. Thus it becomes important when reporting results to be as precise as possible with regard to the conditions under which the measurements were made.

If the monolayer and the subphase-induced polarization are treated as a plane parallel-plate capacitor then the change in SP on going from the clean water surface

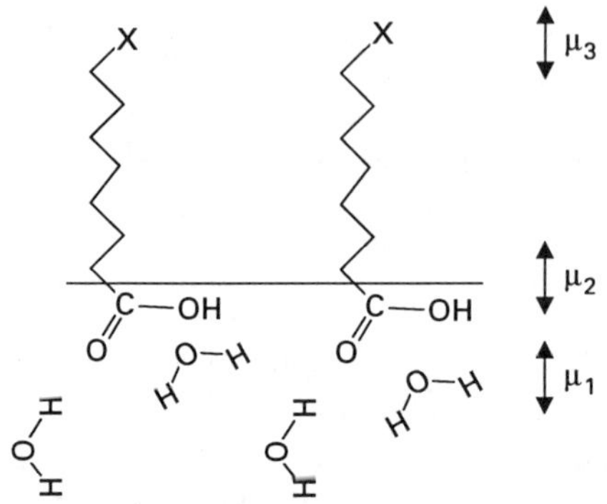

Figure 6.9 **Monolayer at the air–water interface. The dipole contribution to the total net dipole in the surface normal direction is μ_1 from the subphase (water) molecules, μ_2 from the head group, and μ_3 from the terminal group.**

to one coated with a monolayer is given by Equation 6.13. Assuming that dipoles in the vertical direction are additive in a vectorial fashion (Figure 6.9), Davies and Rideal split the μ_n term into three components: μ_1, representing the contribution from the subphase; μ_2, from the head group; and μ_3, from the terminal group. Since the μ_1 term cannot be measured independently and since it is probably dependent on the type of head group anyway, these terms may be combined into one.

Thus, as the film is compressed, ΔV will change if either the head or the tail alters its orientation or conformation with respect to the surface normal. By using molecules that had different ω-terminal groups (at the monolayer–air interface) but the same α-terminal groups (subphase–monolayer interface), and hence the same contribution from μ_1 and μ_2, Davies and Rideal were able to estimate differences in the contributions of different ω-functional moieties to the SP. However, all such calculations led to lower estimates of group dipoles than were expected from bond dipole summations. A couple of factors are thought to account for this: (1) the molecules in the monolayer are not isolated and therefore there will be induced dipolar interactions between them, and (2) the relative permittivity in a condensed monolayer is unlikely to be unity.

Demchak and Fort went some way toward correcting for these problems by suggesting that the relative permittivity of the various regions should not be assumed to be unity, but should instead be assigned some "effective" value.[21]

$$\frac{\mu_{\perp}}{\varepsilon_{net}} = \frac{\mu_1}{\varepsilon_1} + \frac{\mu_2}{\varepsilon_2} + \frac{\mu_3}{\varepsilon_3} \tag{6.14}$$

In doing this, one automatically takes into account the problem of induced polarizations. The relative permittivities of the various regions were calculated by looking at molecules which had the same α-terminal groups but different ω groups, and vice versa, and then solving the necessary simultaneous equations. Implicit in this approach, however, are a number of assumptions which do not necessarily hold, primarily, that the contribution from the subphase is independent of the head group. In spite of this, the model yielded much better estimates of the calculated dipole moments, and the authors went on to show how these could then be applied to distinguish between different possible molecular conformations at the air–water interface.

A comparison of the Demchack–Fort approach to one by Vogel and Möbius, who used a novel partial dipole compensation technique as a means of calculating group dipole moments, leads to the conclusion that the Demchack–Fort model gave a better description of monolayers at interfaces.[22, 23]

One point not implicitly included in the models given here is the dependency of SP on the molecular length. Increasing the molecular chain, say by two CH_2 units, but maintaining the molecular orientation (and hence bond dipole contributions) will cause a change in the SP.[24] From a bond dipole calculation this would not be expected, since the two CH_2 groups introduced should give no contribution to the net normal component of the molecular dipole. This could possibly be explained if each CH_2 unit is considered to be polarizable and serves to shield the dipole. Such a variation must be recognized and can be incorporated into the model through a suitable choice of relative permittivity.

Electrically charged monolayers It is often the case that the monolayers may be charged and that the subphase contains counter-ions. In such situations, a double diffuse layer is set up beneath the monolayer and an additional term must be added to the Helmholtz equation, Equation 6.13, to account for its contribution to the measured SP. There are a variety of models to choose from for calculating the contribution to the double layer, and we shall briefly consider the two main ones.

Gouy proposed that if the monolayer consisted of an impenetrable plane-layer of positive charge spread uniformly over the surface, then the negative counter-ions within the subphase (which were considered to be point-like) would display a Boltzmann distribution (Figure 6.10) and could be represented by

$$\psi_0 = \frac{2kT}{e}\sinh^{-1}\left(\frac{\sigma}{c_i^{1/2}}\frac{500\pi}{\varepsilon RT}\right) \tag{6.15}$$

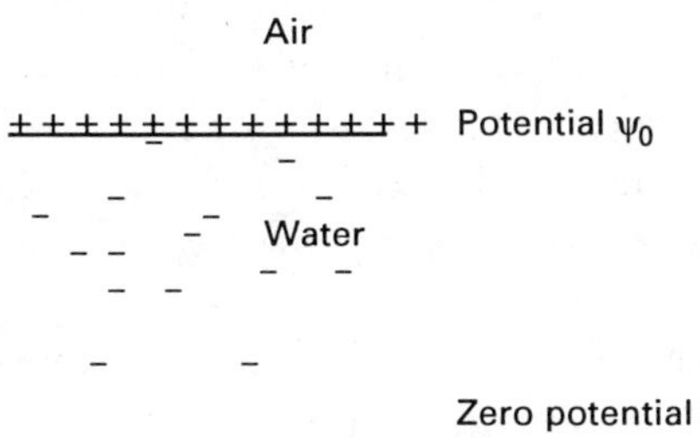

Figure 6.10 Distribution of ions in the Gouy Double Layer.

where e is the charge on an electron, ε is the relative permittivity of the water, R is the gas constant, k is Boltzmann's constant, and $c_i^{-1/2}$ is the total electrolyte concentration in moles per liter.[25]

At 20 °C, this simplifies to

$$\psi_0 = 50.4 \sinh^{-1} \frac{134}{Ac_i^{1/2}} \tag{6.16}$$

where A is the area per charged group (in the surface layer) in square angstroms. The Gouy equation is found to hold as long as the area per charged group is large, that is, for expanded or gaseous monolayers, where the interaction between neighboring head groups is weak. An alternative approach was suggested by Donnan,[26, 27] where the monolayer head group is assumed to be fixed in an ion-permeable membrane, but is no longer required to be strictly planar. Representing the expression derived by Donnan in form similar to that given above for the Gouy equation, we have, at 20 °C,

$$\psi_0 = 25.2 \sinh^{-1} \frac{2 \times 134}{Ac_i^{1/2}} \tag{6.17}$$

This equation is useful for high ionic concentrations in the subphase and for high charge densities in the monolayer (i.e., for more condensed phases).

In cases where there is specific adsorption of one particular type of ion at a surface, the potential is modified, and the model suggested by Stern is usually used, often in conjunction with the Gouy theory. A recent treatment of ionized monolayers of stearic acid was reported by Oliveira et al., in which a modified form of the Gouy equation was used in conjunction with the Demchak–Fort three-layer model.[28]

From these discussions, it is clear that the SP should yield information about the double layer and the degree of ionization within the monolayer, and should enable us to follow either chemical reactions, which change the dipole of the molecules in the film, or specific binding events.

Air–Solid Interface: Surface Potential of L–B and Related Films

In spite of the large number of studies using the SP technique to look at monolayers at air–water interfaces, relatively few have been performed on transferred L–B and related films. There are a number of reasons why this might be so. For instance, unless careful precautions are taken, the results tend to be difficult to reproduce due to the sensitivity/dependency of the method, which may be strongly affected by temperature, humidity, light, surface charges, stray capacitances, adventitious contaminants, etc. Notwithstanding this, the technique may be used to study a wide variety of phenomena, and in the following sections we illustrate how the method may used in the study of L–B and self-assembled films.

Film structure The first experiments looking at the SP of L–B films were made by Porter and Wyman in the late 1930s.[29–31] In these seminal papers they investigated the variation of SP with increasing numbers of layers for stearate films with different structures. For noncentrosymmetric X-type films, they showed that the SP increased to a rather dramatic 8.6 V in 170 layers, while in Y-type films, the potential remained constant, at 0.12 V, after the first 10 or so layers had been deposited. The interpretation of their results should probably be treated with some caution, although their use of the technique illustrates its usefulness in the investigation of L–B films. Langmuir suggested that these results might be interpreted as being due to the incorporation of "surface charges" during the deposition process rather than being directly associated with the molecular (salt) dipoles.[32] With such high current interest in obtaining noncentrosymmetric film structures, the SP technique readily lends itself as a quick method for evaluating film structure and quality. In the second of these papers, Porter and Wyman looked at films with what might be regarded as novel architectures—architectures containing regions of X- and Y-type structures. Further, they showed that the SP can be used to follow the aging and thermal stability of films.

Interpretation of SP results: Schottky model Tredgold and Smith studied the variation of the SP for stearic acid (and chlorophyll) films deposited on GaP, as a function of film thickness. For Y-type deposition, they found that the SP converges asymptotically, with little change after the first 10 or so layers. They pointed out that there are a number of contributions to the observed behavior: (1) the initial step due to the dipole in the first monolayer, (2) the effect of defects existing in the multilayer films, which contribute to band bending, and (3) effects due to the substrate.[33, 34] The conclusions from their studies were that multilayers of stearic acid (Cd salt) contain p-type defects and thus exhibit a depletion region rather like that in a semiconductor.

In a simplified analysis, let us consider an abrupt metal–semiconductor (L–B film) junction (see Figure 6.11). It is clear that at equilibrium the Fermi level in the metal will be brought into coincidence with that in the film (providing it is of sufficient thickness) by electrons transferring from the metal to the p-type defects

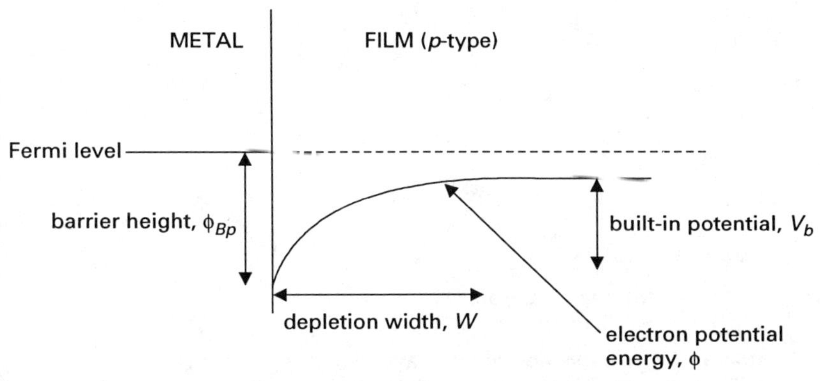

Figure 6.11 Energy-band diagram of a metal-*p*-type semiconductor contact.

(acceptor states) within the film. However, for thin film structures, it is unlikely that the acceptor density will be great enough for this to be realized, and depletion will occur throughout the film. Surface potential measurements made on films with thicknesses less than that of the depletion width should thus explore the band bending within this region (n.b., films of different thicknesses would represent different stages of "frustrated" depletion, and thus the profile obtained does not have precisely the same definition as for band bending in a semiconductor). For films (with Y-type structures) of thickness greater than the depletion width, there should be no further changes in the SP as the number of layers is increased.

From Poisson's equation, in one dimension, we can obtain expressions for the variation of the potential, $V(x)$, for the depletion width, W, and for the built-in potential, V_b:

$$V(x) = \frac{qN_A}{\varepsilon_r \varepsilon_0}\left(Wx - \tfrac{1}{2}x^2\right) - \phi_{Bp}$$

$$W = \left(\frac{2\varepsilon_r \varepsilon_0 V_b}{qN_A}\right)^{1/2} \tag{6.18}$$

$$V_b = \frac{qN_A W^2}{2\varepsilon_r \varepsilon_0}$$

where N_A is the density of acceptor sites in the film.

This model has also proven useful in understanding the electronic properties of a variety of films, for example, in assessing the effects of different electrode materials on charge injection[35, 36] (see Figure 6.12).

Noncentrosymmetric film structures Tredgold and Evans have extended this interpretation to noncentrosymmetric film structures (see Figure 6.13). Here, an

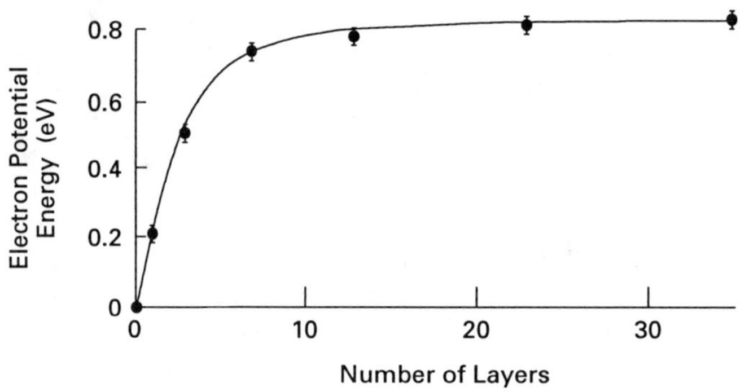

Figure 6.12 Variation of SP as a function of multilayer thickness for films of Ag(II) mesoporphyrin IX diol on aluminium. (From R. Jones et al.[35])

additional term is needed to account for the net dipole associated with each repeat unit (the type of repeat unit depends on whether an X, Z, or ABAB alternating layer structure is being modeled).[37, 38]

The ability of the SP technique to distinguish between different multilayer-type structures probably represents one of its most powerful attributes.[39]

Uniformity of monolayers/multilayers Recently, the technique has been used to study the uniformity of deposited films. Taylor and co-workers placed samples on a stepper-motor controlled *x–y* stage which could be raster-scanned beneath the vibrating probe. Using this setup, they have been able to probe monolayer uniformity (with a resolution of 6 mm) and study stepped structures. Further, they have shown that the technique can be used to follow binding events at the monolayer surface[40–42] (see Figure 6.16). By using a much smaller probe, Heckl et al. have demonstrated that it is possible to map the SP at high lateral resolution (40 μm) for studying the heterogeneity of monolayer films, while maintaining an accuracy of less than ±5 mV.[43]

Chemical composition and surface modification Though it is not possible to perform the chemical identification of surface species using this technique, Ulman et al. have demonstrated that it may be used in a quantitative way to determine the relative surface concentrations of specific functional groups. In studying mixed self-assembled monolayers of alkyl thiols on gold, they have illustrated that the SP varies in a well-defined manner as the ratio of the terminal OH/CH_3 groups is varied (Figure 6.14). Thus, once calibrated against a more absolute technique, such as XPS, SP can be used to make accurate determinations of relative surface concentrations.[44]

Such measurements may also shed some light on how the different molecular species are arranged, that is, whether the distribution of molecules is random or whether

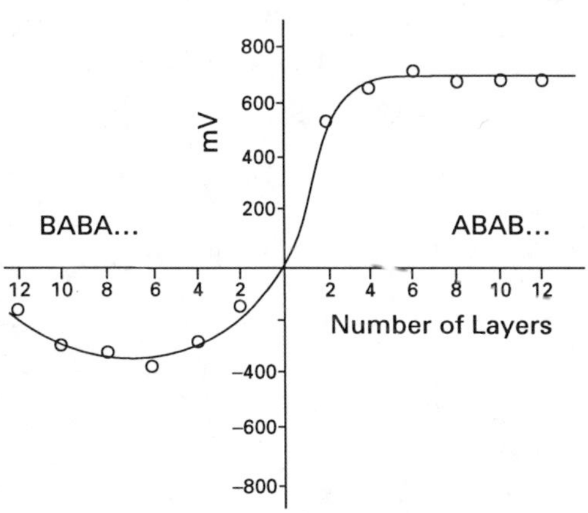

Figure 6.13 **Illustration of the variation of SP for a noncentrosymmetric multilayer structure (A is stearic acid; B is stearylamine). (From R. H. Tredgold et al.[37])**

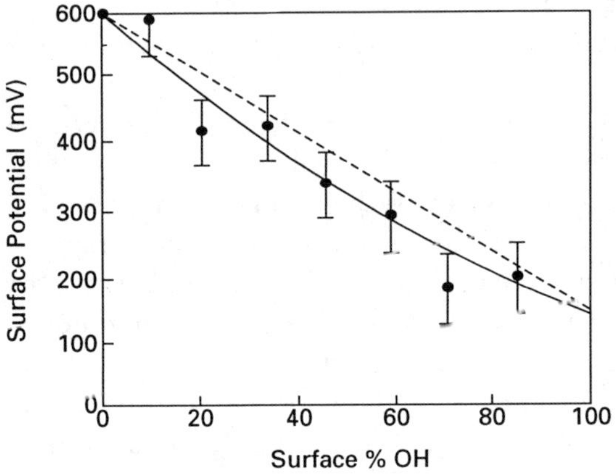

Figure 6.14 **SP for mixed self-assembled monolayers containing molecules with different ω-terminal groups (CH_3/OH). (From Ulman et al.[44])**

macroscopic patches exist, since these would be expected to display different net polarizations and permittivities. However, it is by no means clear how the permittivity of even a random mixture should be calculated. (Tareev gives at least six different equations for estimating it.)[45] In a separate study of self-assembled thiol monolayers, Evans and Ulman also showed that the SP varied linearly as a function of chain length, and that this variation implied that the relative permittivity within the monolayer varied as a function of the molecular length (as a result of the changing molecular polarizability), as shown in Figure 6.15.[46]

It is also apparent that the SP technique should be able to detect modifications to surfaces, either surface chemical reactions or binding events. An example of the latter is provided by Taylor et al. in their study of the biotin–avidin–bisbiotin system (Figure 6.16).[22]

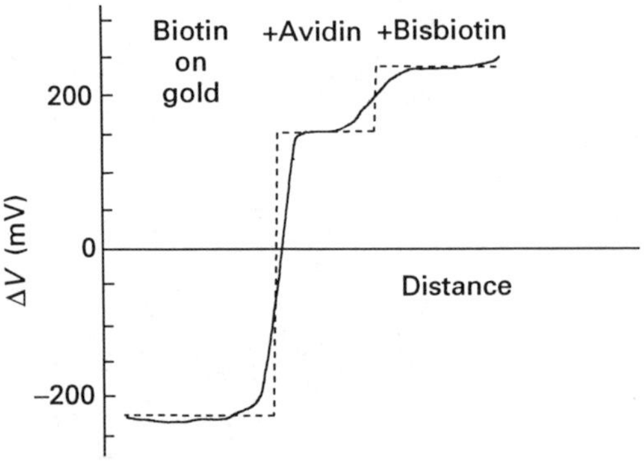

Figure 6.15 **SP profile across a gold substrate coated with a stepped structure of *d*-biotin, avidin, and bisbiotin. (From D. M. Taylor et al.[22])**

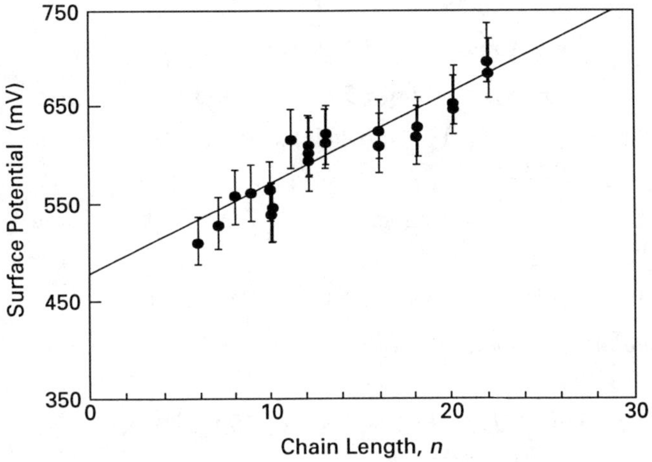

Figure 6.16 **SP variation as a function of alkyl chain length for self-assembled alkylthiol monolayers on gold. (From S. D. Evans and A. Ulman.[46])**

6.5 Conclusions

SP measurements can be extremely useful in understanding the properties and behavior of adsorbed molecular assemblies for both monolayers and multilayers. The technique is not without its problems, however, the most prominent of which is the ambiguity involved in intepreting changes. In practice, this effectively means that SP results should probably be analyzed in conjunction with information obtained from alternative techniques. Having said this, the technique does provide a rich variety of information not readily obtainable from other techniques.

References

1 I. Langmuir. *J. Am. Chem. Soc.* **54**, 2798, 1932.

2 J. C. Rivière. In *Solid State Surface Science.* (M. Green, Ed.) Dekker, New York, 1969, p. 179.

3 G. A. Haas and R. E. Thomas. In *Techniques of Metal Research.* (E. Pasgalia, Ed.) Interscience, New York, 1972, p. 91.

4 C. D. Kinloch and A. I. McMullen. *J. Sci. Instr.* **36**, 347, 1959.

5 W. Thomas. *Phil. Mag.* **46**, 82, 1898.

6 W. A. Zisman. *Rev. Sci. Instrum.* **3**, 367, 1932.

7 K. A. Macfadyen and T. A. Holbeche. *J. Sci. Instru.* **34**, 101, 1957.

8 W. R. Harper. *Proc. Roy. Soc.* **A205**, 83, 1957.

9 Y. L. Yousef, A. Miskriki, S. Aziz, and H. Mikhail. *J. Sci. Instru.* **424**, 873, 1965.

10 R. E. Simon. *Phys. Rev.* **116**, 613, 1959.

11 T. A. Delchar and G. Ehrlich. *J. Chem. Phys.* **42**, 2686, 1965.

12 Y. Petit-Clerc and J. D. Carette. *Rev. Sci. Instr.* **39**, 933, 1968.

13 J. L. M. Venselaar, A. J. Kruger, L. M. H. Verbakel, and J. A. Poulis. *J. Colloid Int. Sci.* **70**, 149, 1979.

14 J. C. Jacobs, A. J. M. Buuron, P. J. M. Renders, and A. F. M. Snik. *J. Colloid Int. Sci.* **84**, 270, 1981.

15 J. H. Schulman and Hughes. *Proc. Roy. Soc.* **A138**, 430, 1932.

16 G. L. Gaines, Jr. In *Insoluble Monolayers at Liquid–Gas Interfaces.* Interscience, New York, 1966.

17 J. H. Schulman and E. K. Rideal. *Proc. Roy. Soc.* **A130**, 259, 270, and 284, 1931.

18 N. K. Adam and J. B. Harding. *Proc. Roy. Soc.* **A138**, 411, 1932.

19 J. T. Davies and E. K. Rideal. In *Interfacial Phenomena.* Academic, New York, 1963.

20 A. Ulman. *Ultrathin Organic Films from Langmuir Blodgett to Self-Assembly.* Academic, New York, 1991.

21 R. J. Demchak and T. Fort, Jr. *J. Colloid Int. Sci.* **46**, 191, 1974.

22 D. M. Taylor, O. N. Oliveira, Jr., and H. Morgan. *J. Colloid Int. Sci.* **139**, 508, 1990.

23 V. Vogel and D. Möbius. *Thin Solid Films.* **132**, 205, 1985.

24 Y. K. Kuchhal, S. S. Katti, and A. B. Biswas. *J. Colloid Int. Sci.* **49**, 48, 1974.

25 J. Gouy. *Phys. Radium.* **9**, 457, 1910.

26 Z. Donnan. *Elektrochem.* **17**, 572, 1911.

27 Z. Donnan. *Chem. Rev.* **1**, 73, 1924.

28 O. N. Oliveira, Jr., D. M. Taylor, and H. Morgan. *Thin Solid Films.* **210/211**, 76, 1992.

29 E. F. Porter and J. Wyman, Jr. *J. Am. Chem. Soc.* **59**, 2746, 1937.

30 E. F. Porter and J. Wyman, Jr. *J. Am. Chem. Soc.* **60**, 1083, 1938.

31 E. F. Porter and J. Wyman, Jr. *J. Am. Chem. Soc.* **60**, 2855, 1938.

32 I. Langmuir. *J. Am. Chem. Soc.* **60**, 1190, 1938.

33 R. H. Tredgold and G. W. Smith. *J. Phys. D: Appl. Phys.* **14**, L193–5, 1981.

34 R. H. Tredgold and G. W. Smith. *Thin Solid Films.* **99**, 215, 1983.

35 R. Jones, R. H. Tredgold, and A. Hoorfar. *Thin Solid Films.* **123**, 307, 1985.

36 M. Iwamoto and A. Fukuda. *Jap. J. App. Phys.* **29**, l638–1640, 1990.

37 R. H. Tredgold, P. Hodge, Z. Ali-Adib, and S. D. Evans. *Thin Solid Films.* **210/211**, 4, 1992.

38 S. D. Evans. M.Sc. thesis, University of Lancaster, 1985.

39 P. Christie, C. A. Jones, M. C. Petty, and G. G. Roberts. *J. Phys. D.* **19**, L167, 1986.

40 D. H. Taylor, H. Morgan, and C. D'Silva. *J. Phys. D: Appl. Phys.* **24**, 1443, 1991.

41 O. N. Oliveira, Jr., D. M. Taylor, C. J. M. Striling, S. Tripathy, and B. Z. Guo. *Langmuir.* **8**, 1619, 1992.

42 P. Christie, G. G. Roberts, M. C. Petty. *Appl. Phys. Lett.* **48**, 1101, 1986.

43 W. M. Heckl, H. Baumgartner, and H. Möhwald. *Thin Solid Films.* **173**, 269, 1989.

44 A. Ulman, S. D. Evans, Y. Shnidman, R. Sharma, and J. E. Eilers. *Adv. Colloid Int. Sci.* **39**, 175, 1992.

45 B. Tareev. *Physics of Dielectric Materials.* Mir Publishers, Moscow, 1975.

46 S. D. Evans and A. Ulman. *Chem. Phys. Letts.* **170**, 462–466, 1990.

7

X-Ray Diffraction

P. DUTTA

Contents

7.1 Introduction

An organic thin film is an artificial structure, at least in the general sense that one of its dimensions is microscopic and so the structure *should* differ in some way from that of the bulk material, but more specifically because the film has been made in order to generate a property that depends upon the expected molecular arrangement. A fundamental question, then, is "how are molecules actually arranged?" This is not always the same as the intended or "logical" arrangement. For example, many people lulled by diagrams of Langmuir–Blodgett films being lifted off the water surface ideal layer by ideal layer have later discovered that life does not always imitate art.

Recently, new microscopies (atomic force microscopy in particular) have proved capable of producing vivid real-space pictures of organic thin film surfaces. X rays do not provide real-space pictures, except indirectly. On the other hand, they provide information, such as correlation lengths, that is hard to obtain from a real-space picture, which has either limited range or limited resolution; they can be used to study films on practically any substrate; and they provide structural information about the entire film, not just the outer surface. Reciprocal-space (diffraction) studies and real-space (microscopy) studies will therefore remain complementary for the foreseeable future.

The use of X-ray diffraction in the laboratory to study reasonably well ordered bulk materials is by now routine, and the technique and its terminology are familiar to most scientists. Studies of thin films differ from bulk studies in only a few respects. First, organic materials scatter X rays weakly, and except for the simplest studies, thin organic films require the very bright radiation produced by synchrotron sources. (Deuterated molecules scatter neutrons more strongly, but synchrotron beams are much brighter than neutron beams, and in most cases synchrotron radiation offers the net advantage.) Second, the simple Bragg picture based on an infinite ideal lattice does not fit a thin film, because at least one of its dimensions is quite small; also, many organic monolayers and thin films are not ideally crystalline even within the film plane, but rather form intermediate-range–ordered mesophase structures. Unlike simple atomic lattices, structural information required for molecular films includes the orientation of the molecule relative to the lattice and the location of the thin film lattice relative to the structure of the substrate surface (if any).

This chapter does not aspire to be a complete scientific review of thin film studies using X-ray diffraction. In the following sections I will discuss some selected techniques and illustrate their use with a few examples. I will also mention related X-ray scattering techniques that have been applied to organic thin films, such as reflectivity, standing-wave fluorescence, and resonance X-ray diffraction. Many other topics are discussed in more detail elsewhere (see Reference 1 and references therein). A description of the basic principles of X-ray diffraction and its application to thin films in materials science can also be found in the lead volume of this Series, *Encyclopedia of Materials Research.*

7.2 Basic Principles

When an X-ray beam of wave vector $\mathbf{K}$ (magnitude $2\pi/\lambda$, incident beam direction) is elastically scattered in a material and then detected such the new wave vector is $\mathbf{K}'$ (same magnitude, scattered direction), the intensity I in this direction depends on $\mathbf{k} \equiv \mathbf{K} - \mathbf{K}'$ as follows:

$$I(\mathbf{k}) \propto \left| \int \rho(\mathbf{r}) e^{i\mathbf{k}\cdot\mathbf{r}} d\mathbf{r} \right|^2 \tag{7.1}$$

where $\rho(\mathbf{r})$ is the local electron density, and the integral is over the dimensions of the sample or the part of it that is illuminated by the X rays. In an ideal point-particle crystalline solid, $\rho(\mathbf{r})$ is nonzero only at discrete values of $\mathbf{r}$ corresponding to allowed lattice sites; the Fourier transform is then also zero, except at discrete values of $\mathbf{k}$; in other words, the diffraction pattern $I(\mathbf{k})$ consists of the familiar Bragg spots. On the other hand, in a completely disordered system, the Fourier transform has no features at all—that is, there is no diffraction pattern.

In an organic thin film, these ideal limits do not apply. Organic molecules cannot be approximated as point particles; moreover, the structure may not be either perfectly periodic or perfectly disordered, but somewhere in between. This is certainly true normal to the film plane, assuming that these "thin" films are not thick enough to be treated as periodic systems (see Section 7.3). The in-plane order can also be intermediate between that of solids and liquids; for example, it is known that Langmuir monolayers on water can form "mesophases" (see Section 7.4).

The goal of X-ray diffraction is to learn about $\rho(\mathbf{r})$, which is the formal answer to our question "how are the molecules actually arranged?" Since it is usually impractical to determine the function $\rho(\mathbf{r})$ completely, one explores those aspects of $\rho(\mathbf{r})$ that are important to the problem at hand. In Section 7.3, we look at studies of the structure normal to the film plane; in Section 7.4, we discuss studies of structure within the film plane, including the determination of tilt angles and directions.

7.3 Structure Normal to Film Plane

When the incident and scattered X rays are in a plane normal to the film plane, and the incident and scattered angles measured from the film plane are the same (θ), the specular condition obtains, and the vector $\mathbf{k} = \mathbf{K} - \mathbf{K}'$ is normal to the film plane (in the z-direction) with a magnitude of $2K \sin\theta$. (A schematic diagram can be found in Figure 7.3, later in this chapter.) It can then be seen by rewriting Equation 7.1 that the scattering depends only on the Fourier transform of the plane-averaged density, $\rho(z) \equiv \int \rho(\mathbf{r})\, dx\, dy$, and so is insensitive to the in-plane structure. If the thin film in question is a multilayer with a large number of layers, $\rho(z)$ is periodic, and so $I(k_z)$ also shows strong, Bragg-like diffraction maxima from whose positions the plane spacing is easily calculated. In Langmuir–Blodgett (L–B) films of saturated fatty salts (soaps), for example, 50 or more bilayers are not uncommon, and many orders of diffraction peaks can be obtained using only a tube X-ray source.[2]

A few-layer system, however, does not show sharp Bragg peaks as a function of k_z, but broader oscillations. It is in fact possible to obtain more information from such data than from the Bragg peaks of an essentially infinite system, since nontrivial data are obtained between, as well as at, the Bragg points. One example is the work of Fischetti et al.,[3] who studied five L–B bilayers of arachidic acid (AA), a disubstituted diacetylene monomer (DAM), and the blue and red polymeric forms of DAM (DaPb, DaPr). The scattering intensities with **k** along the normal to the film planes are shown in Figure 7.1; the oscillations are not Bragg peaks but broad interference effects. From these data, and using a few reasonable assumptions, they obtained $\rho(z)$ for each type of bilayer (Figure 7.2). It can be seen that three or four density regions are distinctly visible: carboxyl head group (high density, center), methyl end groups (low density, two ends) and hydrocarbon chain (intermediate density; slightly higher density between diacetylene and carboxyl groups than between diacetylene and methyl groups). Fischetti et al. concluded that in all the diacetylene-containing materials, the chains are better ordered between the diacetylene and the carboxyl groups than between the diacetylene and the methyl groups; also that the DAPr methyl end groups are relatively disordered, because of the shallower troughs.

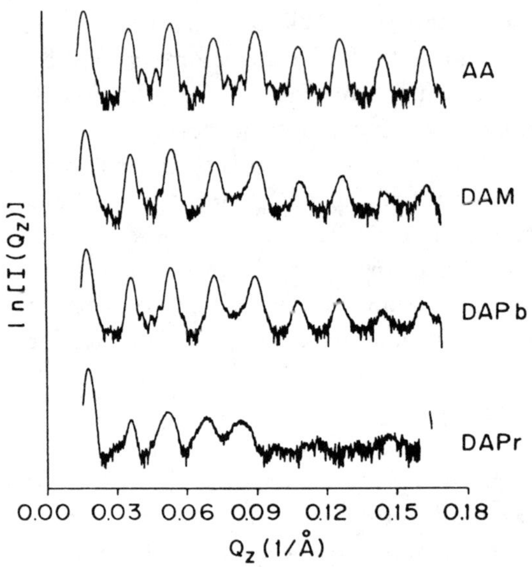

Figure 7.1 **Corrected intensity functions for five-bilayer L–B films of AA, DAM, DAPb, and DAPr. The functions are plotted on a semi-logarithmic scale and normalized to the same peak intensity. (From Fischetti et al.[3])**

In a single monolayer or bilayer, where there is no periodicity at all, the basic principles are the same (see Equation 7.1), but the experiments are known as reflectivity measurements. When θ is small enough, X rays suffer total external reflection from the film surface, because at X-ray wavelengths most materials have a refractive index just under 1.0. At higher angles, the reflected intensity falls off in a way that depends on the form of $\rho(z)$. For example, data obtained by Tidswell et al.[4] from uncoated silicon and from self-assembled (SA) films of different thicknesses deposited on silicon may be seen in Figure 7.3. Loosely, it can be said that the minima in the reflectivity are due to destructive interference between the part of the X-ray beam that is reflected from the film–air interface and the part reflected from the film–silicon interface; this is why the minima move to the left as the thickness increases. More rigorously, one fits the data to a model for $\rho(z)$ and varies the model parameters to get the best fit. Unfortunately, getting good fits usually requires models with a large number of parameters. While simple information such as overall film thickness is provided quite reliably by these fits, the credibility of other computer-generated information by these fits, such as interface widths, is less clear. In some cases, for example, Tidswell et al. *constrained* the electron density within the film to the expected value (i.e., used an estimated value as input) in order to get an acceptable fit to the data, and we have found that this procedure is frequently necessary for our own data as well.

Although some reflectivity data can be obtained using laboratory X-ray sources, the k_z range is extremely limited because the reflectivity drops off so rapidly (note

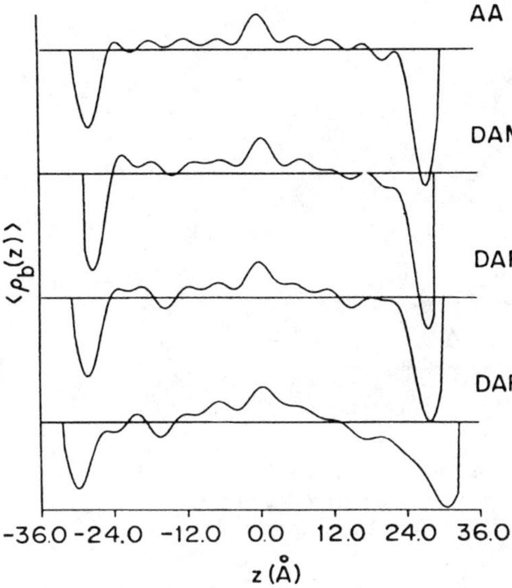

Figure 7.2 Density distribution of an average bilayer for the four types of films for which data are shown in Figure 7.1. The arrows indicate the positions of the diacetylene groups, where present. (From Fischetti et al.[3])

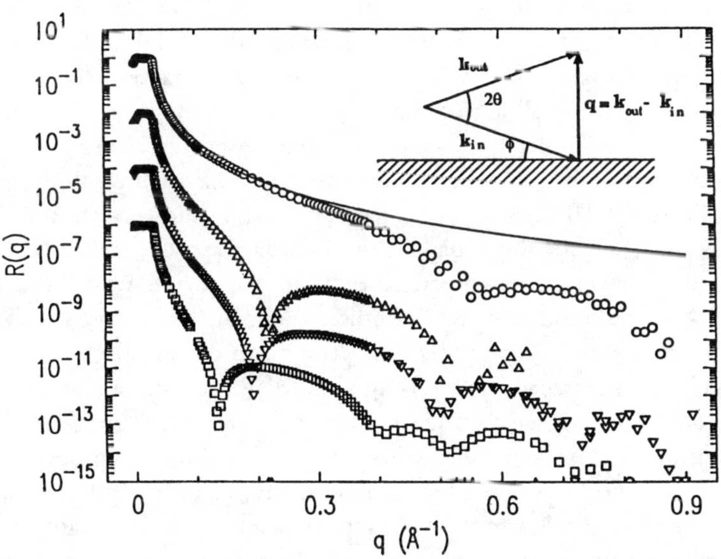

Figure 7.3 Normalized reflectivity data from various samples; successive data sets are shifted vertically by two orders of magnitude for clarity. The line is the theoretical reflectivity from an ideal step-function interface between bulk silicon and air; (○) actual silicon surface, no monolayer; (△) with SA monolayer of C_{10} alkylsiloxane; (▽) C_{12} alkylsiloxane; (□) C_{18} alkylsiloxane. (From Tidswell et al.[4])

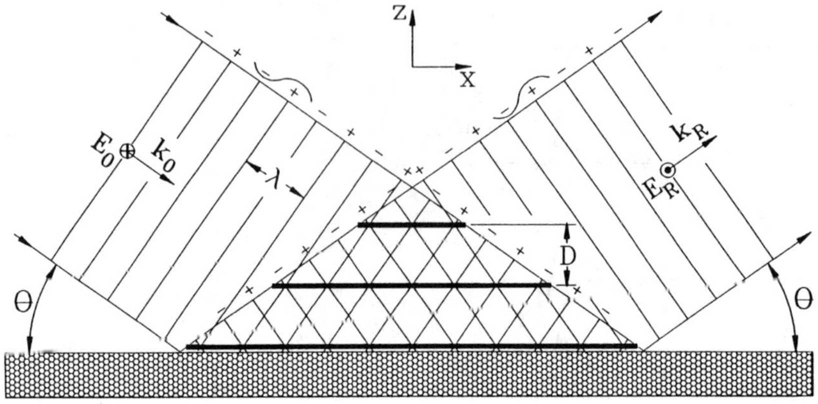

Figure 7.4 **Schematic diagram showing how a standing-wave field is formed by interference between the incident and scattered plane waves above a mirror surface. The spacing between antinode planes is $D = \lambda/2\sin\theta$. (From Bedzyk et al.[6])**

that the *y*-axis in Figure 7.3 is a log scale). Getting better data over a wider range, thus allowing more precise fits, requires synchrotron radiation. Because synchrotron radiation is intense and parallel, and because it contains a wide continuous range of frequencies, a number of other techniques become practicable. Applications of these techniques to organic films are mentioned only briefly here. By tuning the X-ray wavelength to the absorption edge of an atomic species in the film, and also slightly above and slightly below it, and performing diffraction scans at each of these wavelengths, it is possible to obtain information on the distribution of this atomic species. For example, Pachence et al.[5] have used resonance X-ray scattering to locate the iron atom within a cytochrome *c* monolayer bound to the surface of an arachidic acid L–B film. Another technique, standing-wave fluorescence, utilizes the fact that standing waves parallel to the surface are set up as a result of interference between incident and reflected beams in the region where they overlap (see Figure 7.4). The standing-wave periodicity depends on the scattering angle; it is possible to set up small spacings by diffracting from the substrate lattice (at relatively large angles), larger spacings by using a multilayer substrate and thus a smaller angle, and even larger spacings using total external reflection (very small angles). When an antinode plane coincides with a layer of a fluorescing atomic species inside the organic film, one observes maximum fluorescence, and so this technique can be used to locate this atomic species in the *z* direction. For example, Bedzyk et al.[6] studied the zinc fluorescence from an L–B structure consisting of a cadmium arachidate bilayer followed by four layers of ω-tricosanoic acid and by one bilayer of zinc arachidate (Figure 7.5). To get spacings between antinodes comparable to the large spacings in an L–B film, total reflection from the surface was used (as shown by the reflectivity data in Figure 7.5*a*—the critical angle is indicated). Figure 7.5*b* shows the dependence of the zinc fluorescence on incident angle; the

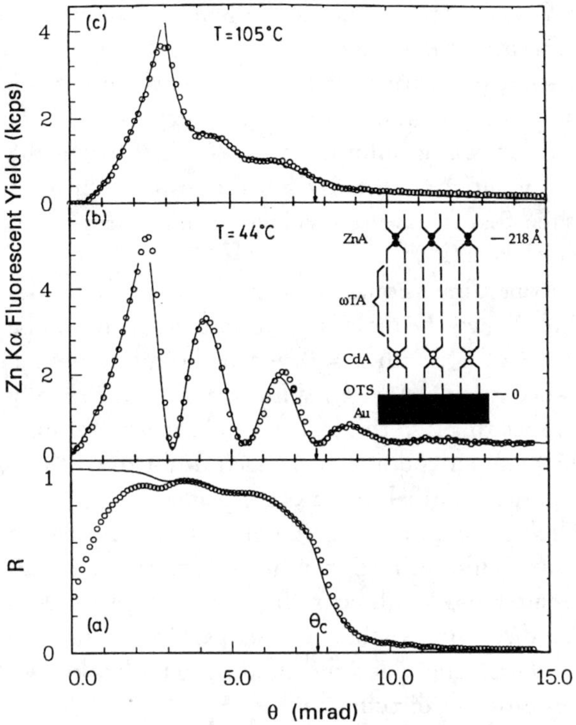

Figure 7.5 **Angular dependence of: (*a*) reflectivity for the L–B structure on an Au mirror surface (schematically depicted in the inset); (*b*) Zn Kα fluorescence count rate at 44 °C; (*c*) Zn Kα fluorescence count rate at 105 °C. The circles are experimental data and the solid lines are calculated. In the inset, circles represent heavy atoms and line segments represent hydrocarbon chains. (From Bedzyk et al.[6])**

oscillations are caused by successive antinode planes passing through the zinc layer. These data indicate that the zinc layer is quite well-defined; Bedzyk et al. calculated that the layer is at a mean distance of 218 Å from the substrate surface with a full width at half maximum (FWHM) of 24 Å. However, upon heating to 105 °C, the zinc fluorescence curve became much broader (Figure 7.5*c*), corresponding to a poorly defined zinc layer (mean distance from substrate ≈ 174 Å, FWHM ≈ 110 Å).

7.4 Structure Within the Film Plane

In this section, we will discuss all studies in which **k** has a component in the film plane. If the substrate is thick enough to be opaque to X rays (as is almost always true), it is not possible to have **k** entirely in the film plane, because this would require a substantial number of X rays to penetrate long distances laterally through the very thin film. However, one can come very close. As mentioned earlier, X rays undergo total external reflection at small incidence angles. When this occurs, the X rays are scattered near the surface rather than from the bulk of the substrate; thus, in this geometry the X rays are surface sensitive (or thin-film sensitive). If the scattered beam is also detected at grazing angles, **k** is almost entirely in the film plane. However, some of the most interesting information is obtained when the beam is incident at grazing angles for surface sensitivity, but the scattered beam is detected in arbitrary directions such that **k** has non-negligible components both normal to the plane (k_z) and parallel to it ($\mathbf{k}_p$).

Consider, as a simple example, a one-dimensional monolayer array of molecules thought of as rigid rods (Figure 7.6). When the molecules are vertical, there will be diffraction maxima at specific values of $\mathbf{k}_p$, with $k_z = 0$; when the molecules are tilted, the diffraction maxima will occur at nonzero k_z. While this follows formally from Equation 7.1, a simple way to see this is to think of the molecules as forming narrow "Bragg planes"—when the plane orientation is changed, **k** for the diffraction peak will also change ($\mathbf{k}_{peak}$ must be normal to the Bragg planes).

In two dimensions, both the tilt angle and the tilt direction relative to the lattice are relevant. If the vector **k** for a specific diffraction peak is normal to the tilt direction, the tilt angle does not matter and **k** will be in the horizontal plane (k_z = 0), but diffraction peaks in all other directions will have nonzero k_z. In general, one can determine the lattice structure, the tilt angle, and the tilt direction if **k** is known for at least two diffraction peaks in different directions.

As an example, Figure 7.7 shows diffraction data of Shih et al.[7] from a Langmuir monolayer of heneicosanoic acid spread on a 0.5 mM $CaCl_2$ solution at zero pressure (5 °C). Perhaps unexpectedly, the monolayer has some degree of order even at zero pressure; two diffraction peaks can be seen, plotted as intensity contours in the

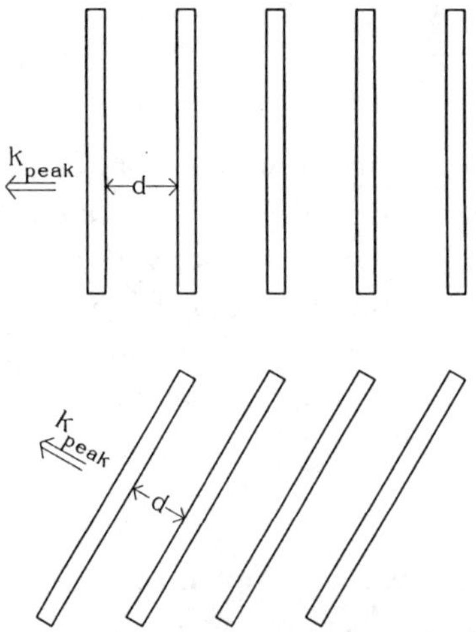

Figure 7.6 **Schematic diagram showing *d* spacing and direction of $\vec{k}$ at diffraction peak for one-dimensional arrays of tilted and untilted molecules.**

$\mathbf{k}_{xy}$–k_z plane. (The monolayers are powders in the plane, thus the x and y components of $\mathbf{k}$ cannot be distinguished.) It is clear that one peak has $k_z = 0$, the other has $k_z \neq 0$; from the data shown, it can be determined that the molecules form a distorted-hexagonal lattice in the horizontal plane, with centered-rectangular two-molecule unit cell dimensions $a \approx 5$ Å and $b \approx 8.75$ Å. Moreover, the tilt direction is towards a near neighbor in the direction of a.

The peak widths in k_z are a function of the thickness of the film and are essentially the same in all three cases because the thickness is insensitive to small changes in tilt angle. The width of the in-plane peak in $\mathbf{k}_{xy}$ is limited by instrumental resolution, but the width of the off-plane peak is considerably broader than the resolution. In other words, this structure, while ordered enough to produce diffraction peaks, is not a solid but an anisotropic mesophase, a two-dimensional analog of a liquid crystal.

It can also be seen in Figure 7.7 that as pH is increased, the in-plane peak hardly moves at all, while the off-plane peak moves to higher $\mathbf{k}_{xy}$ and lower k_z. This means that the unit cell becomes smaller as pH increases, but only in one direction; b remains essentially constant while a changes (from 5.19 Å at pH = 3.4 to 4.96 Å at pH = 7.3). Since the tilt is in the a direction, one might expect that this contraction would be correlated with a decrease of the tilt angle, and indeed the tilt angle decreases from 29° to 24°. Spink and Sanders[8] reported in 1955, using isotherm

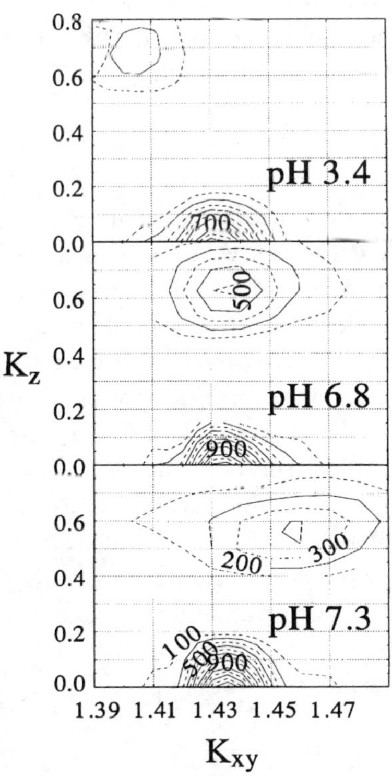

Figure 7.7 **Diffracted intensities, plotted as contours in the k_z–$\vec{k}_{xy}$ plane, for monolayer of heneicosanoic acid on $CaCl_2$ solution (5 °C, 0 dyn/cm) at various pH values, as indicated. Intensity units are arbitrary; each line differs from adjacent lines by the same amount. (From Shih et al.[7])**

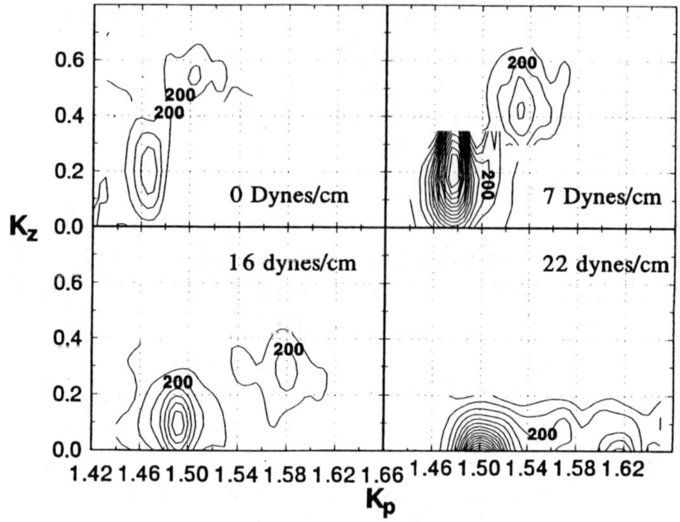

Figure 7.8 **Diffracted intensities, plotted as contours in the k_z–$\vec{k}_{xy}$ plane, for a monolayer of heneicosanol on water (20 °C, pH = 2) at various pressures, as indicated. Intensity units are arbitrary; each line differs from adjacent lines by the same amount. (From Shih et al.[9])**

studies, that the effect of increasing pH was to "condense" the monolayer; these data show that indeed the microscopic area/molecule decreases, but also that the "condensation" is uniaxial, presumably due to an anisotropic interaction of the divalent calcium ions with the carboxyl groups.

As an example of data from a different structure, we show in Figure 7.8 the diffracted intensity contours[9] from a monolayer of heneicosanol on water (pH = 2) at 20 °C. In this case, there are again two peaks, and these peaks are visible even at zero pressure, but now both peaks have $k_z \neq 0$ at lower pressures. From this it can be deduced that the lattice is distorted-hexagonal, and that the tilt direction is towards a next-nearest neighbor (the *b* direction in terms of the rectangular unit cell mentioned earlier). As a function of pressure, both peaks move to higher $\mathbf{k}_{xy}$ and lower k_z; in other words, the unit cell shrinks in both directions as a function of pressure and the tilt angle decreases. At 22 dyn/cm, both peaks have $k_z = 0$, that is, the tilt angle is zero. The isotherm for this monolayer shows a "kink" at about 21 dyn/cm; in other words, these X-ray data reveal that the phase transition in the isotherm is between a tilted phase and an untilted phase. The higher pressure phase is less compressible (steeper isotherm) than the lower pressure phase; the X-ray data show that this is because the tilt angle changes in the lower pressure phase. Incidentally, while peak widths are not very clear from these data, other data show both peaks to be slightly wider than the instrumental resolution—these phases are not solids but partially-ordered mesophases.

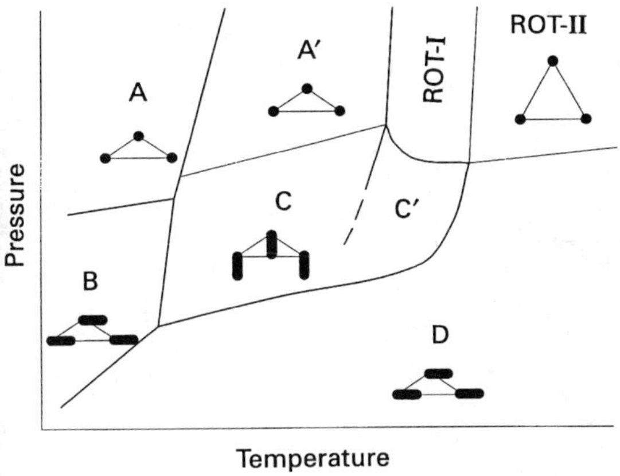

Figure 7.9 **Generalized qualitative phase diagram for Langmuir monolayers of saturated fatty acids, showing structures of each phase as determined by X-ray diffraction (the temperature and pressure origins depend on the chain length). In the schematic diagrams, the molecules are indicated by the "shadows" of cylinders on the water surface, so a circle indicates that the molecules are vertical, while elongation indicates the direction of tilt. Nonequilateral triangles indicate that the structure is distorted-hexagonal, but the distortion has been exaggerated. The phases C and C′ have the same qualitative structure, with a small difference in the unit cell dimensions; the "Rotator-I" phase has a slightly-distorted structure intermediate between the distorted-hexagonal A′ and hexagonal Rotator-II phases. (From Shih et al.[13])**

With this kind of data, much more is known today about Langmuir monolayer structures than a few years ago. Saturated fatty acids, for example, have been known since 1945 to have very complex phase diagrams, but this information has been ignored until recently because of the dogma that monolayer phases are either "solid," "liquid condensed," "liquid expanded," or gaseous. More recently, it has been determined that all fatty acid monolayers have the same phase diagram,[10] with a chain-length dependent shift in the temperature and pressure axes, and that the structures of the various phases[11–14] are, qualitatively, as shown in the schematic diagrams (Figure 7.9). The lowest temperature phases (A, B) have resolution-limited diffraction peaks, but the other phases are less well ordered. There are also curious anomalies in the peak widths[15]: the "Rotator-I" phase, for example, shows wider diffraction peaks (smaller correlation lengths) than the phases surrounding it, including the higher temperature Rotator-II and lower pressure C′ phases. In 1942, Copeland et al.[16] reported an anomalous minimum in the monolayer viscosity exactly in this region of the phase diagram; we now have correlation length information consistent with this finding.

Similar studies can be performed on monolayers transferred to substrates (Langmuir–Blodgett films); in fact such studies are made easier by the fact that solid substrates, unlike water, can be oriented in the X-ray beam. For example, Shih et al.[17]

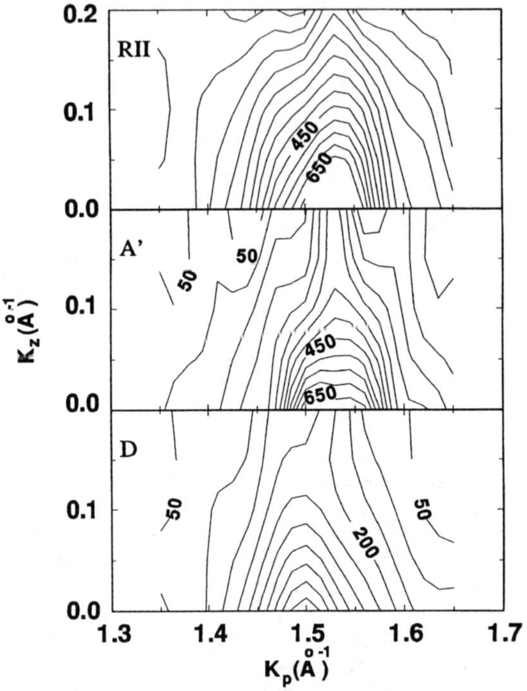

Figure 7.10 **Diffracted intensities, plotted as contours in the k_z–$\vec{k}_{xy}$ plane, for L–B films deposited from three different Langmuir monolayer phases. The label on each panel refers to the phase of the monolayer on water (compare Figure 7.9). Intensity units are arbitrary; each line differs from adjacent lines by the same amount. (From Shih et al.[17])**

transferred monolayers to glass from each of three phases (A′, D, and Rotator II—see Figure 7.9). These phases were chosen because A′ and D have distorted-hexagonal lattices while Rotator-II is hexagonal; also, A′ and Rotator-II have untilted (vertical) molecules, while D has tilted molecules. Figure 7.10 shows diffracted intensity contours from the transferred monolayers, labeled by the "parent" phase. In all cases, there is only one diffraction peak, and this peak is in the horizontal plane ($k_z = 0$). In other words, the L–B film has little "memory" of the structure of the parent system.

Our final example involves self-assembled monolayers of alkylsilanes adsorbed on single-crystal gold substrates, and studied recently by Fenter et al.[18] They found that the molecules attached to specific sites on the gold surface lattice, as shown in real space in Figure 7.11*a* and in reciprocal space in Figure 7.11*b*. The data for scans along the direction of **k** are shown in Figure 7.11*c* for $C_{12}H_{25}SH$, both immediately after deposition (broader peak) and after annealing to 90 °C (sharper peak). From data not shown here, they determined that the chains were tilted by about 32.5° in a direction about 14° away from a near-neighbor direction. However, when a longer

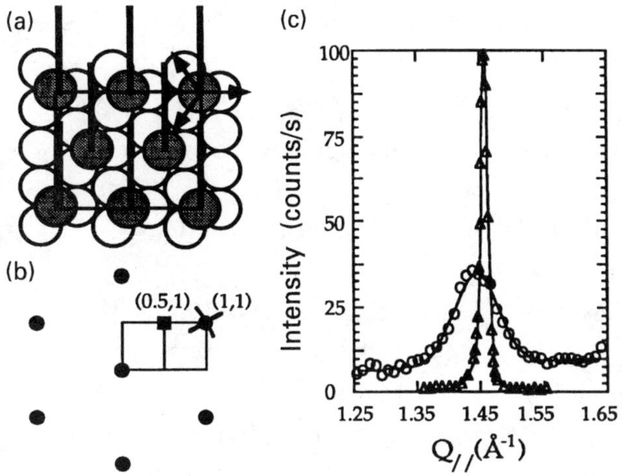

Figure 7.11 (*a*) Top view of SA film surface, indicating Au atoms (*open circles*) and S atoms (*shaded circles*), a next-nearest-neighbor tilt of the hydrocarbon chains (*vertical solid lines*), the unit cell, and the direction of the lateral displacement of the top layer of Au atoms (*arrows*). (*b*) Reciprocal-space structure, showing only one (0.5,1) peak of the monolayer structure (*square*). (*c*) Radial scan through (1,1) diffraction peak of C_{12} monolayer: as-deposited (*circles*) and after anneal (*triangles*). (From Fenter et al.[18])

molecule ($C_{18}H_{37}SH$) was used, the tilt direction was 52° away from the nearest-neighbor direction, while the tilt angle was almost the same, about 30.5°.

7.5 Summary

We have given examples of how various kinds of structural information may be obtained using X rays. These examples illustrate that structure is not just a question of, for example, measuring the contact angle and deciding that the monolayer is quite densely packed and therefore "ordered." Considerably more information can be and should be extracted; the functionality of an organic film may depend critically on these details.

References

1 A. Ulman. *An Introduction to Ultrathin Organic Films.* Academic, New York, 1992.

2 M. Prakash, J. B. Peng, J. B. Ketterson, and P. Dutta. *Chem. Phys. Lett.* **128**, 354, 1986.

3 R. F. Fischetti, M. Filipkowski, A. F. Garito, and J. K. Blasie. *Phys. Rev.* **B37**, 4714, 1988.

4 I. M. Tidswell, B. M. Ocko, P. S. Pershan, S. R. Wasserman, G. M. Whitesides, and J. D. Axe. *Phys. Rev.* **B41**, 1111, 1990.

5 J. M. Pachence, R. F. Fischetti, and J. K. Blasie. *Biophys. J.* **56**, 327, 1989.

6 M. Bedzyk, G. M. Bommarito, and J. S. Schildkraut. *Phys. Rev. Lett.* **62**, 1376, 1989.

7 M. C. Shih, T. M. Bohanon, J. M. Mikrut, P. Zschack, and P. Dutta. *J. Chem. Phys.* **96**, 1556, 1992.

8 J. A. Spink and J. V. Sanders. *Trans. Faraday Soc.* **51**, 1154, 1955.

9 M. C. Shih, T. M. Bohanon, J. M. Mikrut, P. Zschack, and P. Dutta. *J. Chem. Phys.* To be published.

10 A. M. Bibo and I. R. Peterson. *Adv. Materials.* **2**, 309, 1990.

11 P. Dutta. In *Phase Transitions in Surface Films 2.* (H. Taub, G. Torzo, H. J. Lauter, and S. C. Fain, Eds.) Plenum, New York, 1991, p. 183.

12 B. Lin, M. C. Shih, T. M. Bohanon, G. E. Ice, and P. Dutta. *Phys. Rev. Lett.* **65**, 191, 1990.

13 M. C. Shih, T. M. Bohanon, J. M. Mikrut, P. Zschack, and P. Dutta. In *Surface X-Ray and Neutron Scattering.* Springer Proceedings in Physics, Vol. 61, (I. K. Robinson and H. Zabel, Eds.) Springer-Verlag, Berlin, 1992, p. 151.

14 R. M. Kenn, C. Böhm, A. M. Bibo, I. R. Peterson, H. Möhwald, K. Kjær, and J. Als-Nielsen. *J. Phys. Chem.* **95**, 2092, 1991.

15 M. C. Shih, T. M. Bohanon, J. M. Mikrut, P. Zschack, and P. Dutta. *Phys. Rev.* **A45**, 5734, 1992.

16 L. E. Copeland, W. D. Harkins, and G. E. Boyd. *J. Chem. Phys.* **10**, 357, 1942.

17 M. C. Shih, J. B. Peng, K. G. Huang, and P. Dutta. To be published in *Langmuir.*

18 P. Fenter, P. Eisenberger, and K. S. Liang. Submitted to *Phys. Rev. Lett.*

8

High Resolution EELS Studies of Organic Thin Films and Surfaces

LAWRENCE H. DUBOIS

Contents

8.1 Introduction

The casual reader might consider the scattering of low-energy electrons from polymer surfaces or organic thin films to provide little useful information due to charging and electron-induced damage. This is in fact *not* the case; I will demonstrate in this chapter that by choosing appropriate materials and equipment, high quality vibrational spectra can be obtained by the inelastic scattering of low-energy electrons. A representative spectrum is shown in Figure 8.1. The sample consists of a multilayer of tetracyanoquinodimethane (TCNQ) adsorbed on a Cu(111) single crystal substrate at 100 K.[1] The strong peak at the left is due to elastically scattered electrons at an energy of 5.4 eV. A small fraction of these incident electrons lose energy (i.e., scatter inelastically) and these are shown to the right. Peak frequencies are in wavenumbers and are referenced to the elastic peak at 0 energy loss. Each peak corresponds to a vibrational mode in the TCNQ film.[1]

The technique used to record this spectrum is called high resolution electron energy loss spectroscopy or HREELS. (The basic principles are described in the lead

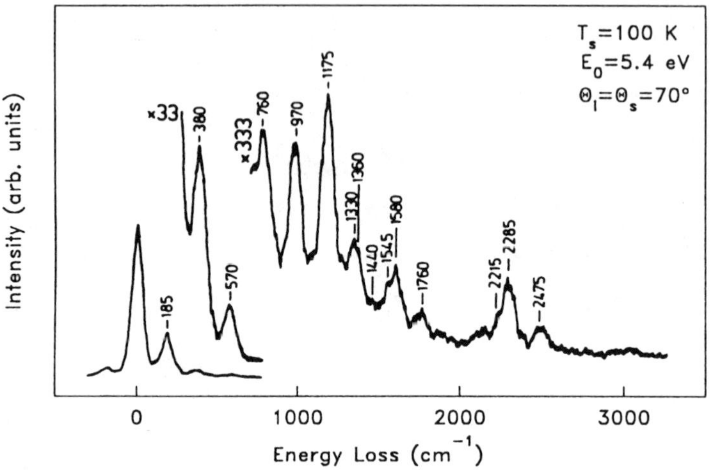

Figure 8.1 HREELS spectrum of a multilayer of tetracyanoquinodimethane (TCNQ) physisorbed on a Cu(111) single crystal substrate at 100 K. Mode assignments are discussed in Reference 1.

volume of the Series, *Encyclopedia of Materials Characterization.*) Historically, EELS has been employed to study the structure and bonding of small molecules on single crystal metal and semiconductor surfaces.[2] Only recently has it been used to develop an understanding of the properties of organic thin films. This includes a knowledge of polymer chain configurations and the relative orientation of chemical groups at an interface. In this chapter, I concentrate on why this is a valuable technique for the study of these materials and will use several examples from the literature to illustrate these points.

8.2 The Scattering Mechanism

When an electron is incident on the surface of a material, one of several processes can take place. For example, an electron can penetrate the surface of a conducting substrate and eventually reach earth potential. Inelastic scattering can take place during this process, and information about the surface layer or a thin film can be gained by measuring the electron current absorbed as a function of the incident energy.[3]

If the material is an insulator, however, the electron can be absorbed into the surface layer, causing the material to charge negatively. This process occurs until the surface potential gets too high and all further electrons are repelled. Alternatively, the incident electron can produce secondary electrons, which exit into the vacuum, causing the material to charge positively. When the surface charges, little information is gained, since the energy of the scattered electrons is continuously changing. By "flooding" the surface with the correct flux of electrons, the positive or negative charging problem can be compensated for and useful spectra recorded.[4]

The final possibility is that the incident electron scatters from the surface. This scattering process can be elastic, which, if the surface is ordered, can provide useful structural information through a diffraction process. In addition, inelastic scattering can occur. Energy can be lost to either electronic (from 1 to several electron-volts) or vibrational (≲500 meV or ≲4000 cm^{-1}) degrees of freedom of the substrate, and this has proven to be a powerful method for obtaining information on the structure, bonding, and orientation of molecules at surfaces (see, for example, Figure 8.1 and Reference 2).

This chapter concentrates only on the excitation of surface vibrational modes using the HREELS technique. This inelastic scattering process can occur in one of three ways, depending on the details of how the incident electron interacts with the surface layer. In order of importance, they are dipole scattering, impact scattering, and resonance scattering.[5]

Dipole Scattering

The dipole-scattering mechanism involves the interaction of the electric field of the incoming electron with the long-range (≳100 Å) *dipole* field of the sample. Thus, only infrared active modes (i.e., those modes having a dipole derivative) will be excited. The scattering can be thought of as a two-step process—a simple reflection by the surface followed and/or preceded by energy loss. Since the interaction can occur at a long distance, the momentum transfer *parallel* to the surface is small and scattering is approximately specular (≲±1°). If the surface is rough or disordered, as in the case of most organic thin films, scattering may appear diffuse on a macroscopic scale. An example of this is shown by the triangles in Figure 8.2, where the angular distribution of electrons scattered from a polystyrene sample is shown.[5] Although the angular width is ~30°, the scattering is still peaked in the specular direction (arrow), indicating that dipole scattering must be occurring. This broad angular distribution will also tend to limit the spectral resolution.

On *conducting* substrates, an image charge of the incident electron is formed, and this effectively cancels the electric field parallel to the surface. This gives rise to a surface selection rule, which, simply stated, is that only those modes with a component of the dipole derivative perpendicular to the surface will be excited.[2, 5] This is the same selection rule which applies to infrared spectroscopy.[2] Thus, under certain specialized conditions, the orientation of the surface layer or molecules adsorbed on the surface can be determined.

Impact Scattering

If dipole scattering was the only mechanism by which electrons could interact with organic thin films, there would be little advantage to using EELS for probing the vibrational modes of a surface. Electrons may also lose energy by interacting directly with the atomic potentials of the sample. This short-ranged phenomenon (~1 Å) results in a large momentum transfer and leads to very diffuse scattering with no clear specular lobe (see, for example, the circles in Figure 8.2[6]). Because of this

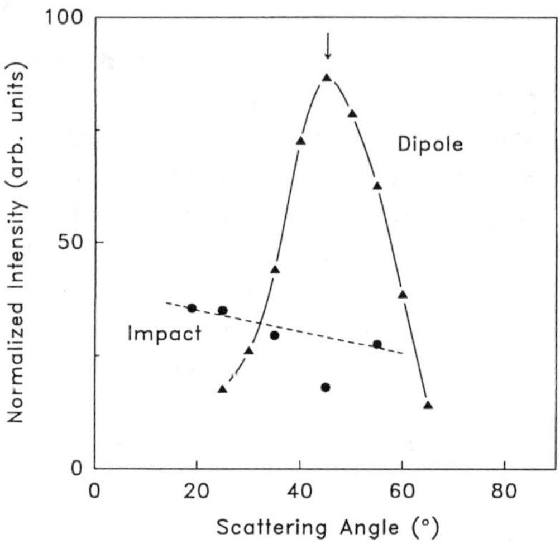

Figure 8.2 **Angular distribution of low-energy electrons inelastically scattered from polystyrene (C–H bend, triangle)[4] and polyimide (C–H stretch, circles)[6] thin films. Dipole interactions dominate the scattering from polystyrene since the intensity is peaked in the specular direction (arrow, 45°). The lobe here is much broader than that seen on well-ordered substrates.[2] The absence of a peak in the angular distribution of the polyimide data is strong evidence for impact scattering. The data is normalized to the elastic scattering intensity in both cases.**

broad angular distribution, impact scattering is typically 1–2 orders of magnitude less intense than dipole scattering *in the specular direction.* The short interaction distance implies that an image potential of the incident electron is not formed, and therefore there are effectively no selection rules for impact scattering† (see Reference 5). Thus, both infrared and Raman active modes are observed. An additional consequence of this short interaction distance is the strong dependence of the scattering intensity on the incident electron energy.[4, 5] Along with the differing angular dependences, this can be effectively used to discriminate between dipole and impact scattering regimes. Furthermore, this energy dependence can also be used to "tune" the penetration depth of the interaction by changing the mean free path of the incident beam in the material. In practical terms, the effective sampling depth can be varied from ~3 to 25 Å.[7]

Resonance Scattering

The third scattering mechanism, resonance scattering, is well known in gas phase EELS experiments and has recently been observed in polymer thin films. Examples

† There is actually a weak selection rule on metal surfaces due to the geometry of the scattering, but this is not the case for organic thin films (see, for example, Reference 4).

are presented in References 8–10. Here, a short lived (~10^{-12}–10^{-15} s) negative ion is formed by trapping the incident electron in a shape resonance of a molecule in the film. Ejection of the electron leaves the resulting neutral molecule in an excited vibrational state. The transition is governed by the energies of the ion and ground state, and the Franck–Condon overlap factors. As expected, the capture cross section is a strong function of the incident electron energy and depends intimately on the *local* electronic structure of the material.[5] The angular distribution of the ejected electrons is also quite broad. Unlike dipole and impact scattering, resonance scattering typically results in the formation of intense overtone bands. In combination, these three scattering modes provide tremendous detail about the structure and bonding of the surfaces of organic materials.

8.3 The Spectrometer

HREELS is, in principle, a straightforward technique. All that is required is a narrow, monochromatic beam of low-energy electrons, a surface to scatter them off, and an energy analyzer and detector for the scattered electrons. Generating a high quality incident beam is not easy, however, and numerous articles and books have been written on the subtleties of this process.[2, 11] Since this is an inelastic scattering process, the spectral width is determined either by the energy resolution of the incident electron beam (typically 30–60 cm^{-1}) or by the linewidth of the transition (typically ≤20 cm^{-1}). Despite the large number of potentially overlapping transitions in many organic materials and the inhomogeneous nature of most polymer surfaces, experiments are generally limited by the resolution of the spectrometer. However, recent advances in both hardware and software have helped to alleviate this problem.

A schematic diagram of a typical HREELS spectrometer is shown in Figure 8.3. Electrons are boiled off a hot tungsten or LaB_6 filament and focused onto the entrance slit of an electrostatic energy analyzer by means of a simple lens system. Due to the short mean free path of electrons in air, a high vacuum (≤10^{-5} torr) is required. A monochromator is needed at this point, since the energy spread of thermally generated electrons is typically 200–250 meV (1600–2000 cm^{-1}). In this case, the electrons are monochromatized at a low pass energy in a 127° cylindrical electrostatic sector. They are then accelerated and focused onto the sample by another set of lenses. Typical incident beam energies and currents are 5 eV and 2×10^{-10} A, respectively, though these values can be varied by an order of magnitude higher or lower. These low-energy incident beam energies and currents ensure minimal sample damage.

Only a small fraction (typically <1%) of the incident electrons actually scatter from the sample (this includes both elastic and inelastic processes), the remainder being absorbed into the substrate. The few scattered electrons are decelerated by another set of lenses and focused onto the entrance slit of the analyzer sector. The

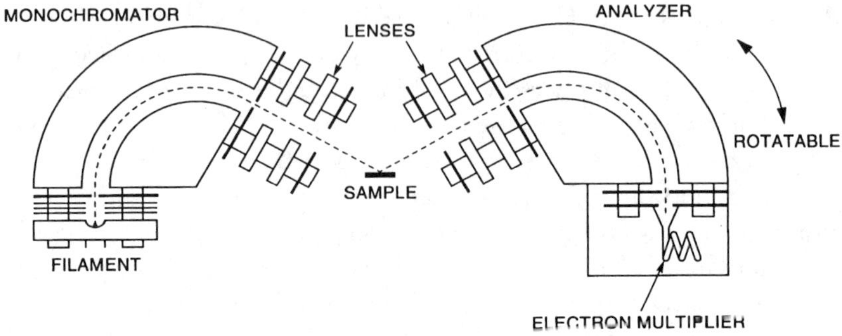

Figure 8.3 **Schematic diagram of a typical HREELS spectrometer using 127° cylindrical sectors for the energy dispersing elements. The main features of this system (including an electron source, monochromator, lenses, analyzer, and detector) are found in all HREELS spectrometers. The dashed line shows the path of the electrons through the different elements. The analyzer is rotated to measure the angular distribution of the scattered electrons. More details can be found in Reference 2.**

energy distribution of these electrons is then measured by scanning the analyzer in a retarding potential ("window") mode. The electrons are finally detected, and the signal is amplified, counted, and recorded outside of the vacuum system. The angular distribution is measured by rotating the analyzer sector (see Figure 8.3). Though milliamperes of current ($>10^{15}$ e/s) are emitted by the filament, only 10^{-17} A (~10–100 e/s) of inelastically scattered electrons actually reach the detector. The overall spectrometer resolution is determined by a convolution of the resolution of the monochromator and analyzer, and is typically $\sqrt{2}$ times that of the individual components (30–80 cm^{-1}).

Clearly, the limiting feature in the usefulness of HREELS is the resolution of the spectra (see, for example, Figure 8.1). Recent advances allow an improvement by a factor of approximately two. The first is instrumental. In a recent publication, Ibach and co-workers[11] demonstrated an enhanced spectrometer with a resolution of 16 cm^{-1}. These results were obtained on a pristine metal single crystal surface. Corresponding studies on disordered organic materials have not been reported, but based on this work, it is expected to improve the resolution there as well. This is not a complete cure, however, since most conventional HREELS spectrometers yield spectra with significantly better spectral resolution on single crystal substrates than on disordered samples. The reason for this lies in the broad angular distribution of the scattered electrons (see, for example, Figure 8.2), which, because of the finite slit width and angular resolution of the analyzer, decrease the spectral resolution. The other tack is to use one of several available resolution-enhancing signal-processing algorithms to treat the stored data. This is a common technique in infrared spectroscopy and is now becoming popular for the deconvolution of HREELS data.[7, 12] An example of this is discussed later.

8.4 EELS Versus Other Techniques: Advantages and Disadvantages

It is clear from the preceding discussion that HREELS is no ordinary technique—it requires highly specialized equipment and a high-vacuum environment. Samples must not outgas significantly and must be able to withstand bombardment by low-energy electrons. Does EELS have any advantages that other vibrational spectroscopies do not? The answer is a definitive yes. First of all is its greater inherent surface sensitivity. Unlike photons, electrons, which eventually scatter from the surface, do not penetrate deeply into the bulk of the sample (≤25 Å), producing a greater surface sensitivity. As previously noted, this sensitivity can be tuned somewhat by varying either the incident electron energy or the scattering angle. The ultimate sensitivity is on the order of 0.1% of a monolayer. Furthermore, by appropriately choosing both the incident electron energy and scattering angle, either dipole, impact, or resonance scattering can be enhanced, thus allowing modes of all symmetries to be observed. The broad spectral range (including the near, mid, and far infrared) allows a complete vibrational spectrum to be collected in a single instrument. The range can be extended easily to include electronic transitions (for an example, see Reference 13).

No single technique can provide all of the information necessary to obtain a complete picture of the structure and bonding of organic materials. Though HREELS has its place, it is not the perfect tool. In the following sections I will summarize results from the literature that demonstrate some of the advantages of EELS and, in so doing, compare the results of these studies with the results of complementary experiments using both infrared and Raman spectroscopies.

8.5 Examples

Resolution Enhancement

As noted above, one of the major problems with HREELS is that by optical standards it is *not* a high resolution technique. Even the best resolution is still an order of magnitude worse than with either infrared or Raman spectroscopies. Good resolution is essential for correctly assigning vibrational peaks, separating closely spaced modes, and observing subtle changes in peak shapes and/or frequencies. Some of this need for enhanced resolution is obviated in the case of organic thin films and surfaces, since most linewidths are greater than 20 cm^{-1}. This is caused by both the number of modes (due to the complex structure of many polymers) and the inhomogeneous broadening of these bands[†]. Nevertheless, improved resolution would go a long way toward making EELS a more generally useful technique. Resolution-

† This is not the case for well-ordered monolayers, where peak splittings on the order of 1–2 cm^{-1} can be observed by infrared spectroscopy. (See Reference 14.)

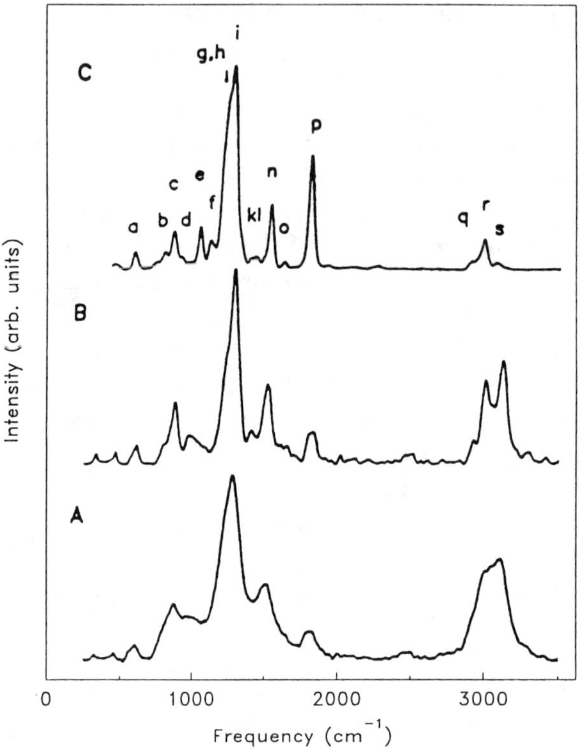

Figure 8.4 **Comparison of HREELS and infrared spectra from a polycarbonate thin film.[10] The EELS spectrum of the film adsorbed on a graphite substrate is shown in A. The incident energy was 5 eV, and the background was subtracted. In B, the resolution of this spectrum was enhanced using the maximum likelihood method, taking the peak shape from the elastic scattering peak. These results are compared to a transmission infrared spectrum (part C) that had been artificially broadened to a 30 cm^{-1} resolution. Mode assignments for peaks "a" through "s" are discussed in Reference 10.**

enhancement methods, such as Fourier deconvolution and maximum likelihood, have proven successful in many areas and have recently been applied to the study of polymer surfaces by HREELS.[7, 12]

Figure 8.4 is an example of the application of the maximum likelihood method to improve the resolution of the vibrational spectrum of a polycarbonate thin film adsorbed on a graphite substrate.[10] Maximum likelihood is a statistical approach which uses the known instrumental lineshape (determined by the shape of the elastic scattering peak) and a knowledge of the type of data being restored to calculate a "most probable" estimate of the correct spectrum.[12] The application of this technique is straightforward, transforming the rather broad, ill-defined spectrum in **A** (in Figure 8.4) to the much more detailed spectrum in **B**. Each of these peaks can be assigned by a comparison with the transmission infrared spectrum of bulk polycarbonate, shown in **C**. For example, peaks **b** and **c**, attributable to out-of-plane carbonate-group vibrations and aromatic deformations, respectively, are successfully resolved in **B**. Similarly, peaks **k** and **l**, which are due to methyl CH bending modes, and peak **o**, due to the aromatic C–C skeletal ring modes, are also distinguishable. Complete mode assignments are discussed in Reference 10. The maximum likelihood method is superior to standard linear deconvolution schemes in many ways.[12] First, it does not produce negative-going artifacts, nor does it leave strong side lobes which may mask weak features. Second, it does not require a high signal-to-noise ratio, and it can deal with asymmetric peak shapes. Finally, no a priori guesses need be made about filter functions or filter shapes. Thus, despite the limited resolution of EELS, high quality vibrational spectra can be recorded. An improved picture of the bonding and orientation of the polycarbonate film can now be obtained.

Linearity

With or without resolution enhancement, EELS would be a relatively useless technique if it could not be made quantitative. Though one must always be careful, because of changes in the scattering mechanism (see below), under certain circumstances EELS intensities can be used to quantitate surface coverage. Figure 8.5*a* shows an example of this using mixed monolayers of Cd-stearate with different ratios of perdeuterated and undeuterated stearic acid.[15] Samples were formed by Langmuir–Blodgett techniques on a Ag(111) single crystal substrate. Traces (i) and (v) are for the reference phases of pure deuterated and pure hydrogenated stearic acids, respectively. The strong peaks at 2210 cm^{-1} in (i) and 2952 cm^{-1} in (v) result from excitation of C–D and C–H stretching vibrations, respectively. The stretching vibrations of the carboxylate group are observed at 1419 and 1533 cm^{-1} in (i) and (v), respectively. The nature of the weaker features is discussed in Reference 15.

Traces (ii) through (iv) in Figure 8.5*a* show a series of HREELS spectra recorded as a function of the concentration in solution of different mixtures of perdeuterated and undeuterated stearic acid. For ideal mixtures, the signal intensities for the C–H and C–D stretching vibrations should be proportional to the corresponding concentration ratios. This is clearly shown by the plot in Figure 8.5*b*, which summarizes data recorded from a more extensive data set.[15] Here, the *ratio* of the peak areas of the C–H and C–D stretching vibrations is plotted, since absolute intensities vary from sample to sample due to the subtleties of film preparation and to surface roughness.[8,9] The plot yields a straight line with a slope of 1.0 ± 0.02, demonstrating that no large domain formation and/or demixing is occurring in the films. The data also show that the EELS intensities are linearly proportional to the coverage. Another interesting feature of these data is that the intensity ratio at a surface coverage of 1:1 perdeuterated-to-undeuterated stearic acid is 1.0, implying that the cross sections for the excitation of C–H and C–D vibrations are equal. This observation is not a general phenomenon and is only a consequence of the incident energy used (5 eV).[15]

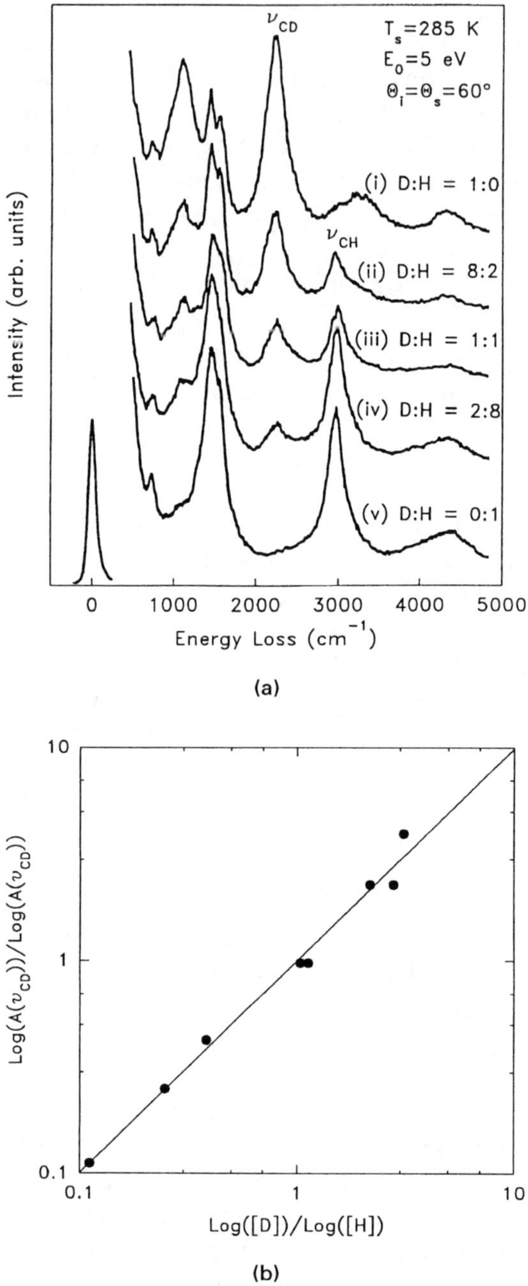

Figure 8.5 **(*a*) HREELS spectra of several mixtures of undeuterated and perdeuterated Cd-stearate thin films formed by Langmuir–Blodgett techniques. Mode assignments are discussed both in the text and in Reference 15. The ratio of the peak area of the C–D and C–H stretching vibrations versus the corresponding concentrations of perdeuterated and undeuterated stearate in the films is plotted in a log–log fashion in (*b*).**

Depth Sensitivity

The selective deuteration of monolayers can also be used to test the depth sensitivity of the incident electron beam. An example of this is shown by the data in Figure 8.6. Here, a thick layer of deuterated polymethylmethacrylate (PMMA) was covered by successive layers (~12 Å thick) of hydrogenated PMMA, and the vibrational spectrum was recorded with an incident energy of 6 eV.[7] The circles are the intensity of the C–D stretching vibrations at 2175 cm^{-1}, and they decrease dramatically with increasing layer thickness. The increase in the absolute intensities of the C–H stretching vibrations (2985 cm^{-1}, triangles) is not linear and may be a consequence of the absolute scattering efficiency of the individual film (films were deposited by Langmuir–Blodgett techniques). Nevertheless, without any detailed analysis the rapid attenuation of the C–D stretch is clear. These data allow the electron probe depth to be bracketed between 24 and 36 Å for the C–D stretch at an electron impact energy of 6.0 eV. A similar experiment with normal PMMA covered with multiple layers of deuterated PMMA gave an identical result.[7]

These observations are consistent with the results of Wandass and Gardella.[8, 9] They show that the carboxyl stretch of fatty acid monolayers (~1650 cm^{-1}) is *not* observed by HREELS despite its large oscillator strength in the infrared. This result implies that 25 Å of pure hydrocarbon can completely attenuate this mode. Incident energies varied between 1.5 and 6.5 eV in those experiments.

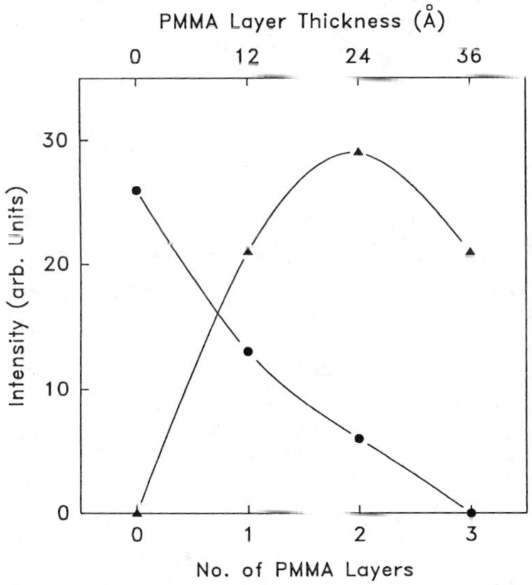

Figure 8.6 **By covering a film of perdeuterated PMMA with a series of 12 Å thick layers of undeuterated polymer, the depth penetration of low-energy electrons can be measured. The circles correspond to the intensity of the C-D stretch at 2175 cm^{-1}; these disappear with only a 24–36 Å thick overlayer. A corresponding increase in the intensity of the C-H stretch (2985 cm^{-1}, triangles) is also observed (the decrease in the last data point may be due to surface disorder).[8]**

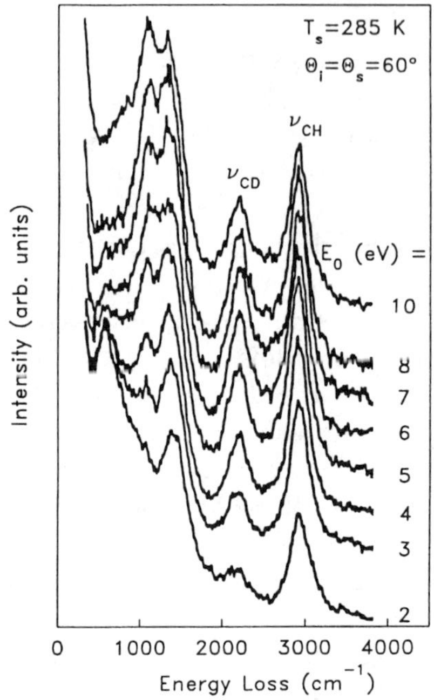

Figure 8.7 **HREELS spectra of a monolayer of the Cd-salt of octadecanoic-18,18,18,-d_3 acid at different primary energies.[15] The film was deposited on a Ag(111) single crystal substrate using Langmuir–Blodgett techniques. The vibrational spectra change dramatically in going from top to bottom, showing both the depth sensitivity of EELS and a change in the scattering mechanism at low incident energies.**

As noted above, changing the incident energy can be used to control the effective probe depth of the incident electrons. Figure 8.7 shows an example of this for a monolayer of the Cd-salt of octadecanoic acid deposited on a Ag(111) substrate using Langmuir–Blodgett techniques.[16] The adsorbate is selectively deuterated at the last carbon (18,18,18-d_3) so that the relative intensity of the tail can be studied independently. The chain axis is approximately normal to the surface and extends ~25 Å away from the substrate. As the energy is tuned from 2 to 10 eV, the contribution to the spectrum of the C–D stretch at 2218 cm^{-1} grows, while that of the C–H stretch at 2912 cm^{-1} decreases. Above approximately 8 eV both bands decrease in intensity. Thus, by changing the incident energy, the vibrational spectrum of the outermost surface of the film can be probed exclusively.

Perhaps of greater interest is the fact that at low incident energies the carboxylate bands at ~1500 cm^{-1} become visible.[16] These are buried ~25 Å below the surface at the metal–acid interface, and their observation implies that the incident electron can effectively penetrate the bulk of the sample. The ν_{CD}/ν_{CH} intensity ratio also decreases, consistent with this notion. These observations are in good agreement with the work of Cartier et al., who showed that at very low incident energies the penetration depth of electrons increases.[17] A change in scattering mechanism is also

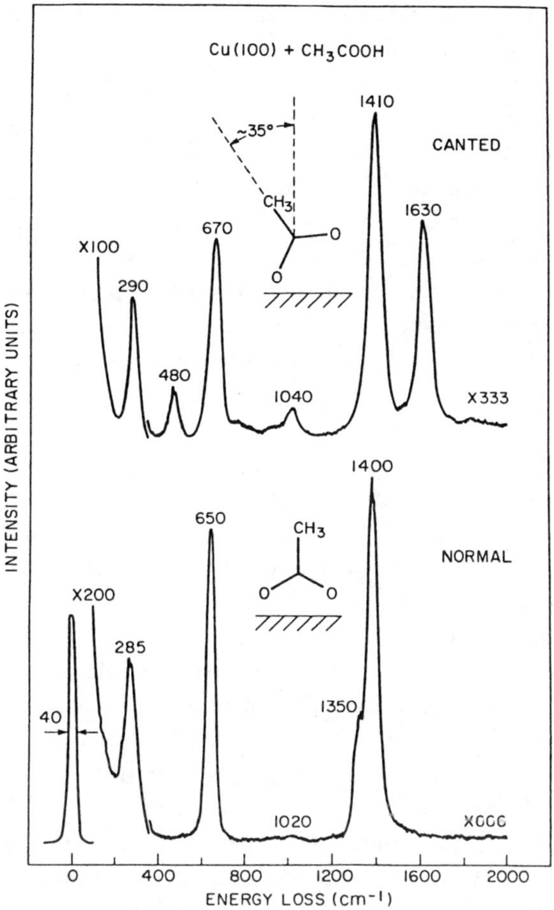

Figure 8.8 HREELS spectra from two acetic acid monolayers of differing structure on a Cu(100) substrate. The bottom trace is the spectrum of a low-coverage, symmetrically bonded carboxylate phase. The top trace shows the spectrum that develops when additional acetic acid is added. This latter monolayer corresponds to a mixed carboxylate/carboxylic acid phase exhibiting significant intermolecular hydrogen bonding and therefore asymmetric bonding of the carboxyl head group to the substrate. Preparative methodologies and mode assignments are discussed in Reference 18.

observed here since the deeper penetration into the solid favors close electron approach and therefore impact scattering.[16]

Molecular Orientation

If the dipole selection rule is applicable (i.e., if the properties of a thin organic film on a conducting substrate are being studied), then the orientation of adsorbed molecules potentially can be determined. Figure 8.8 shows two HREELS spectra of monolayers of acetic acid adsorbed on a Cu(100) single crystal substrate at different densities.[18] While this does not constitute an organic surface or thin film in the context of this chapter, it does provide a well-defined model system on which to test the sensitivity of HREELS to molecular orientation. The two modes to concentrate on in this example are the symmetric (1410 cm^{-1}) and asymmetric (1630 cm^{-1}) carboxylate stretching vibrations, which have their dipole derivatives parallel and perpendicular to the C–C axis, respectively. Since *both* bands are observed in the top trace of Figure 8.8, the adsorbate must be canted, although the exact angle is difficult to determine. In contrast, the bottom trace shows only the symmetric carboxylate stretch, implying that the C–C axis must be normal to the surface. These types of observations have been extended to polymer films, where measurements of the orientations of C–H and C=O bonds in polyimide have recently been made.[6]

Local Versus Long-Range Interactions

The scattering mechanism and, therefore, the details about bonding in a polymer film can be probed by measuring the vibrational cross section as a function of the incident electron energy. As noted above, both impact scattering and negative-ion resonance formation should have a strong, nonmonotonic incident energy dependence. This is because they involve highly localized excitations and therefore depend on the microscopic details of the electron–surface interaction. An example of this is shown by the data in Figure 8.9 for electron scattering from a polycarbonate film on a graphite substrate (open circles).[10] Here, the intensity of the aromatic CH stretching vibrations (frequencies $\gtrsim$3000 cm^{-1}) normalized to the intensity of the elastic scattering peak are plotted as a function of the incident electron energy. A clear maximum is observed at 4.5 eV and is attributed to electron capture by an ion which is predominantly phenyl in character.[10] In contrast, dipole scattering is dominated by long-range forces, and as a result, should have a weaker electron energy dependence.[19] An example of this is shown by the solid symbols in Figure 8.9 for electron scattering from a polyethylene surface. The sample was cut from a large piece of bulk polyethylene and charging was minimized using an auxiliary electron flood gun. We note for completeness that the cross section for all three types of scattering (but especially dipole scattering) will also be strongly dependent on the intensity of the elastically scattered electrons.

Still, there are some subtleties to dipole scattering which can be exploited by a detailed energy-dependent measurement. For example, the relative intensity of a vibrational band (normalized to the elastic peak height) will vary as a function of the incident energy, depending on whether the vibration is related to a collective property of the material (e.g., the excitation of a surface phonon) or to the excitation of a more localized vibration. In the former case, the intensity should scale as $E_0^{-1/2}$, while in the latter it should scale as E_0^{-1}.[19] Figure 8.9 is a log–log plot of the normalized intensity of the C–H stretching bands (~2900 cm^{-1}, square) and the bending modes (integrated over the 700–1800 cm^{-1} region, triangle) versus the incident electron energy, E_0. While the data is not fit by a true E_0^{-1} dependence, it

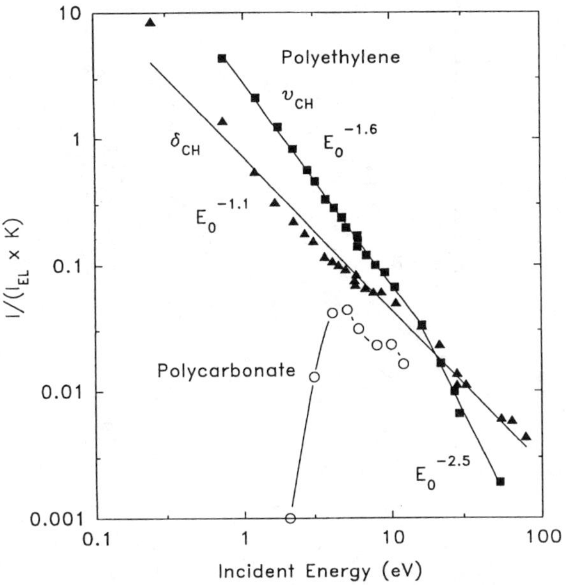

Figure 8.9 **Vibrational excitation cross sections as a function of the incident energy are shown for polyethylene (squares and triangles)[19] and polycarbonate (open circles).[10] The data are normalized to the intensity of the elastic peak and are divided by a constant *k*, which is a function of the spectrometer used and of the exact scattering geometry. The scattering of low-energy electrons from polyethylene is dominated by long-range dipole interactions. The incident energy distribution for electrons scattered from polycarbonate is highly peaked, providing strong evidence for impact scattering.**

is fit more closely by this function. This has been interpreted as being due to the excitation of localized areas of the polymer surface (i.e., by exciting individual methyl, methylene, or carbon skeleton vibrations).[19] This indicates that polymers do not behave like true solid-state materials showing collective excitations such as phonons, but more like condensed gases with weak intermolecular interactions.

Additional subtleties to this data, such as the different slopes in the two curves or the break in the slope of the curve for the C–H stretching vibration above ~20 eV, are not discussed further here.[19] Since the slopes are not unity, however, there must be some contribution from either impact or resonance scattering.

Surface Segregation

Potentially, the most useful aspect of HREELS is its inherent surface sensitivity. Despite its limited resolution, this attribute can be exploited to study surface segregation in blends of miscible polymers. Figure 8.10 is a comparison of infrared (upper curve) and HREELS (lower curve) spectra of thin ($\gtrsim$100 Å) polystyrene films cast onto gold substrates from carbon tetrachloride solution.[20] The solution consisted of mixtures of normal and perdeuterated polystyrene of the same average

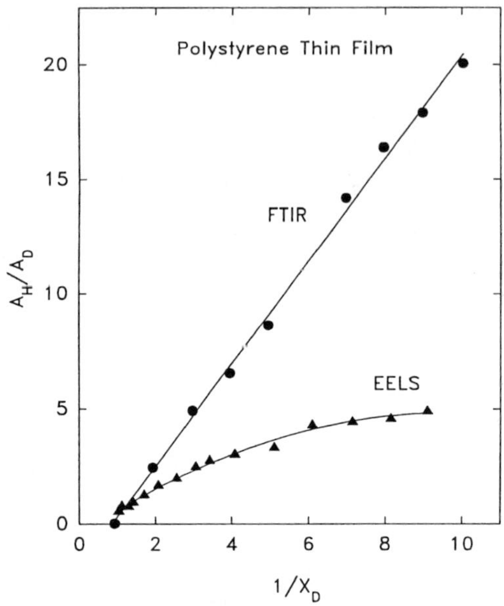

Figure 8.10 **Fourier transform infrared (circles) and HREELS (triangles) analysis of the composition of mixtures of normal and selectively deuterated polystyrenes with different molar ratios (X_D).[20] The intensity ratios of the C–H and C–D stretch bands (A_H/A_D) show a deuterium enrichment at the surface. This conclusion is made possible by the added surface sensitivity of EELS.**

molecular weight (~40 000), but mixed in different molar ratios. HREELS spectra were recorded with an incident energy of 7 eV to ensure high surface sensitivity. Indeed, the data in Figure 8.10 show that the intensity ratios of the C–H to C–D stretching vibrations only change weakly with solution composition, proving a net surface enrichment in deuterated polymer.[20] Fourier transform infrared spectra show a linear relationship between the band intensities and the inverse of the deuterium mole fraction, indicating that the bulk of the films contain the correct composition. The origin of this surface segregation effect is beyond the scope of this chapter and is discussed in more detail elsewhere.[7, 20]

8.6 Conclusions

HREELS has recently expanded our knowledge base about the structure of polymers and their surfaces, but there is still much to be learned. This technique will continue to grow in importance in the future because of improvements to better control the spectral resolution and angular divergence of the scattered beam as well as the use of new, better-understood model systems. Still, no single tool will provide all types of information, and a combination of appropriately chosen techniques will certainly be more useful than any single one. This point is made even stronger by considering the results summarized in other chapters in this volume.

References

1 W. Erley and H. Ibach. *Surface Sci.* **178**, 565, 1986.

2 H. Ibach and D. L. Mills. *Electron Energy Loss Spectroscopy and Surface Vibrations.* Academic Press, New York, 1982.

3 K. Hiraoka and W. H. Hamill. *J. Chem. Phys.* **57**, 3870, 1972.

4 J. J. Pireaux, P. A. Thiry, R. Sporken, and R. Caudano. *Surf. Interface Anal.* **15**, 189, 1990.

5 P. A. Thiry, M. Liehr, J. J. Pireaux, and R. Caudano. *Physica Scripta.* **35**, 368, 1987.

6 J. J. Pireaux, M. Vermeersch, C. Gregoire, P. A. Thiry, R. Caudano, and T. C. Clarke. *J. Chem. Phys.* **88**, 3353, 1988.

7 J. J. Pireaux, C. Gregoire, R. Caudano, M. Rei Vilar, B. Brinkhuis, and A. J. Schouten. *Langmuir.* **7**, 2433, 1991.

8 J. H. Wandass and J. A. Gardella, Jr. *Langmuir.* **2**, 543, 1986.

9 J. H. Wandass and J. A. Gardella, Jr. *Langmuir.* **3**, 183, 1987.

10 G. Apai and W. P. McKenna. *Langmuir.* **7**, 2266, 1991.

11 H. Ibach, M. Balden, D. Bruchmann, and S. Lehwald. *Surface Sci.* **269/270**, 94, 1992.

12 W. P. McKenna and G. Apai. *J. Phys. Chem.* **96**, 5902, 1992.

13 J. J. Pireaux, C. Gregoire, P. A. Thiry, R. Caudano, and T. C. Clarke. *J. Vac. Sci. Technol.* **A5**, 598, 1987.

14 L. H. Dubois, B. R. Zegarski, and R. G. Nuzzo. *J. Chem. Phys.* **98**, 1993.

15 M. Schreck, M. Abraham, A. Lehmann, H. Schier, and W. Göpel. *Surface Sci.* **262**, 128, 1992.

16 M. Schreck, M. Abraham, W. Göpel, and H. Schier. *Surface Sci.* **237**, L405, 1990.

17 E. Cartier, P. Pfluger, J. J. Pireaux, and M. Rei Vilar. *Appl. Phys.* **A44**, 43, 1987.

18 L. H. Dubois, B. R. Zegarski, and R. G. Nuzzo. *Langmuir.* **2**, 412, 1986.

19 J. J. Pireaux, P. A. Thiry, R. Caudano, and P. Pfluger. *J. Chem. Phys.* **84**, 6452, 1986.

20 M. Rei Vilar, M. Schott, J. J. Pireaux, C. Gregoire, R. Caudano, A. Lapp, J. Lopes Da Silva, and A. M. Botelho Do Rego. *Surface Sci.* **212/213**, 782, 1989.

9

Wetting

D. LI and A. W. NEUMANN

Contents

9.1 Introduction

Some of the important foundations of wetting were established in the early 19th century, as in the remarkable work of Thomas Young[1], yet because of the many challenging problems in the field, it is attracting growing research interest from a wide spectrum of disciplines, such as material, biomedical, and environmental science and engineering. Wetting itself is such a broad subject that a complete review of the extensive literature in this field is very difficult. However, although the complexity of wetting phenomena varies from situation to situation, insight can always be gained from the key parameters: the contact angle and the surface tensions involved in a wetting system. It is a common conception that wetting studies concern the measurements of contact angles. However, the interpretation of contact angles is an equally important side of wetting studies. The aim of this chapter is to introduce some novel techniques in contact angle measurement and modern developments in the theories of wetting phenomena.

9.2 Contact Angles

From daily experience we know that a drop of water deposited onto the surface of an organic material will form a sessile drop. Also, we often see that many liquids can climb up to a certain height in a capillary tube and form a meniscus at the top. In these phenomena, the angle formed between the liquid–vapor interface and the liquid–solid interface at the solid–liquid–vapor three-phase contact line is conventionally defined as *contact angle.* The study of wetting, adhesion, and flotation requires a knowledge of the contact angle. A liquid is wetting if it spreads easily across a surface, that is, the contact angle is zero or very close to zero. Wetting liquids make good contact with solid surfaces. For example, insecticides should fully wet the surfaces of waxy leaves to protect the leaves from insects and disease; and animal dips should properly coat the greasy hair of animals to repel vermin and lice. In studies of biological adhesion, the free energy of adhesion is determined using the contact angle. This free energy is a key parameter in the adhesion of cells to foreign surfaces. In the mining industry, flotation plays an important role in separating the minerals from the nonminerals in crushed ores. The mineral particles float to the surface by attaching themselves to air bubbles. If the contact angle at the interface is finite, then the particle may assume a stable position at the interface. Impurities present in or added to a liquid may alter the contact angle considerably. Wetting agents or detergents change the contact angle from a large value, greater than 90°, to a value often much smaller than 90°. Conversely, waterproofing agents applied to cloth cause the contact angle of water in contact with the cloth to be larger than 90°.

It should be noted that at the molecular level the three phases do not meet in a line but within a zone of small but finite dimensions in which the three interfacial regions merge. Recent experimental evidence[2] shows steps on a molecular scale in the merging region of the three interfaces. Therefore, the microscopic contact angles may be different from the contact angles measured by ordinary instruments, such as a goniometer. Discussion of such microscopic contact angles is beyond the scope of this chapter. For this chapter, the macroscopic contact angles will be at the center of attention; the type of system considered here generally consists of three bulk phases—solid, liquid, and vapor—and three interface phases—the solid–liquid, solid–vapor, and liquid–vapor interfaces.

The interest in contact angles is twofold: they play a major role in a number of technological, environmental, and biological phenomena and processes; they are also a manifestation of the surface tension of the solid on which the contact angle is formed. At the present time, an interpretation of contact angles is one of the better prospects for determining the surface tensions of solids. Unfortunately, contact

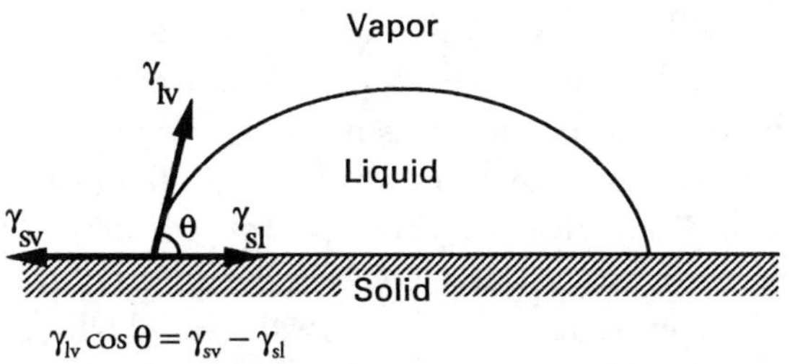

Figure 9.1 A sessile drop on a solid surface in equilibrium with the vapor phase.

angle phenomena are complex, which can be most easily demonstrated by invoking the classical theory of capillarity.[3] Minimizing the overall free energy of a system consisting of a liquid in contact with a solid (Figure 9.1) yields the Laplace equation of capillarity

$$\gamma_{lv}\left(\frac{1}{R_1}+\frac{1}{R_2}\right) = \Delta\rho g z + c = \Delta P \tag{9.1}$$

and Young's equation

$$\gamma_{lv}\cos\theta_e = \gamma_{sv} - \gamma_{sl} \tag{9.2}$$

where γ_{sv} = solid–vapor interfacial tension, γ_{lv} = liquid–vapor interfacial tension, γ_{sl} = solid–liquid interfacial tension, θ_e = equilibrium contact angle, R_1, R_2 = principal radii of curvature at any point of the liquid surface, $\Delta\rho$ = density difference between the liquid and vapor phases, ΔP = capillary pressure or pressure of curvature, z = the ordinate of the point of the liquid surface at which the principal radii of curvature are R_1 and R_2, c = a constant, and g = acceleration due to gravity.

The derivation of these relations assumes that the solid surface in contact with the liquid is smooth, homogeneous, isotropic, and nondeformable. This assumption is of little or no consequence for the range of validity of the Laplace equation, since it essentially describes the shape of the liquid–vapor or liquid–liquid interface, away from the solid–vapor and solid–liquid interfaces. Young's equation, on the other hand, is an equilibrium condition that involves properties which are a function of the solid surface, that is, γ_{sv} and γ_{sl}. Since real surfaces quite often do not meet the above restrictions, the persistence of the controversy over the validity and applicability of Young's equation and the thermodynamic status of contact angles is not surprising.

The above statement that the Laplace equation describes the shape of the liquid–vapor interface requires some further elaboration. To be more specific, the Laplace equation prescribes, for a given liquid of surface tension γ_{lv}, a value of the mean curvature $J = 1/R_1 + 1/R_2$ as a function of the ordinate z at any point on the liquid–vapor interface. In addition, it will be necessary in many cases to introduce, as a boundary condition, what we term the phenomenological contact angle θ. It is the angle between the solid–liquid interface and the tangent to the liquid–vapor interface at the three-phase line measured in a plane normal to the solid–liquid and the liquid–vapor interfaces. The phenomenological contact angle θ may or may not be the equilibrium contact angle θ_e, and it may or may not be a dynamic contact angle. It will in any case represent the appropriate boundary condition which determines the shape of the liquid meniscus.

The equilibrium contact angle θ_e is a unique function of the interfacial tensions γ_{sv}, γ_{sl}, and γ_{lv}, given by Young's equation (Equation 9.2). In the vast majority of cases, however, experimentally observed contact angles are not uniquely determined by the surface tensions of a given solid–liquid pair; there exists normally a range of contact angles, quite often as large as 50° or 90°, over which any one contact angle gives rise to a mechanically stable liquid meniscus. Normally, the largest and the smallest of these angles are recorded and are termed advancing contact angle θ_a and receding contact angle θ_r, respectively. The difference between the advancing and receding contact angles is called contact angle hysteresis. However, these names may be misleading; advancing and the receding contact angles are not dynamic contact angles, as their names may imply. The advancing contact angle is the contact angle of the liquid *tending* to advance across a dry solid surface. The receding contact angle is the contact angle of the liquid *tending* to recede from a solid surface. The dynamic contact angles, however, are the contact angles associated with a moving solid–liquid–vapor three-phase contact line, as opposed to a static contact angle. At low rates of motion of the three-phase line, static and dynamic contact angles tend to be identical on smooth solid surfaces.[4] Wetting kinetics and, specifically, dynamic contact angles are discussed in the literature in terms of a hydrodynamic or, alternatively, a molecular-kinetic description. The most detailed hydrodynamic treatment, elaborated by Cox,[5] leads to an expression relating the contact line velocity and the static and dynamic contact angles. A molecular-kinetic approach was formulated by Blake and Haynes[6] in terms of a molecular adsorption–desorption oscillation frequency and the distance between adjacent adsorption centers. More details, including experimental results, can be found elsewhere.[7]

Contact angle hysteresis may be conveniently observed by first advancing and then receding a liquid drop over a solid surface. If the surface tension of the solid, γ_{sv}, and the interfacial tension, γ_{sl}, change due to the contact with the liquid, we expect a change of the equilibrium contact angle θ_e as an result of this contact. In this case, the fact that observed contact angles are not a unique property for any given solid–liquid system would not necessarily conflict with Young's equation. Although there may exist such a true hysteresis, such as in biological systems, we will see that there are more common causes for contact angle hysteresis.

It was only in 1936 that Wenzel[8] recognized that Young's equation may not be a universal equilibrium condition for the physical interaction between a solid and a liquid. He essentially argued that if the solid surface is rough, the interfacial tensions γ_{sv} and γ_{sl} should not be referred to the geometric area, but to the actual surface area. If we let

$$r = \frac{\text{actual surface area}}{\text{geometric surface area}} \tag{9.3}$$

this leads to the so-called Wenzel equation

$$r(\gamma_{sv} - \gamma_{sl}) = \gamma_{lv} \cos\theta_W \tag{9.4}$$

where θ_W may be called the Wenzel contact angle; it is in fact the equilibrium contact angle θ_e on a rough solid surface.[9] The same type of reasoning was applied to heterogeneous surfaces by Cassie and Baxter.[10] A heterogeneous solid surface is one which contains domains of different surface tension. Examples of heterogeneous solid surfaces are surfaces on which patches of a monomolecular film are adsorbed, or a surface of a polycrystalline material, which exposes different crystallographic planes at its surface. If we consider a solid surface consisting of two types of domains, the intrinsic contact angles with respect to a given liquid being θ_{e1} and θ_{e2}, Cassie[10] obtained

$$\cos\theta_C = a_1 \cos\theta_{e1} + a_2 \cos\theta_{e2} \tag{9.5}$$

where a_1 and a_2 are the fractional surface areas of the two types of surfaces such that

$$a_1 + a_2 = 1 \tag{9.6}$$

and θ_C may be called the Cassie contact angle. It is, like θ_W for a rough surface, the appropriate equilibrium contact angle for a heterogeneous solid surface.[9] However, it will also not be amenable to experimental determination.

One of the difficulties with relations like the Cassie equation or the Wenzel equation is that they seem to conflict with experimental observation: the latter equation predicts that the contact angle on a rough surface should be smaller than the contact angle on a smooth surface if the contact angle on the smooth surface is smaller that 90°. If the contact angle on the smooth surface is larger that 90°, then the contact angle on the rough surface should be larger. Experimentally, however, one finds that with increasing roughness the advancing contact angle increases and the receding contact angle decreases.[11, 12] In addition to these difficulties, the explanations of the relation between roughness and heterogeneity, on the one hand, and the phenomenon of contact angle hysteresis on the other made further investigations necessary.

There have been many studies on contact angle hysteresis phenomena. Even prior to Cassie's work, Shuttleworth and Bailey[13] explained how roughness may produce contact angle hysteresis. The important concept of metastable states was first introduced by Derjaguin[14] in connection with heterogeneous solid surfaces. It is the development of this concept which facilitates a deeper understanding of contact angle phenomena. The more quantitative analyses[9, 15–17] reveal that the surface imperfections, roughness and heterogeneity, produce a large number of metastable states or local minima of the overall free energy of the system and, correspondingly, many possible values of contact angle. For a heterogeneous surface with two different types of domains, it has been shown[9] that the advancing and receding contact angles are equal to the respective equilibrium contact angles that would be observed on ideal surfaces of the two types. Therefore, the advancing and receding contact angles may be used in Young's equation (Equation 9.2) with the understanding that they characterize different types of surfaces.

With the present technology, surface roughness usually can be minimized by very careful surface preparation. Nevertheless, contact angle hysteresis still remains, probably due to heterogeneity. Generally, the receding contact angles are less reproducible than the advancing contact angles. Fortunately, it can be made plausible that[17] the advancing contact angle be representative for the actual, dominant surface material, and the receding contact angle, being only a manifestation of impurities of the solid surface, often may be safely disregarded when the main goal is a characterization of the wettability and energetics of the solid surface.

A direct experimental proof of the thermodynamic status of advancing contact angles is the determination of the heat of wetting by measuring advancing contact angles. The heat of wetting is heat liberated when a clean solid surface is immersed in a liquid. It can be determined calorimetrically by measuring the heat evolved on the immersion of a clean solid or solid powders in a liquid. Using Young's equation (Equation 9.2) the heat of wetting is given by

$$\Delta H_{\text{wetting}} = -\left(\gamma_{lv}\cos\theta - T\frac{d(\gamma_{lv}\cos\theta)}{dT}\right) \tag{9.7}$$

As seen from this equation, a calculation of the heat of wetting requires only data on liquid surface tension and contact angle as well as the respective temperature dependence. Therefore, it is possible to compare calorimetric heats of wetting with

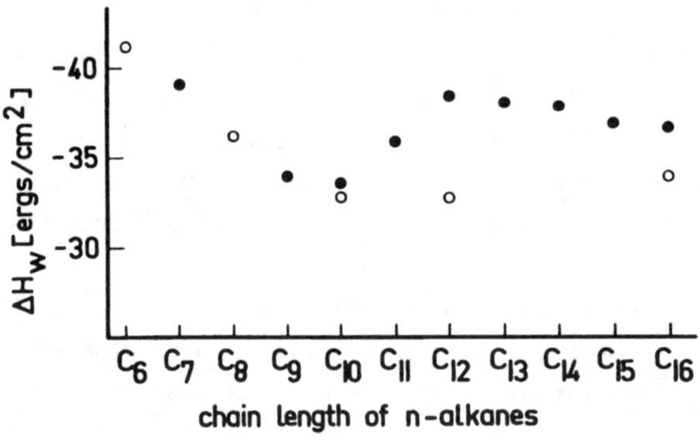

Figure 9.2 **Heats of wetting calculated from advancing contact angles (●) and calorimetric heats of immersion (○) of PTFE in *n*-alkanes.**

that calculated from the temperature dependence of advancing contact angles. Such a comparison was performed[18] with seven n-alkanes against PTFE and good agreement was found (Figure 9.2).

9.3 Techniques for Contact Angle Measurements

A survey of techniques for measuring contact angles can be found elsewhere.[19, 20] Among these, the direct measurement of contact angles from sessile drops or photographs of sessile drops is most widely used. In this technique, the angle is measured by aligning a tangent with the drop profile at the point of contact with the solid surface. The measurement can be performed either using a telescope equipped with a goniometer eyepiece or on a photograph of the drop using a protractor; a precision of $\pm 2°$ is usually claimed. Aligning the tangent is subjective and depends on the experience of the operator. An alternative method uses the capillary rise[21] at a vertical plate (Figure 9.3). The solid surface is brought into contact with the liquid and the capillary rise at the vertical surface is measured. In addition to the capillary rise, the liquid surface tension, the density difference, and the gravitational acceleration need to be known before the contact angle can be computed. The precision obtained with this technique is $\pm 0.1°$, an order of magnitude better than the direct method. The disadvantage of this method is that relatively large volumes of liquid are required. Two other techniques, not frequently used, are interference microscopy and specular reflection. Interference microscopy relies on interference fringe patterns produced by the edge of the drop surface to calculate the contact angle. Specular reflection uses a beam of light reflected from the liquid surface. By pivoting the beam on its axis, the orientation at which the specular reflection disappears is determined, and the contact angle established.

In this chapter, we introduce a more powerful, novel methodology.

Axisymmetric Drop Shape Analysis-Profile (ADSA-P)

ADSA-P[22–24] is a technique to determine liquid–fluid interfacial tensions and contact angles from the shape of axisymmetric menisci, that is, from sessile as well as

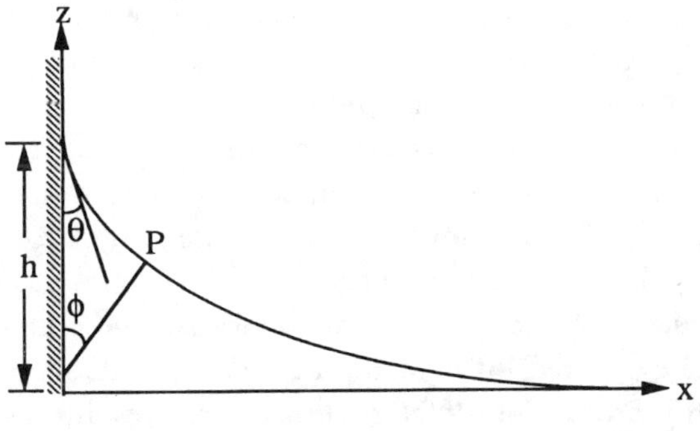

Figure 9.3 Schematic of capillary rise at a vertical solid surface.

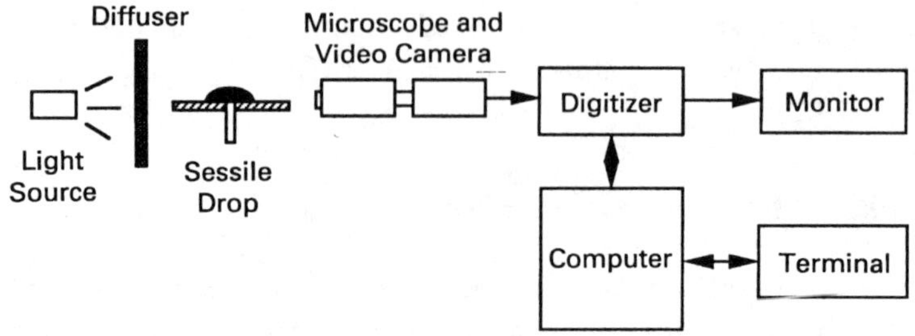

Figure 9.4 Experimental setup for sessile drop contact angle measurement by ADSA.

pendent drops. The strategy employed is to construct an objective function which expresses the deviation of the physically observed profile of a drop from a theoretical Laplacian curve, that is, a curve satisfying the Laplace equation of capillarity. The objective function is minimized numerically, with the liquid surface tension as one of the adjustable parameters.

Apart from local gravity and densities of liquid and fluid phases, the only input information required by ADSA-P is several arbitrary but accurate coordinate points selected from the drop profile. Specifically, it is not necessary to identify the apex of the drop profile or the drop width. To achieve rapid and accurate data acquisition and preprocessing, an automatic digitization technique utilizing recent developments in digital image acquisition and analysis has been used. A FORTRAN program has been developed to implement this method. The output of the ADSA-P program will provide the values of surface tension, contact angle, drop volume, surface area, and radius of the three-phase contact line.

A block diagram of the experimental setup using digital image processing is shown in Figure 9.4. In this setup, a Cohu 4800 CCD monochrome camera is mounted on a Wild Heerbrugg M7S microscope. The video signal of the sessile drop is transmitted to a Quantex DS-30 digital video processor, which performs the frame-grabbing and digitization of the image to 512×512 pixels with 256 gray levels, where 0 represents black and 255 represents white. A DEC MicroVax II computer is used to acquire the image from the Quantex and perform the image analysis and computation. For more cost effectiveness, the processor and the computer can be replaced by an ITI PCVISION plus image processor and a personal computer.

In the measurements, the advancing contact angles of sessile drops are produced by feeding the liquid very slowly from the bottom of the surface through a small hole by a motor-driven syringe device. When the motor is stopped, the sessile drop is allowed to relax for approximately 30 s to reach equilibrium. Then several pictures of this sessile drop are taken successively at certain time intervals. These images of the sessile drop are immediately stored in the computer, and the ADSA-P program automatically finds the drop profile coordinates with subpixel resolution and corrects the optical distortion due to the camera and lenses. More liquid can

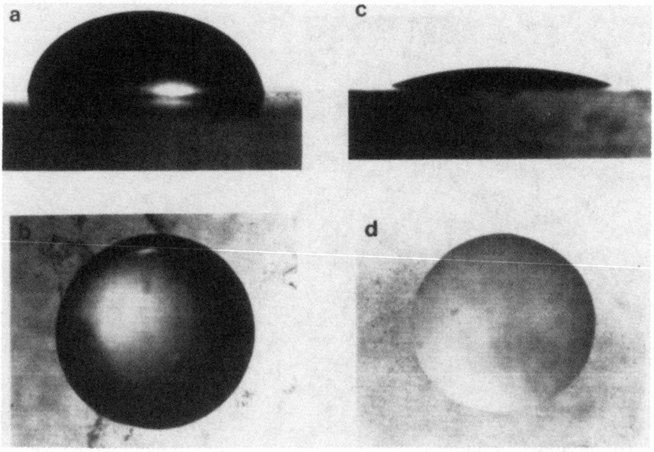

Figure 9.5 (*a*) Side and (*b*) top views of a sessile drop with a contact angle larger than 60°. (*c*) Side and (*d*) top views of a sessile drop with a small contact angle.

then be pumped into the drop to increase the volume of the sessile drop until it reaches another desired size, and the above procedure is repeated. Usually three or more drops are measured on one test surface. For each solid–liquid system, the above measurements are conducted two or three times, and a fresh solid surface is used each time. The accuracy of the contact angles measured by ADSA-P is generally better than 0.2°.

Axisymmetric Drop Shape Analysis-Contact Diameter (ADSA-CD)

For the cases of very low contact angles, say below 20°, it becomes increasingly difficult to measure contact angles with most techniques (see Figure 9.5). The precision of ADSA-P also decreases, since it becomes more difficult to acquire accurate coordinate points along the edge of the drop profile. ADSA-CD was developed for these situations.[24–26] In ADSA-CD, the drop is viewed from above, rather than from the side. The contact angle is determined by minimizing the difference between the prescribed drop volume and the drop volume calculated from the contact diameter of the drop in conjunction with the Laplace equation of capillarity. For ADSA-CD the input consists of (1) the surface tension of the liquid, (2) the density of the liquid, (3) the volume of the drop, and (4) the contact diameter of the drop. Similar to ADSA-P, the contact diameter can be determined by using the method of digital image acquisition and analysis. But more-elementary manual procedures are also feasible. A computer program for ADSA-CD has also been developed. The accuracy of ADSA-CD for small contact angles on smooth and homogeneous surfaces is found to be better than ±0.3°.

An important area of application is the determination of contact angles on biological substrates. Measuring contact angles on biological materials is generally

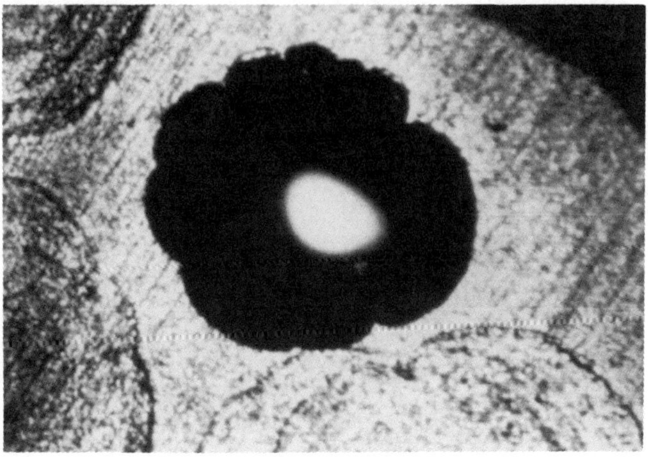

Figure 9.6 Sessile drops of water on a layer of T. ferrooxidans: example of a nonaxisymmetric drop.

difficult due to the fact that such surfaces usually present not only small contact angles but also morphological and energetic imperfections, leading to irregularities of the three-phase contact line (Figure 9.6). It would be difficult to measure contact angles on such drops by, for example, finding the tangent of the drop profile at the three-phase contact point, and the results would be of dubious value. However, ADSA-CD avoids this problem and provides averaging by analyzing to drop contact area. Recently, water contact angles on layers of bacterial cells have been measured.[26] The layers are prepared by depositing the cells by suction filtration onto a cellulose membrane filter. Using ADSA-CD, a video image of the drop contact area is digitized semi-automatically using a "mouse" on the video screen. An average drop contact diameter is then calculated from these values. ADSA-CD then calculates the contact angle based on such a drop contact diameter and the input values for the drop volume, the density, and the surface tension of the liquid.

Drops of water on layers of cells are inherently unstable because the water tends to penetrate between the cells by capillary action. As a result, the contact angle decreases rapidly, so that accurate and reproducible contact angles may be assessed with a goniometer only with much experience and skill. With ADSA-CD, such effects of liquid penetration do not present a major problem; because of contact angle hysteresis the three-phase line is hinged in a fixed position and will not move due to liquid penetration for at least many seconds, allowing more than adequate time for image acquisition.[26]

Recently, ADSA-CD has been adapted to contact angles larger than 90°.[27] In such situations, the maximum, equatorial diameter of the drop is determined and replaces the contact diameter in the calculation with a modified computer program called ADSA-MD (MD = Maximum Diameter).

Capillary Rise Technique

Basically, the capillary rise technique works as follows: It is well known that the contact angle is related to the capillary rise at a vertical plate by

$$\sin\theta = 1 - \frac{\Delta\rho g h^2}{\gamma_{lv}} \tag{9.8}$$

where $\Delta\rho$ is the density difference between the liquid and the vapor, g is the gravitational acceleration, γ_{lv} and θ are the liquid surface tension and the contact angle, respectively, and h is the capillary rise at the vertical plate. Thus, the task of measuring contact angle is reduced to a measurement of a length, which can be performed optically with high accuracy by means of a catheometer.

In the automated capillary rise technique,[28] the conventional catheometer is replaced by a motorized, computer-controlled positioning stage, a microscope (Wild Heerbrugg M5A stereomicroscope), and a CCD scanner (Cohu 4810). The scanner is connected to a digital video processor and a computer, as well as a video monitor. The positioning stage, the microscope, and the scanner are mounted on a vibration-free table. The chamber containing the test liquid cell and the test surface is mounted on another vibration-free table. To identify accurately the position of the undisturbed liquid surface, a reference fiber is mounted on each side of the vertical solid surface. Images of the two reference fibers, their mirror images, and the solid–liquid–vapor three-phase contact line on the solid surface are digitized and stored in the computer. Using modern digital image processing and analysis techniques, the position of the undisturbed liquid surface is determined automatically by the computer from the images of the reference fibers and their reflections; then the capillary rise, h, will be calculated by the computer as the difference between the position of the three-phase line and that of the free liquid surface. Details regarding the setup, procedures of the image digitization and analysis, and autofocusing and calibration can be found elsewhere.[4, 28] With this technique, it has been found that the accuracy in the measurement of h is approximately ± 1.4 μm; the corresponding accuracy in contact angle is approximately $\pm 0.06°$.

The high accuracy, automated capillary rise technique has been employed to study dynamic contact angles,[4] with the result that on a very smooth solid surface the dynamic contact angles at relatively low rates are essentially identical with the corresponding static advancing contact angles. Figure 9.7 gives such an example.

9.4 Phase Rule for Moderately Curved Surface Systems

Utilization of the measured contact angles largely depends upon thermodynamic arguments. Surface thermodynamics is a major theoretical tool in studies of wetting and adhesion phenomena, especially in the interpretation of surface tension and contact angle data. However, there is a less well known component of surface thermodynamics that is fundamental and has a profound impact on interpreting contact angles—the phase rule for capillary systems.[29, 30]

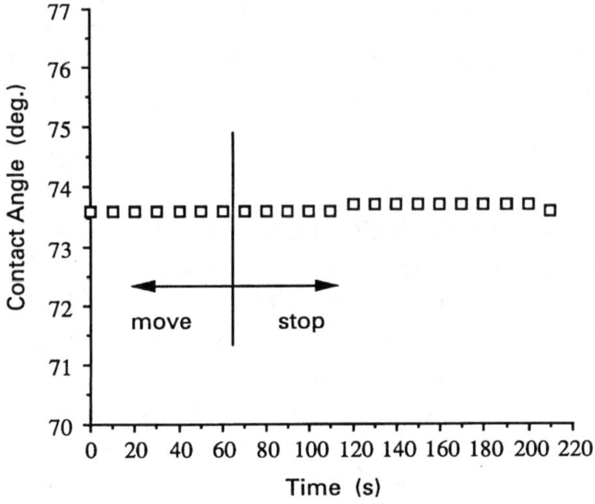

Figure 9.7 Comparison of dynamic contact angles at a velocity of 0.49 mm/min of the three-phase contact line with static contact angles of dodecane on a FC721 surface. Contact angles were measured by an automated capillary rise technique.

Within Gibbsian thermodynamics, the phase rule is a well-known relation employed to predict the number of thermodynamic degrees of freedom, or the number of independent, intensive variables that determine the equilibrium state of a many-component, multiphase system. Essentially, the *number of degrees of freedom* or *variance* of any composite system is evaluated by subtracting the number of equilibrium constraint equations from the number of variables used to describe the composite system. For a multicomponent, multiphase system in equilibrium, the number of independent, intensive variables or the number of degrees of freedom, f, is given by

$$f = r + 2 - n \tag{9.9}$$

Relation 9.9 represents the standard Gibbs phase rule. It simply states that in a system with r independent chemical components and n coexisting phases, the maximum number of independent, intensive variables that may be used to describe the state of the system is f.

However, this phase rule is applicable only for bulk phase systems. This is easily illustrated by considering a one-component capillary system consisting of a liquid droplet surrounded by its vapor. In this case, there are two bulk phases, one liquid–vapor interface but only one component. That is, $n = 3$ while $r = 1$. According to Equation 9.9, the number of degrees of freedom, f, would be zero. This is obviously unreasonable, since it is well-known from experimental work that both temperature, T, and pressure, P, can be chosen as independent variables for such a system; thus the number of degrees of freedom should be two.

The reason why Equation 9.9 is not generally applicable to capillary systems is that it does not consider the effects of curved interfaces or bulk phase boundaries.[30–32] For a composite bulk phase system, the mechanical equilibrium condition is pressure equality through all phases.

$$P_\alpha = P_\beta = \cdots = P_\lambda \tag{9.10}$$

However, for a capillary system, because of the presence of curved interfaces the pressures cannot be equal in all phases. The mechanical equilibrium condition for two bulk phases separated by a curved interface is given by the Laplace equation of capillarity (Equation 9.1), $P_\alpha - P_\beta = \gamma_{\alpha\beta} J_{\alpha\beta}$. The key point is that Equation 9.1 does not function as a constraint relation among P_α, P_β, and $\gamma_{\alpha\beta}$. This is simply because these three parameters, P_α, P_β, and $\gamma_{\alpha\beta}$, may all be freely assigned while Equation 9.1 can still be satisfied by adjusting the value of the mean curvature of the interface, $J_{\alpha\beta}$. Thus, a reduction in the number of constraint conditions is necessary whenever a curved surface separates two bulk phases. However, Equation 9.9 fails to account properly for the reduction in the number of mechanical equilibrium constraint equations in a capillary system.

It is worthwhile to emphasize that whenever using the Laplace equation an additional degree of freedom is operative. For example, for pure water at 1 atm, the curvature or Laplace pressure causes the boiling temperature of a bubble with a 1 μm radius to be 121 °C. In comparison with the boiling temperature in the case of a flat surface (i.e., 100 °C), this is a 21% increase in the boiling temperature of water. As another example, consider several capillary tubes of identical radii but different materials vertically dipping into a bulk liquid. Since the contact angles of the liquid on the surfaces of these tubes are different, the curvatures of the three menisci will be different; hence, the Laplace pressures are different so that there can be large differences in capillary rise between these tubes. Such a phenomenon can be observed for the cases where radii of the capillary tubes are of the order of millimeters. This illustrates the significance of the additional degrees of freedom for systems containing moderately curved interfaces with a radius of curvature of the order of millimeters.

For a multicomponent system consisting of bulk phases, moderately curved interface phases, by counting the number of intensive variables required to describe all phases in the system and the number of equilibrium constraint relations, the phase rule has been derived as[30]

$$f = r + 1 - N \tag{9.11}$$

where r is the number of independent chemical components or species in this system and N is the number of $P_\alpha = P_\beta$ pressure equality relations among the mechanical equilibrium conditions. Generally, the phase rule for systems having moderately curved interfaces predicts a larger number of degrees of freedom than that predicted by the phase rule for bulk phase systems (Equation 9.9).

The phase rule for capillary systems is of particular importance because knowledge of the number of degrees of freedom, as given by Equation 9.11, provides a theoretical justification for functional relationships between interfacial tensions. Let us consider the following two examples.

Example 1: Liquid substrate systems[30] Consider a capillary system consisting of three bulk fluid phases of different densities, as shown in Figure 9.8. In addition to the three bulk phases, there are three surface phases—liquid$_1$–fluid$_3$, liquid$_1$–liquid$_2$, and liquid$_2$–fluid$_3$. If all the surfaces are curved, the mechanical equilibrium conditions related to pressures are three Laplace equations of capillarity. There are no pressure equality relations of the type $P_\alpha = P_\beta$, since all the interfaces are curved. As a consequence, $N = 0$, and the phase rule stated in Equation 9.11 yields $f = r + 1$. For a two-component system, with $r = 2$, $f = r + 1$ gives the number of degrees of freedom as $f = 3$. The equation $f = r + 1$ has been previously derived by Defay and Prigogine[29] by a different type of reasoning.

Example 2: Sessile drop systems with a planar solid substrate[30] Consider a sessile drop system (a two-component, three-phase system), as shown in Figure 9.1, which has three bulk phases—solid, liquid, and the liquid's vapor—and three surface phases—solid–vapor, solid–liquid, and liquid–vapor. Under the simplifying assumption that the solid phase is isotropic such that uniaxial normal stresses and tangential stresses are zero, one may characterize its bulk mechanical state by a single hydrostatic pressure P_s.[32, 33] In order to apply the phase rule for capillary systems (Equation 9.11) one must determine the number, N, of pressure equality relations of the form $P_\alpha = P_\beta$ that exist among the mechanical equilibrium conditions. For the system shown in Figure 9.1, the mechanical equilibrium conditions are given by

$$P_s = P_v; \qquad P_l - P_v = \gamma_{lv} J_{lv}$$

and Young's equation (Equation 9.2) for the three-phase contact line boundary. Only one pressure equality relation exists, so $N = 1$. Accordingly, the phase rule (Equation 9.11) gives the number of degrees of freedom of this system as $f = r$. Realizing that this system has just two components (solid and liquid), $r = 2$, it can readily be seen that the number of degrees of freedom is just *two*. That is, for a two-component, solid–liquid–vapor capillary system, there are two degrees of freedom.

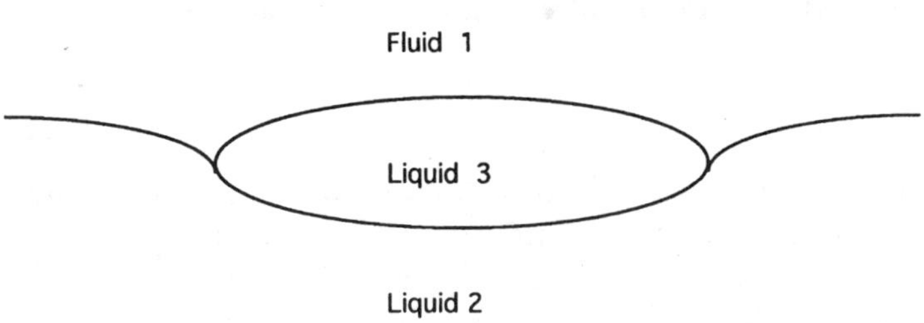

Figure 9.8 A general liquid–liquid lens–fluid system with three curved interfaces.

This result means that any two of the intensive variables used to describe this system are sufficient to characterize the system completely. Alternatively, any two intensive variables may be chosen as the independent variables, and the remaining intensive variables can be expressed as functions thereof. Thus, if one chooses γ_{sv} and γ_{lv} as the two independent variables, then one may express γ_{sl}, the remaining surface tension, as a function of the other two surface tensions in the following functional form

$$\gamma_{sl} = f(\gamma_{sv}, \gamma_{lv}) \tag{9.12}$$

This provides the thermodynamic basis for an equation of state for interfacial tensions.

9.5 Equation of State for Interfacial Tensions of Solid–Liquid Systems

The surface tensions of the solid–vapor interface and the solid–liquid interface, γ_{sv} and γ_{sl}, are important parameters in many areas of applied science and technology. These interfacial tensions are responsible for the behavior and properties of commonly used materials such as paints, adhesives, detergents, and lubricants. Also, many interfacial phenomena, such as the adhesion of biological cells to biomaterials and synthetic materials, are not only governed by specific chemical interactions, but are frequently dominated by changes in the overall free energy of the process of adhesion, which is a function of the surface tensions involving a solid phase.

In order to perform a thermodynamic analysis of interfacial phenomena, such as cell adhesion, one has to know the solid surface tensions, γ_{sv} and γ_{sl}, involved in that system. Generally, the liquid–fluid interfacial tension can be measured directly by many different methods; however, because of the absence of mobility of a solid interface, the solid interfacial tensions cannot be measured directly. Therefore, that determination depends on indirect methods. So far, the most reliable way of estimating the solid surface tension is to interpret contact angle data by an equation of state for interfacial tensions.[34, 35]

For a solid–liquid–vapor capillary system, the classical Young's equation relates the surface tensions of the solid–liquid interface and the solid–vapor interface, γ_{sl} and γ_{sv}, to the liquid–vapor surface tension, γ_{lv}, and the equilibrium contact angle, θ (Equation 9.2). Of the four parameters in Young's equation, only γ_{lv} and θ can be readily measured; hence, Young's equation can only provide the difference between the solid–vapor surface tension, γ_{sv}, and the solid–liquid interfacial tension, γ_{sl}. Nevertheless, Young's equation suggests that the observation of the equilibrium contact angles of liquids on solids may be a starting point for investigating solid surface tensions. The pioneers in this area are W. A. Zisman and his co-workers. Zisman's extensive studies of contact angles and wetting led him to the conclusion that the solid surface tension, γ_s, is approximated by a "critical surface tension," γ_c, which is the surface tension of a liquid forming a zero contact angle on the solid surface. However, as Zisman himself later pointed out,[36] "The empirical nature of γ_c is obvious, and it would be helpful to replace γ_c by parameters having a sound basis in thermodynamic or statistical mechanical considerations. . . . A more precise theory of wetting of low-energy surfaces should at least include γ_{sl}."

The attractiveness of using Young's equation to estimate solid interfacial tensions results from the fact that the liquid surface tension γ_{lv} and the contact angle θ can be measured relatively easily. However, since both γ_{sv} and γ_{sl} in Young's equation are parameters to be determined, an additional equation, such as $\gamma_{sl} = f(\gamma_{sv}, \gamma_{lv})$, providing a relation among the surface tensions in Young's equation is required. Such an equation is referred to as an equation of state for interfacial tensions. Combining Equation 9.12 with Equation 9.2 yields

$$\gamma_{lv} \cos\theta = \gamma_{sv} - f(\gamma_{sv}, \gamma_{lv}) \tag{9.13}$$

As seen from Equation 9.13, the solid–vapor surface tension γ_{sv} can be determined when γ_{lv} and θ are known, provided that f is an explicitly known function. Once knowing γ_{sv}, the solid–liquid interfacial tension γ_{sl} can be determined either from the equation of state, Equation 9.12, or from Young's equation, Equation 9.2.

As demonstrated previously, the existence of the equation of state $\gamma_{sl} = f(\gamma_{lv}, \gamma_{sv})$ for two-component solid–liquid–vapor systems is a direct result of the phase rule, Equation 9.11. The remaining problem then is how to formulate an explicit form of the equation of state. Based on a wide variety of contact angle data on low-energy polymeric surfaces,[34, 35] an empirical equation of state was derived over 25 years ago:

$$\gamma_{sl} = \frac{\left(\sqrt{\gamma_{lv}} - \sqrt{\gamma_{sv}}\right)^2}{1 - 0.015\sqrt{\gamma_{lv}\gamma_{sv}}} \tag{9.14}$$

So far, this equation of state approach has been used successfully to predict many phenomena, such as cell adhesion, the sedimentation of particles, and the behavior of particles at solidification fronts.

However, there are two considerations which have lead to a recent reformulation of this equation. First, the explicit form of the equation of state, as shown in Equation 9.14, has an impediment in its mathematics; that is, when the product $(\gamma_{lv}\gamma_{sv})$ is large enough, the denominator of Equation 9.14 will tend to be zero, and the function becomes discontinuous. Although this mathematical difficulty has been remedied for all practical purposes by using a computer program,[35] published tables,[37] or a hand-held calculator,[38] it has nevertheless deterred potential users of the equation of state for interfacial tensions. Therefore, the equation of state must be reformulated in terms of a function which is continuous and represents physical reality. Second, the contact angle data used to formulate the previous equation of state were obtained over 30 years ago. Although these were the best available at that time, their thermodynamic status and especially accuracy (about $\pm 3°$) certainly do not satisfy today's standards.

Recently, a new form of an equation of state for interfacial tensions has been formulated, as summarized below. The free energy of adhesion per unit area of a solid–liquid pair is equal to the work required to separate a unit area of the solid–liquid interface, that is,

$$W_{sl} = \gamma_{lv} + \gamma_{sv} - \gamma_{sl} \tag{9.15}$$

The free energy of adhesion, W_{sl}, was proposed[39] as a modified geometric mean of the free energy of cohesion of the solid pair, W_{ss}, and the free energy of cohesion of the liquid pair, W_{ll}. Using the definitions $W_{ll} = 2\gamma_{lv}$ and $W_{ss} = 2\gamma_{sv}$, a modified geometric mean for the free energy of adhesion of a solid–liquid pair is given by

$$W_{sl} = 2\sqrt{\gamma_{lv}\gamma_{sv}}\, e^{-\beta(\gamma_{lv} - \gamma_{sv})^2} \tag{9.16}$$

In the above, β is as yet an unknown constant. Combining Equation 9.16 with Equation 9.15, a new equation of state of interfacial tensions can be written as[39]

$$\gamma_{sl} = \gamma_{lv} + \gamma_{sv} - 2\sqrt{\gamma_{lv}\gamma_{sv}}\, e^{-\beta(\gamma_{lv} - \gamma_{sv})^2} \tag{9.17}$$

Combining Equation 9.17 with Young's equation yields

$$\cos\theta = -1 + 2\sqrt{\gamma_{sv}/\gamma_{lv}}\, e^{-\beta(\gamma_{lv} - \gamma_{sv})^2} \tag{9.18}$$

Recently, more accurate data for γ_{lv} and θ of various two-component solid–liquid systems have been obtained using the ADSA-P technique.[40] By finding the best fit of Equation 9.18 to these data, the constant β was determined to be 0.0001247 $(m^2/mJ)^2$.

The new equation of state does not have the difficulty of a discontinuity. The solid–vapor surface tension γ_{sv} and the solid–liquid interfacial tension γ_{sl} can be determined from Equations 9.17 and 9.18 when the measured liquid surface tension γ_{lv} and the contact angle θ are given. Table 9.1 gives such an example.[40] As seen from Table 9.1, the equation-of-state approach yields a constant γ_{sv} value (11.90 ± 0.21 mJ/m^2) for the given solid surface (FC721) from the contact angle data of 16 different liquids measured by ADSA-P.[40]

It is worthwhile to emphasize the applicability of this equation of state. First, as proven by the phase rule, this type of equation generally does not exist for sessile drop–liquid substrate systems; therefore, either of the above-mentioned explicit forms of the equation of state for interfacial tensions should not be applied to sessile drop–liquid substrate systems. Second, this type of equation of state is applicable to solid–liquid systems with two independent chemical components only, a solid and a pure (single-component) liquid. For two-component solid–liquid systems with no significant vapor adsorption on the solid surfaces, the above equation-of-state approach predicts a constant γ_{sv} value from the contact angle data for one and the same solid surface in contact with various liquids. This result can be expected,

Liquid	γ_{lv} (mJ/m²)	θ (degrees)	γ_{sv} (mJ/m²)
decane	23.43	65.97	11.98
dodecane	25.44	69.82	12.03
tetradecane	26.55	73.31	11.63
hexadecane	27.76	75.32	11.63
trans-decalin	29.50	76.71	12.03
cis-decalin	31.65	79.87	12.04
ethylcinnamate	38.37	88.20	12.11
dibenzylamine	40.63	92.06	11.64
DMSO	43.58	94.47	11.90
1-bromonaphthalene	44.01	95.29	11.75
diethylene glycol	45.04	96.84	11.56
ethylene glycol	47.99	99.03	11.82
thiodiglycol	54.13	103.73	12.20
formamide	57.49	107.32	11.90
glycerol	63.11	111.38	12.17
water	72.75	119.05	12.06

Table 9.1 Solid surface tensions γ_{sv} for an FC721 surface calculated by the equation of state (Equation 9.17) at 21 °C.

as the solid–vapor surface tension γ_{sv} in this case should be equal to γ_s, the surface tension of the solid in contact with a vacuum; the latter is a property of the given solid only. However, if the liquids are mixtures, say binary liquid mixtures, the above equation of state is not applicable simply because the number of degrees of freedom is more than two for such a system, and hence the $\gamma_{sl} = f(\gamma_{lv}, \gamma_{sv})$ type of relation does not exist. It was demonstrated[41] that if contact angles of binary liquid mixtures were interpreted by using the above equation of state, the calculated "γ_{sv}" of a given solid surface will show a strong dependence on the surface tensions of the binary liquid mixtures. This is actually a reflection of the third degree of freedom of such three-component systems, and the "γ_{sv}" values are artifacts.

9.6 Drop Size Dependence of Contact Angle and Line Tension

An important phenomenon in wetting is the drop size dependence of the contact angle. Usually, contact angles on low-energy solid surfaces vary only a few degrees for a change of a few millimeters in drop size. Thus, conventional contact angle measurement techniques with an accuracy of one or two degrees would probably fail to detect such a dependence. The higher accuracy ADSA-P will, therefore, be particularly useful in such cases. However, for very small drops, such a drop size dependence will have much more significant effects on contact angles, as indicated by the equations that follow.

In a generalized theory of capillarity,[42] the Gibbsian approach of capillarity was extended to include three-phase (solid, liquid, and vapor) contact lines, or intersection lines of surfaces. Analogous to a surface or an interface phase, a three-phase line is considered a linear phase. The line tension, analogous to the surface tension, is defined as the specific free energy of the three-phase contact line or, mechanically, as a tensile force operating in the three-phase line. The drop size dependence of contact angles can be explained in terms of line tension through a modified Young equation derived in the generalized theory of capillarity.

Consider a sessile drop resting on a horizontal, ideal solid surface, that is, the solid surface is smooth and homogeneous. For only this case, the three-phase contact line is a smooth circle. The mechanical equilibrium condition for any point at such a smooth circle has been derived as[42]

$$\gamma_{lv} \cos\theta = \gamma_{sv} - \gamma_{sl} - \frac{\sigma}{R} \tag{9.19}$$

where γ_{lv}, γ_{sv}, γ_{sl}, and θ are the liquid–vapor, solid–vapor, and solid–liquid interfacial tensions and the contact angle; σ is the line tension, or the tension of the three-phase contact line; and R is the radius of the three-phase contact circle. Equation 9.19 is called the modified Young equation; it considers the contribution of line tension to the equilibrium of the three-phase line.

When $R \to \infty$, Equation 9.19 will be replaced by the classical Young equation

$$\gamma_{lv} \cos\theta_{\infty} = \gamma_{sv} - \gamma_{sl} \tag{9.20}$$

Combining Equation 9.19 with Equation 9.20 yields

$$\cos\theta = \cos\theta_{\infty} - \frac{\sigma}{\gamma_{lv}}\frac{1}{R} \tag{9.21}$$

Equation 9.21 shows that the contact angle will vary with drop size, or more explicitly, with the radius of the three-phase contact circle, and that line tension can be related to the measurable quantities: contact angle, liquid surface tension, and radius of the three-phase contact circle. Therefore, it should be possible to determine the mean value of line tension by measuring the dependence of contact angle on the radius of the three-phase contact circle. As indicated by Equation 9.21, for a given solid–liquid system, the cosine of the contact angle will be a linear function of $1/R$. If the line tension is positive, then as $1/R$ increases, $\cos\theta$ will decrease. According to Equation 9.21, the slope of the straight line will be equal to the ratio $-\sigma/\gamma_{lv}$; thus the line tension σ can be calculated from these slopes, since we know the values of the liquid surface tension.

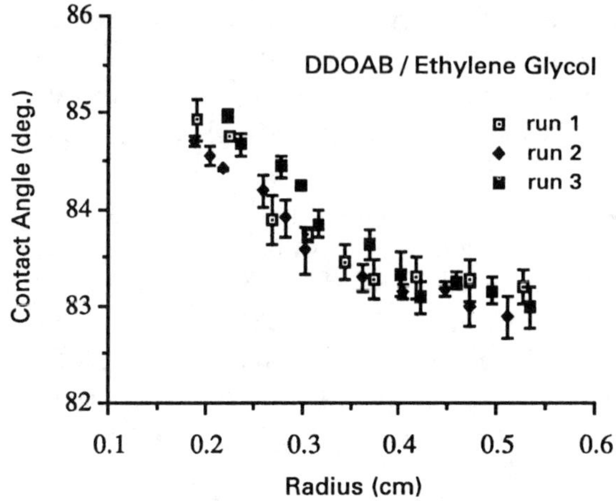

Figure 9.9 **An example of contact angle drop size dependence.**

Recently, the drop size dependence of the advancing contact angle of different liquids on several hydrophobic solid surfaces was studied.[43–45] It was found that contact angles decrease 3° to 5° as the contact radius increases from approximately 1 to 5 mm. Thereafter, the contact angles tend to reach a final constant value. Figure 9.9 gives such an example. This drop size dependence of contact angles is interpreted in terms of line tensions through the modified Young equation (Equation 9.21). The values of θ_∞ and line tension, σ, for these systems can be calculated using the experimental data and the modified Young equation. The line tensions were found to be positive, ranging from 2 to 4 μJ/m. It should be mentioned that there are also reports in literature showing an increase of contact angles with drop size. This kind of phenomenon can be attributed to solid surface imperfections (heterogeneity or roughness).[46, 47]

9.7 Contact Angles in the Presence of a Thin Liquid Film

In the study of contact angle phenomena, it is common knowledge that the adsorption of the liquid component on the surface of a solid substrate may affect the magnitude of the contact angle. Generally, there are two ways to consider adsorption on a solid surface. The first is the classical surface thermodynamics approach established by Gibbs,[3] which takes adsorption into account by considering a solid–vapor interface with a surface tension γ_{sv}, rather than a pure solid–vacuum interface with a surface tension, γ_s. The difference between γ_s and γ_{sv} is defined as the equilibrium spreading pressure, $\pi = \gamma_s - \gamma_{sv}$. Generally, the adsorption will decrease the solid surface tension, hence, $\pi > 0$. The classical Young equation, in principle, includes the effect of adsorption on the equilibrium contact angle in terms of the solid–vapor surface tension, γ_{sv}.

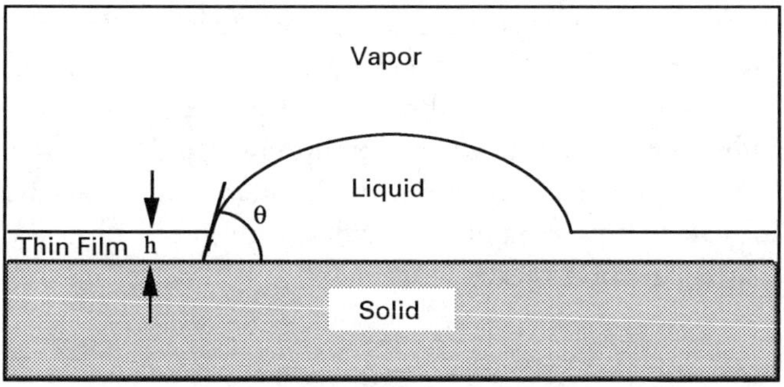

Figure 9.10 **Illustration of a sessile drop on a solid surface in equilibrium with the liquid's vapor and a flat thin liquid film.**

A second way to consider the adsorption of the liquid component on solid surfaces is given in the theory of thin liquid films. When the adsorbate on the surface of a solid substrate forms a thin film with a thickness ranging from ten to a few hundred angstroms, the above interface model may become untenable. On the one hand, the thickness of the thin film is now much greater than the thickness of a heterogeneous interfacial region between two uniform bulk phases, which is usually not more than a few molecular diameters; on the other hand, although such a film is quite thick, the intermolecular interactions of the two heterogeneous boundary regions of the thin film overlap, causing the adsorbate inside this film to have different properties from both that of a bulk phase and that of an interface.

Thin liquid films have been studied extensively. The modern theory of thin liquid films has been developed by Derjaguin, Churaev, and many other researchers. An excellent summary can be found in Derjaguin, Churaev, and Muller's monograph *Surface Forces.*[48] To account for the intermolecular forces in a thin liquid film, Derjaguin introduced "disjoining pressure" as a distinctive property of the thin liquid film. In this section, however, we look at the thin liquid film–contact angle phenomena from a thermodynamic point of view.[49]

Consider a sessile drop of liquid resting on an ideal solid substrate in equilibrium with the liquid's vapor and a thin liquid film, as illustrated in Figure 9.10. As usual, the solid surface is assumed to be isotropic, homogeneous, smooth, rigid, and insoluble in the liquid. Physically, between the flat film and the free (without disturbance of the film) liquid–vapor meniscus, there is a transition zone. For the sake of simplicity, such a small transition zone will not be considered in the present model, that is, only a flat film intersecting with the liquid–vapor meniscus governed by the Laplace equation will be considered, as shown in Figure 9.10. In this model, the film is considered to be a simple thermodynamic phase, as are all the other phases in the system. That is, all the phases are assumed to be homogeneous and not subject to any chemical reactions and interactions with external fields. As the main difference in the physical nature of the thin film and that of an interface separating two bulk phases stems from the two overlapping regions of interfacial interactions of the thin liquid film, the thickness of the film is a characteristic parameter, and hence the thin liquid film is considered a three-dimensional phase. Because the thickness of the thin liquid film makes a significant contribution to the film energy, the film thickness, h, must be considered as a geometric variable, in addition to the film area A_f. Applying Gibbsian thermodynamics to such a thin liquid film, the energy form of the fundamental equation can be written as

$$U_f = U_f(S_f, A_f, h, N_{1f}, \ldots, N_{rf}) \tag{9.22}$$

where U_f is the total internal energy of the thin liquid film phase; S_f is the total entropy of the thin film; and N_{kf} is the mole number of the kth component of the film.

Microscopically, as implied by such a fundamental equation, this thin liquid film model considers only the van der Waals type long-range intermolecular forces and the reorientation effects of molecular structure due to these long-range forces. The differential form of the fundamental equation is

$$dU_f = TdS_f + \gamma_f dA_f - A_f \Pi\, dh + \sum_{i=1}^{r} \mu_i dN_{if} \tag{9.23}$$

Other than the temperature T and the chemical potentials μ_i, the two new intensive variables here are the film tension and the disjoining pressure:

$$\gamma_f = \left(\frac{\partial U_f}{\partial A_f}\right)_{S_f, h, N_{if}} \quad \text{and} \quad \Pi = -\frac{1}{A_f}\left(\frac{\partial U_f}{\partial h}\right)_{S_f, A_f, N_{if}} \tag{9.24}$$

Generally, the film tension, γ_f, depends on the molecular nature of both the liquid and the solid, and is different from both the liquid–vapor interfacial tension, γ_{lv}, and the solid–liquid interfacial tension, γ_{sl}, since the intermolecular interactions of the two "interfacial regions" (the film–vapor and the solid–film) overlap. The disjoining pressure, Π, is the force required to separate a unit area of the two "interfaces" of the film. In other words, in order to compress the film thickness by dh, an amount of work, $dW = A_f \Pi\, dh$, must be done on the film.

It has been demonstrated[49] that for the considered thin film system, the mechanical equilibrium conditions are given by the Laplace equation of capillarity (Equation 9.1) and

$$\Pi = P_v \tag{9.25}$$

$$\gamma_{lv}\cos\theta = (P_l - \Pi)h + \gamma_f - \gamma_{sl} \tag{9.26}$$

Equation 9.1, the Laplace equation, governs the equilibrium shape of the liquid–vapor meniscus. Equation 9.25 is the mechanical equilibrium condition between the flat film and the bulk vapor phase. It clearly states that the disjoining pressure in the flat thin film must balance the external compressive force exerted by the vapor. It should also be noted that, as indicated by Equation 9.25, the disjoining pressure, Π, is a positive quantity, as is the vapor pressure, P_v, for such a simple system (i.e., without considering any electric effects). For a nonflat thin film, a Laplace equation type of relation between Π and P_v can be expected. Equation 9.26 is the mechanical equilibrium condition at the intersection of the solid–liquid interface, the liquid–vapor interface, and the thin liquid film. Although Equation 9.26 has a similar form to the classical Young equation, here γ_{sv} is replaced by $(P_l - \Pi)h + \gamma_f$. Equation 9.26 directly accounts for the effects of a thin liquid film on the equilibrium contact angle in terms of the disjoining pressure Π, the film thickness h, and the film tension γ_f. It should be noted that γ_f, intrinsically, is also a function of h. When $h \rightarrow 0$, that is, the thin liquid film vanishes from the solid surface, the first term, $(P_l - \Pi)h$, drops off Equation 9.26, and the second term γ_f can be expected to revert to the solid surface tension γ_s. That is, Equation 9.26 becomes Young's equation for the case of no adsorption on the solid surface. In light of Equations 9.1 and 9.25, Equation 9.26 can be rewritten equivalently as

$$\gamma_{lv} \cos\theta = \gamma_{lv} Jh + \gamma_f - \gamma_{sl} \tag{9.27}$$

where J is the mean curvature of the liquid–vapor meniscus at the intersection of the liquid–vapor interface and the thin film. Obviously, Equation 9.27 shows an explicit dependence of contact angle on the curvature of the liquid–vapor meniscus and on the film thickness.

9.8 Effects of Elastic Liquid–Vapor Interfaces on Wetting

There is another important category of wetting systems: solid–liquid–fluid systems with liquid–fluid interfaces that exhibit elasticity. For such a system, an immediate concern is the effect of the elasticity of the liquid–fluid interface on the basic thermodynamic relations of such a system, such as the Laplace equation of capillarity and Young's equation.

Generally speaking, surface tension is the work to create a unit area of new surface, and surface stress is related to the work of deforming a surface. These two quantities are numerically equal when atomic mobilities are sufficiently high to preserve the microscopic configuration of the surface following the deformation. This is true for many liquid–fluid interfaces whose surface tensions are constant, independent of surface deformation. However, there are many other liquid–fluid systems where the value of the surface tension of the liquid–fluid interface depends on the deformation of the surface. For example, there are some polymer or protein solution systems where molecules adsorbed at the solution–fluid interface may form an elastic "skin." This is also the case of spread monolayers of insoluble substances (e.g., certain kinds of surfactants) on liquid–fluid interfaces.

For an elastic liquid–fluid interface, since the degree of deformation and hence the surface strain generally changes from place to place, the fundamental equation of the liquid–fluid interface phase has to be written in density form with respect to an element surface area dA_{lf}. Under the assumption of moderate curvature and absence of chemical reactions and electrical and magnetic interactions, the energy form of the fundamental equation for such an element can be written as[50]

$$u_{lf} = u_{lf}(s_{lf}, \varepsilon_{11}, \varepsilon_{12}, \varepsilon_{21}, \varepsilon_{22}, n_{1,lf}, n_{2,lf}, \ldots, n_{r,lf}) \tag{9.28}$$

where the subscript lf denotes the liquid–fluid interface, u_{lf} is the area density of the internal energy of the liquid–fluid interface, s_{lf} is the entropy density, $n_{k,lf}$ ($k = 1, 2, \ldots, r$) is the mole density of the kth component, and ε_{ij} ($i,j = 1, 2$) is the surface strain tensor component. As u_{lf} is defined as the energy density with respect to an element of surface area, dA_{lf}, the total internal energy of this elastic liquid–fluid interface phase is the integration of u_{lf} over the entire liquid–fluid interface. The differential form of the fundamental equation can be written as[50]

$$du_{lf} = T ds_{lf} + \sum_{i,j=1}^{2} \tfrac{1}{2} g_{ij} d\varepsilon_{ij} + \sum_{k}^{r} \mu_k dn_{k,lf} \tag{9.29}$$

where the surface stress tensor is given by

$$g_{ij} = 2\left(\frac{\partial u_{lf}}{\partial \varepsilon_{ij}}\right)_{s_{lf}, \varepsilon_{xy \neq ij}, n_{k,lf}} = \gamma\delta_{ij} + \frac{\partial \gamma}{\partial \varepsilon_{ij}}; \qquad i,j = 1,2 \tag{9.30}$$

The last term in Equation 9.30 represents the surface elasticity effects; that is, the surface tension varies with various kinds of surface deformation: stretching, compressing, twisting, bending, etc. Equation 9.30 gives a relationship between surface stress g_{ij} and surface tension γ. Generally, surface stress and surface tension will be numerically equal only if γ is unaffected by the deformation, that is, when surface atomic mobilities are high and there is no long range correlation in atomic positions. In these cases, $(\partial\gamma/\partial\varepsilon_{ij}) = 0$, so that $g = \gamma$. However, in cases where surfactant or biological molecules adsorbed at a liquid–fluid interface form a network, the long-range correlation in atomic or molecular positions and the low molecular mobilities result in a surface deformation dependence of the local configuration around any particular molecule in the interfacial region. Therefore, $(\partial\gamma/\partial\varepsilon_{ij}) \neq 0$.

The mechanical equilibrium conditions for a solid–liquid–fluid system (e.g., a sessile drop system) with an elastical liquid–fluid interface are given by[50]

$$P_l - P_f = \sum \tfrac{1}{2} g_{ij}\left(\frac{1}{R_1} + \frac{1}{R_2}\right) \tag{9.31}$$

$$\sum \tfrac{1}{2} g_{ij} \cos\theta = \gamma_{sf} - \gamma_{sl} \tag{9.32}$$

Although Equations 9.31 and 9.32 have a similar form to the classical Laplace equation (Equation 9.1) and Young's equation (Equation 9.2), respectively, γ_{lv} in

Equations 9.1 and 9.2 now is replaced by $\Sigma\frac{1}{2}g_{ij}$. Equations 9.31 and 9.32 predict that the equilibrium shape of the liquid–fluid interface and the contact angle are controlled by both the surface tension and surface elasticity. This implies that in the presence of elasticity the "surface tension" of the liquid–fluid interface measured by drop shape analysis techniques (such as ADSA) is actually a summation of the true surface tension and the elasticity effects. As seen from Equation 9.30, for ideal liquid–fluid surfaces where $(\partial\gamma_{lf}/\partial\varepsilon_{ij}) = 0$, $\Sigma\frac{1}{2}g_{ij} = \gamma_{lf}$, then Equations 9.31 and 9.32 will revert to the classical Laplace equation and Young's equation, respectively. For the cases of $(\partial\gamma_{lf}/\partial\varepsilon_{ij}) \neq 0$, one has to obtain the expressions for ε_{ij} in order to show what Equations 9.31 and 9.32 imply explicitly.

For simplicity, let us consider a spherical sessile drop of radius R on a solid surface. In this case, the elasticity may result from a monolayer of surfactant on the liquid–fluid interface. Assuming that the strain tensor does not contain any shear components, Equations 9.31 and 9.32 for such a system can be written as[50]

$$P_l - P_f = \frac{2}{R}\left(\gamma_{lf} + R\frac{\partial\gamma_{lf}}{\partial R} + \frac{1}{2}\theta\frac{\partial\gamma_{lf}}{\partial\theta}\right) \tag{9.33}$$

$$\left(\gamma_{lf} + R\frac{\partial\gamma_{lf}}{\partial R} + \frac{1}{2}\theta\frac{\partial\gamma_{lf}}{\partial\theta}\right)\cos\theta = \gamma_{sf} - \gamma_{sl} \tag{9.34}$$

As seen from the classical Laplace equation (Equation 9.1) and Young's equation (Equation 9.2), the curvature of the liquid–fluid interface affects only the Laplace pressure $(P_l - P_f)$, and not the contact angle. This, however, is not the case for the system discussed here. Apparently, $(\partial\gamma_{lf}/\partial R)$ in Equations 9.33 and 9.34 reflects the curvature (drop size) or the bending effects of an elastic liquid–fluid interface on both the Laplace pressure and the contact angle, that is, the contact angle is no longer independent of the curvature of the interface. With respect to the term $(\partial\gamma_{lf}/\partial\theta)$ in Equation 9.33, we point out that, unlike a system with an ideal liquid–fluid interface, the Laplace pressure $(P_l - P_f)$ across an elastic liquid–fluid interface also depends on the boundary condition of this interface, that is, the equilibrium contact angle, θ, at the three-phase contact line. This special feature of surface elasticity may be understood as the long-range dependence of the local configuration around any molecule on the deformation of the entire interface. For a given solid–liquid system γ_{sf}, γ_{sl} and hence the right hand side of Equation 9.34 should be constant. Therefore, the left hand side of Equation 9.34 shows a dependence of contact angle on the drop size due to the presence of surface elasticity.

It has also been demonstrated[50] that in the presence of an elastic liquid–fluid interface the conventional capillary rise technique based on Equation 9.8 is no longer valid to determine θ by measuring the height of capillary rise; and the contact angle in a capillary rise system will generally be different from that in a sessile drop system for the same solid–liquid pair. This is because the surface tension of the liquid–fluid interface and the contact angle depend on the geometric configurations of the elastic liquid–fluid interface. Generally, surface elasticity introduces more complications into surface tension and contact angle measurements. As the existing techniques may provide information only about the simultaneous effects of surface tension and surface elasticity, one has to find other means to separate the elasticity contribution from the results of the conventional experiment methods.

References

1 T. Young. *Phil. Trans. Royal Soc. (London).* **95**, 65, 1805.

2 F. Heslot, N. Fraysse, and A. M. Cazabat. *Nature.* **338**, 640, 1989.

3 W. D. Gibbs. *The Scientific Papers of J. Willard Gibbs*, Vol. 1. Dover, New York, 1961.

4 C. J. Budziak. Ph. D. thesis, University of Toronto, Canada, 1991.

5 R. G. Cox. *J. Fluid Mech.* **168**, 169, 1986.

6 T. D. Blake and J. M. Haynes. *J. Colloid Interface Sci.* **30**, 421, 1969.

7 R. V. Sedev, C. J. Budziak, J. G. Petrov, and A. W. Neumann. Accepted for publication in *J. Colloid Interface Sci.*

8 R. N. Wenzel. *Ind. Eng. Chem.* **28**, 988, 1936.

9 A. W. Neumann and R. J. Good. *J. Colloid Interface Sci.* **38**, 341, 1972.

10 A. B. D. Cassie and S. Baxter. *Trans. Faraday Soc.* **40**, 546, 1944.

11 A. W. Neumann, D. Renzow, R. Reumuth, and I. E. Richter. *Fortschr. Kolloide Polym.* **55**, 49, 1971.

12 J. J. Jasper and E. V. Kring. *J. Phys. Chem.* **59**, 1019, 1955.

13 R. Shuttleworth and G. L. Bailey. *Discussions Faraday Soc.* **3**, 16, 1948.

14 B. V. Derjaguin. *C. R. Acad. Sci., USSR.* **51**, 361, 1946.

15 R. E. Johnson, Jr., and R. H. Dettre. *J. Phys. Chem.* **68**, 1744, 1964.

16 R. E. Johnson, Jr., and R. H. Dettre. *Advan. Chem. Ser.* **43**, 112, 1964.

17 D. Li and A. W. Neumann. *Colloid Polym. Sci.* **270**, 498, 1992.

18 A. W. Neumann. *Adv. Colloid Interface Sci.* **4**, 105, 1974.

19 J. F. Padday. In *Surface and Colloid Science*, Vol. 1. (E. Matijevic, Ed.) Wiley, New York, 1968, p. 101.

20 A. W. Neumann and R. J. Good. In *Surface and Colloid Science*, Vol. 11. (R. J. Good and R. S. Stromberg, Eds.) Plenum, New York, 1979, p. 31.

21 A. W. Neumann. *Adv. Colloid Interface Sci.* **4**, 105, 1974.

22 Y. Rotenberg, L. Boruvka, and A. W. Neumann. *J. Colloid Interface Sci.* **93**, 169, 1983.

23 P. Cheng, D. Li, L. Boruvka, Y. Rotenberg, and A. W. Neumann. *Colloids and Surfaces.* **43**, 151, 1990.

24 D. Li, P. Cheng, and A. W. Neumann. *Adv. Colloid Interface Sci.* **39**, 347, 1992.

25 F. W. Skinner, Y. Rotenberg, and A. W. Neumann. *J. Colloid Interface Sci.* **130**, 25, 1989.

26 W. C. Duncan-Hewitt, Z. Policova, P. Cheng, E. I. Vargha-Butler, and A. W. Neumann. *Colloids and Surfaces.* **42**, 391, 1989.

27 E. Moy, P. Cheng, Z. Policova, D. Kwok, D. R. Mack, P. M. Sherman, and A. W. Neumann. *Colloids and Surfaces.* **58**, 215, 1991.

28 C. J. Budziak and A. W. Neumann. *Colloids and Surfaces.* **43**, 279, 1990.

29 R. Defay and I. Prigogine. *Surface Tension and Adsorption.* (A. Bellemans, collab. with D. H. Everett, trans.) Longmans, Green & Co., London, 1966.

30 D. Li, J. Gaydos, and A. W. Neumann. *Langmuir.* **5**, 1133, 1989.

31 J. G. Kirkwood and I. Oppenheim. *Chemical Thermodynamics.* McGraw-Hill, Toronto, 1961.

32 A. Münster. *Classical Thermodynamics.* (E. S. Halberstadt, trans.) Wiley, Toronto, 1970.

33 A. G. McLellan. *The Classical Thermodynamics of Deformable Materials.* Cambridge University Press, New York, 1980.

34 O. Driedger, A. W. Neumann, and P. J. Sell. *Kolloid-Z.Z. Polymere.* **201**, 52, 1965.

35 A. W. Neumann, R. J. Good, C. J. Hope, and M. Sejpal. *J. Colloid Interface Sci.* **49**, 291, 1974.

36 W. A. Zisman. *Contact Angle, Wettability and Adhesion.* Advances in Chemistry Series, Vol. 43. American Chemical Society, Washington, D.C., 1964.

37 A. W. Neumann, D. R. Absolom, D. W. Francis, and C. J. Van Oss. *Separation and Purification Method.* **9** (1), 69, 1980.

38 C. P. S. Taylor. *J. Colloid Interface Sci.* **100**, 589, 1984.

39 D. Li and A. W. Neumann. *Adv. Colloid Interface Sci.* **39**, 299, 1992.

40 D. Li and A. W. Neumann. *J. Colloid Interface Sci.* **148**, 190, 1992.

41 D. Li, C. Ng, and A. W. Neumann. *J. Adhesion Sci. and Technology.* **6**, 601, 1992.

42 L. Boruvka and A. W. Neumann. *J. Chem. Phys.* **66**, 5464, 1977.

43 J. Gaydos and A. W. Neumann. *J. Colloid Interface Sci.* **120**, 76, 1987.

44 D. Li and A. W. Neumann. *Colloids and Surfaces.* **43**, 195, 1990.

45 D. Duncan, D. Li, J. Gaydos, and A. W. Neumann. Submitted to *J. Colloid Interface Sci.*

46 D. Li, F. H. Y. Lin, and A. W. Neumann. *J. Colloid Interface Sci.* **142**, 224, 1991.

47 F. H. Y. Lin, D. Li, and A. W. Neumann. Accepted for publication in *J. Colloid Interface Sci.*

48 B. V. Derjaguin, N. V. Churaev, and V. M. Muller. *Surface Forces.* Plenum, New York, 1987.

49 D. Li and A. W. Neumann. *Adv. Colloid Interface Sci.* **36**, 125, 1991.

50 D. Li and A. W. Neumann. Accepted for publication in *Langmuir.*

10

Secondary Ion Mass Spectrometry As Applied to Thin Organic and Polymeric Films

JIAN-XIN LI, ROBERT W. JOHNSON, JR.,
and JOSEPH A. GARDELLA, JR.

Contents

10.1 Introduction and Background

In this chapter, we review and outline the application of secondary ion mass spectrometry (SIMS) to the study of thin organic and polymeric films. Much of the work that has been accomplished, and is discussed in this chapter, utilizes Langmuir–Blodgett and self-assembly methods to construct *model systems* for the development of the technique. The SIMS method of surface analysis has the potential to solve chemical questions about the synthesis and chemistry of thin films. It has very low detection limits and, as a mass spectrometry technique, the ability to analyze mixtures and solve molecular structures. We shall demonstrate that the SIMS method can answer important questions about thin organic film structure and composition at the present state of the art.

Overview of the SIMS Method and Experiment

Secondary ion mass spectrometry (SIMS) is a simple and very powerful surface analytical technique.[1] Experimentally, the SIMS technique involves focusing a primary

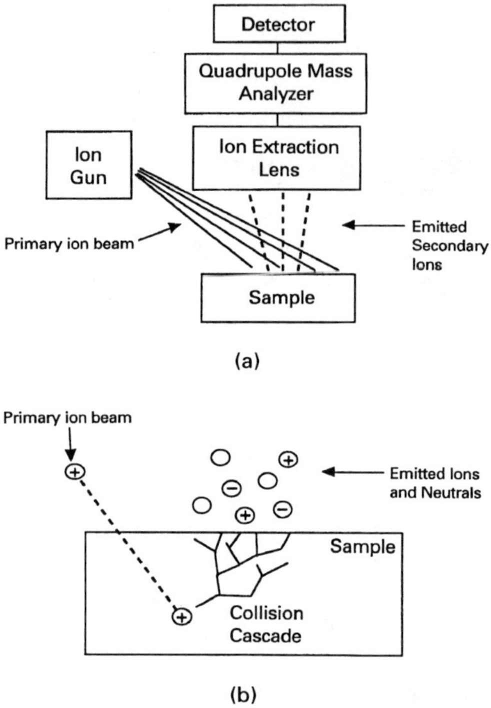

Figure 10.1 Schematic diagrams of (*a*) SIMS instrument and (*b*) secondary ion emission process.

ion beam of 0.5–30 keV energy onto a sample surface and then extracting the secondary ions that are generated. These secondary ions are separated by mass-to-charge ratio in a mass analyzer[2, 3] and next come in contact with an ion multiplier, where a signal is generated. This process is shown schematically in Figure 10.1. In SIMS, the primary ions usually consist of ionized Ar, Xe, Kr, O, Cs^+, or Ga atoms. Provisions are usually made for focusing/defocusing and rastering the primary ion beam, and the secondary ions are extracted by an electrostatic focusing lens arrangement. Ultra high vacuum conditions (pressures less than 1×10^{-8} torr) are necessary to ensure the efficient generation and detection of secondary ion species. In some instances, the focusing lenses are set up to provide a kinetic energy filter so that organic secondary ions may be collected more efficiently. Mass analysis is commonly accomplished using quadruple mass filter (QMF), time-of-flight (TOF), or magnetic sector mass analyzers. Instrumentally, the development and application of TOF-SIMS is an area of great current interest, since TOF mass analysis gives the ability to detect ions of very high mass and therefore analyze high molecular weight samples (10 000 to 20 000 amu).[4, 5] TOF-SIMS also has very low limits of detection due to the quasisimultaneous detection of all masses.

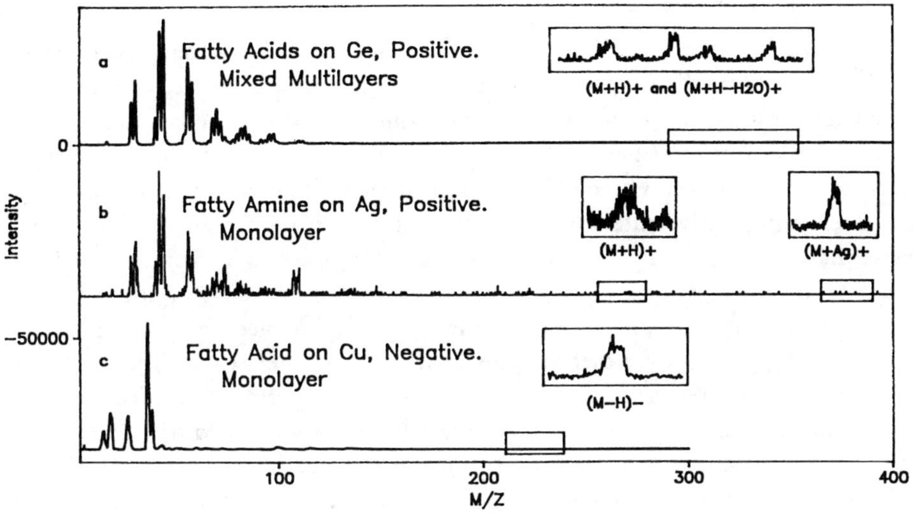

Figure 10.2 **SIMS of L–B films: (*a*) 1 layer behenic/4 layers arachidic acid on Ge, (*b*) 1 layer octadecylamine on Ag, and (*c*) 1 layer myristic acid on Cu.**

The SIMS experiment can be performed in one of two different modes—dynamic or static.[1] In dynamic SIMS, high primary ion current densities, $<1\ \mu A/cm^2$, are used to gain elemental information as a function of sputter erosion depth into the bulk. In addition, elemental maps or images can be obtained by focusing the ion beam to small (1–3 μm) spot sizes (the so-called ion microprobe) or by illuminating the sample with a large-diameter ion beam and using ion optics to preserve the spatial resolution (the ion microscope). In the dynamic mode, virtually all molecular information is lost as a result of the high current densities used. In the static mode, low (primary ion) current densities and ion dosages (less than 1 nA/cm^2 and 1×10^{12} ions/cm^2, respectively) yield a very small probability that a primary ion will strike the same molecule or surface region twice in a typical analysis time. Therefore, molecular structure and bonding information can be preserved. Static SIMS was first described in 1970 by Benninghoven.[6] With current density lowered to less than 10 nA/cm^2 and defocused ion beams, molecular secondary ions were obtained for organic carboxylic acid compounds. In 1976–1977, static SIMS spectra of 15 amino acids[7, 8] displaying parent-like molecular ions were published, causing increased interest in the technique. From that time, static SIMS has been used to analyze a wide variety of synthetic polymers,[4, 9–15] biomolecules,[16] and pharmaceuticals.[17] The focus of a majority of these studies was to investigate secondary ion emission mechanisms and/or show the ability of static SIMS to provide mass specta of large, thermally labile organic compounds. Figure 10.2 shows typical data for organic molecules on surfaces: high intensities of low mass ions from fragmentation mechanisms and low intensity molecular (or quasimolecular) ions at higher masses.

Ion Formation Mechanisms

The mechanism of secondary ion formation is complex. As the primary ion or atom impinges upon a surface it becomes implanted in the sample lattice. The primary ion transfers momentum to the solid surface, thus perturbing the lattice structure and causing the emission (desorption) of ions (Figure 10.1*b*), electrons, and neutral species from the surface. The mechanisms of desorption of atomic and molecular ions are not completely understood. Theoretical models have been developed that suggest that the sputtering of atomic species results from a "collision cascade"[18] process. Some workers have suggested that this mechanism is also responsible for the desorption of molecular species,[19] while others[20, 21] suggest that organic molecules cannot be desorbed by mechanisms in which binary elastic collisions are the predominant events. Garrison and Winograd and co-workers have studied collision cascades induced in organic overlayers. They have shown that direct desorption cannot be completely responsible for molecular ion formation and that recombination is also an important mechanism.[22] They postulate that collision cascades distribute energy to the molecular degrees of freedom of lattice molecules. This causes the formation of a vibrationally excited region prior to desorption. Formation of this region would be a common step in the desorption–ionization techniques that use kiloelectronvolt (SIMS) or megaelectronvolt (plasma desorption mass spectrometry) primary ion bombardment. More detailed reviews of the SIMS technique can be found in recent review articles[2, 3, 23] and books,[24, 25] and a description of the basic principles as well as applications in materials science and technology may be found in the lead volume of this Series, *Encyclopedia of Materials Characterization.*

Comparisons to Other Surface Analysis Techniques

As a surface sensitive technique, SIMS has several advantages over X-ray photoelectron spectroscopy [XPS; also known as electron spectroscopy for chemical analysis (ESCA)] and Auger electron spectroscopy (AES).[1, 26] Such benefits include the ability to detect all elements and their isotopic distributions, low detection limits, molecular secondary ions, which are directly related to the molecular structure, and surface compositional sensitivity because of the matrix dependence of secondary emission. Major difficulties that limit routine analysis using SIMS include sample damage due to sputtering and a lack of understanding of the relationship between matrix dependent secondary emission and surface composition. Also, the application of SIMS is limited by the great difficulty in obtaining reproducible, accurate quantitative molecular information.

The Motivation for Thin Organic Films as Model Systems

Langmuir–Blodgett (L–B) films have been now extensively used as model systems to study the SIMS sampling process for large organic molecules, to investigate secondary molecular ion formation mechanisms, and to develop a quantitative method for static SIMS.[24–41] The use of L–B films as well-characterized model systems for SIMS analysis provides several advantages: (1) a highly regular and pure model system can be prepared, (2) molecular orientations for large molecules are precisely determined, and (3) modifications such as substrate changes, added cations, and most importantly, the "physical state" of the monolayer can be changed by varying the applied surface tension during film production.[42–45] In addition, techniques such as optical ellipsometry and reflection–adsorption infrared spectroscopy can corroborate the nature of the film.[46] Most work[27–37] has employed static primary ion conditions because the limitations on primary ion beam current density, energy, and total ion dose reduce sample disruption and decrease fragmentation, thus benefitting the production of structure-related molecular and quasimolecular ions. Much static SIMS analysis has been accomplished on L–B films of classical long chain alkanoic (fatty) acids[27–29, 31, 37] or acid salts.[33–36, 38–41] Molecular and quasimolecular ions are reported in most of the later studies.[27–37] The detection of molecular ions from a single (quantified) monomolecular layer demonstrates the promise of the L–B technique for producing a homogeneous system of large molecules. These experiments illustrate the sensitivity of static SIMS to the chemistry of the molecular species and substrate, the "pre-ionized state," and the final emitted molecular ion. Some of the studies have focused on multilayer films produced by L–B methods, a distinct advantage of the L–B approach, mainly to address sampling depth issues.[28, 29, 33–41] Methods to obtain quantitative molecular ion data from static SIMS analyses of L–B films have also been published.[31, 32, 34] Applications of SIMS to the study of self-assembled monolayer (SAM) films have been more recent and have focused on using SIMS to solve the chemistry of the films, rather than the use of the films as model systems.[47, 48]

The following sections detail the application of SIMS based on studies that focus on determining surface chemistry (qualitative aspects), using thin films as models to understand sampling depth in SIMS, and the development of quantitative applications.

10.2 Qualitative Information: Mechanisms of Secondary Molecular Ion Formation

Structure–Ion Formation Relationships

A series of (saturated and unsaturated) alkanoic (fatty) acid and acid-salt (using divalent barium counter-ions) monolayers and multilayers have been studied.[27–29] Table 10.1 summarizes the molecular ions detected from both positive and negative SIMS analyses. "M" signifies a molecular mass of the fatty acid. The data can be analyzed by focusing on the surface chemistry that governs the formation of three typical molecular ions: $(M+H)^+$, $(M-H)^-$, and $(M+Subs)^{+\text{ or }-}$, for which examples are $(M+Ag)^+$ and $(M+Au)^-$.

From the monolayers, the substrate metal plays an important role in governing the formation of ions. For silver substrates, $(M+H)^+$ ions are not generally observed in the positive ion spectra. Yet these signals, although weak, are detected from less active surfaces, such as gold and germanium. Clearly, as in Benninghoven's work,[7, 8]

Sample/Substrate[a]	Signal Detected[b] (M+H)$^+$	(M–H)$^-$	(M+Subs)$^+$	(M–H+2Subs)$^+$	(M–H+Ba^{2+})$^+$
Bu-1/Cu	no	yes (s)	no	no	na
M-1/Cu	no	yes (s)	no	no	na
O-1/Ag	no	yes	yes (s)	ns	na
L-1/Ag	no	yes	yes (s)	ns	na
S-1/Cu	no	yes	no	no	na
S-1/Ag	no	yes	yes	yes	na
S-5/Ag	yes	yes	yes	yes	na
BaS-1/Ag	no	yes	yes	ns	yes
BaS-15/Ag	ns	yes	yes	ns	yes
S-1/Ge	yes	yes	no	no	na
S-15/Ge	yes	yes	no	no	na
BaS-1/Ge	ns	yes	no	no	yes
BaS-15/Ge	ns	yes	no	no	yes
S-1/Au	yes	yes	no	no	na
S-15/Au	yes	yes	no	no	na
BaS-1/Au	ns	yes	no	no	yes
BaS-15/Au	ns	yes	no	no	yes
A-1/Ag	no	yes	yes	yes (w)	na
A-5/Ag	yes	yes	no	no	na
A-1/Ge	yes	yes	no	no	na
A-5/Ge	yes	yes	no	no	na
BaA-1/Ge	no	yes	no	no	yes
BaA-5/Ge	yes	yes	no	no	yes
B-1/Cu	no	yes	no	no	na
B-1/Ag	no	yes	yes	yes (w)	na
B-5/Ag	yes	yes	no	no	na
B-1/Ge	yes	yes	no	no	na
B-5/Ge	yes	yes	no	no	na
BaB-1/Ge	no	yes	no	no	yes
BaB-5/Ge	yes	yes	no	no	yes

[a] Samples are listed as acid-# L–B layers/substrate. Ba = barium salt of fatty acid, Bu = butanoic, M = myristic, S = stearic acid, A = arachidic acid, B = behenic acid, L = linolenic, O = oleic, Ag = silver, Ge = germanium, Cu = copper, and Au = Gold.

[b] Yes = signal present, no = no signal present, w = weak signal, na = not applicable, and ns = not scanned.

Table 10.1 Detection of molecular and quasimolecular ions from alkanoic acid L–B systems.

the active silver surface prevents protonation from the monolayer. From monolayers, (M–H)$^-$ is always detectable because of the stability of the carboxylate anion. (M+Ag)$^+$ and (M–H+2Ag$^+$)$^+$ are typical cationized molecular ions detectable because of the attachment of a substrate ion (or ions). It is strongest from the unsaturated fatty acid monolayers, indicating a preference for attachment of the silver to a double bond.[27] Monolayers of acid-salt films yield very interesting results, both preformed ion $(M{-}H{+}Ba^{2+})^+$ and $(M{+}Ag)^+$, indicating—as is known for barium—only partial deprotonation of the monolayer by the barium.[45] Obviously, no $(M{+}H)^+$ and a strong $(M{-}H)^-$ are detectable from all acid-salt monolayer films. Multilayer acid and acid-salt films produce interesting differences, the most important being the presence of a strong $(M{+}H)^+$ ion signal from the multilayer acid films. The head-to-head configuration in the Y deposition produces the appropriate geometry to encourage intermolecular proton transfer, leading to $(M{+}H)^+$ and the complementary $(M{-}H)^-$ ions. The production of multilayers attenuates the silver ion signal and thus decreases the formation probability of $(M{+}Ag)^+$, since the presence of Ag^+ and M are both necessary for the formation of this ion from attachment. The multilayer acid-salt films produce higher intensities of $(M{-}H{+}Ba^{2+})^+$ ions and attenuate the substrate signals to a greater degree due to the higher crystallinity of the films and because fewer defects are present for substrate ions to move through.

With these results, it is clear that static SIMS can be used to probe slight changes in the interfacial chemistry between a thin film and substrate. The formation of protonated and deprotonated ions is governed by acid–base chemistry, and can be used as a guide for the surface environment at the interface. In addition, it is clear that the ion formation mechanisms are quite sensitive to small changes in the chemistry, since the incomplete deprotonation of the barium counter-ions[33, 34, 36] is clearly represented by the formation of $(M{+}H)^+$ and $(M{+}Ag)^+$ ions. This shows great promise in the sense that simple preexisting surface chemistry governs the formation of molecular ions. Thus, SIMS can be used as a surface sensitive probe of such chemistry.

Applications to Self-Assembled Film Chemistry

Self-assembled monolayers (SAMs), like L B films, enable the experimenter to construct a highly oriented organic monolayer.[46] However, very few SIMS studies of SAM chemistry have appeared in the literature. Tarlov and Newman have performed static SIMS experiments on a series of thiol monolayers on Au surfaces.[47] These experiments have shown that a wide variety of molecular and quasimolecular secondary ions appear in the negative spectrum for these systems (summarized in Table 10.2). From their results, they inferred that the thiols are bound as thiolates at the Au surface, and that these thiolates convert to sulfonates upon prolonged exposure to air. Li et al. obtained somewhat similar results on thiol/Au systems using laser desorption Fourier transform mass spectrometry.[48] Their results seem to suggest that the tendency for sulfonates to form decreases as alkyl chain length increases. These studies have shown that, as in the case of L–B films, static SIMS is useful as a very sensitive tool for obtaining information about the surface chemistry and reactivity of SAMs.

10.3 The Study of Sampling Depth in the SIMS Experiment

The depth from which secondary ions can originate depends on the extent and intensity of the energy transfer from the primary ion to the sample lattice. In general, the sampling depth of static SIMS is thought to be on the order of a monolayer to greater than 20 Å.[49] Since SIMS involves the sputtering of material from the surface, depth profiling is possible. The ability of the L–B film technique to allow the construction of highly oriented multilayers containing one or more "marker" layers has provided an excellent method for examining the depth dependence of SIMS.

Dynamic SIMS has been used to perform depth-profile analyses of L–B films. Laxhuber et al.[38, 39] have investigated the sputtering characteristics of arachidic acid multilayers on Au backings under 2.5 keV Ar^+ primary ion bombardment using dynamic SIMS conditions. Ion beam compositional analysis was feasible to a depth of 25 Å. In addition, cadmium and magnesium substituted arachidate L–B films were studied.[40] Depth-profiling experiments were conducted in which the divalent cations served as depth markers. Only the emission of atomic ions was examined because of the high-damage cross section of molecular species. The intensity of the Mg^{2+} was found to be greater when placed at the near surface region, and it was found that dynamic SIMS could determine the position of the metal counter-ions in the matrix to an accuracy of better than 10 Å. The results were complicated by selective sputtering and a thickness-dependent erosion rate. Additional dynamic sputter depth-profile studies on built-up layers of cobalt(II) stearate have also shown this ability to gain positional information on the metal counter-ion.[41] Sputter depth-profiling experiments using lower beam currents (static SIMS) provided inconclusive information.[40]

In a related study, Hakansson and Sundqvist[50] examined the sputtering characteristics of L–B films during "electronic" sputtering processes using megaelectronvolt ^{16}O and ^{12}C ions. Electronic sputtering occurs when a surface is bombarded

Thiols on Au

Positive Ions	Negative Ions
$(M+Au)^+$	$(M-H)^-$
—	$(M+Au)^-$
—	$(MS+Au)^-$
—	$(M-H+2Au)^-$
—	$([M-H]^2+Au)^-$
—	$(MSO^3)^{-a}$

[a] Detected only after several days exposure to air.

Table 10.2 Molecular and quasimolecular ions detected from thiolate SAMs.

by "fast ions," that is, ions with an incident energy in the megaelectronvolt range. The fast ions interact with electrons of the material, and the relaxation of the energy deposited by the incident ion is slow. The results indicated the creation of a crater approximately 150 Å deep, having a radius of 80 to 200 Å. Hakansson and Sundqvist suggested that the molecular ions originated from this sputter volume. Additional experiments utilized megaelectronvolt ions of various energies to examine multicomponent L–B film multilayers.[51] These studies showed unambiguously that the sampling depth, and the secondary ion yield, varied from approximately 100 to 200 Å, and increased as primary ion energy increased. Studies that examined L–B film multilayers showed that the secondary ion yield increased, for ions of a particular energy, for samples with thicknesses up to approximately 150 Å. For thicker films, the yields remained constant.[52] Studies done by Bolbach and coworkers showed that the secondary ion yield increases as the energy of the primary ion increases.[53–55] The emission of molecular ions in static SIMS is also dependent upon the energy density of the primary ion. However, in contrast to electronic sputtering, molecular ions in static SIMS are thought to originate at some distance away from the point of primary ion impact.

In summary, Wittmaack suggested a sampling depth of 25 Å for 2.4 keV Ar^+. Hakansson and Sundqvist reported a sampling depth of 80 Å for megaelectronvolt ^{16}O and ^{12}C ions. The increased sampling depth for the high-energy ions seems logical. However, Wittmaack was only able to observe atomic ion emission from the L–B film. Furthermore, the damage induced by a high-energy ion in the Hakansson work might be of concern.

Studies from the authors' laboratory[29, 33, 34] have focused on the production of molecular ions from model systems constructed by alternating L–B multilayer systems. A typical series is illustrated by the schematic in Figure 10.3. An analyte layer (arachidic acid or barium arachidate) is placed in an array of four other layers (behenic acid or its barium salt). The goal is to monitor the intensity and detection of molecular ions from the analyte layer. The results, summarized in Table 10.3, show that the sampling depth in SIMS varies depending on the molecular or quasi-molecular ion being monitored, the substrate metal present, the integrity of organic overlayers, and the state of the L–B film molecule. For instance, the detection of the highest stability molecular ion, the $(M-H)^-$, is clear, even when the analyte

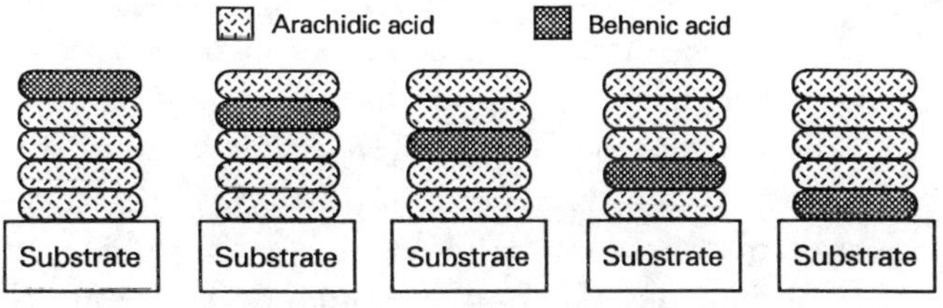

Figure 10.3 **Schematic diagram of L–B film systems on which sampling depth studies have been performed.**

Sample/Substrate	Signal Detected[a]			
	$(M+H)^+$	$(M-H)^-$	$(M+Ag)^+$	$(M+Ba)^+$
	SIMS of Fatty Acids on Silver			
4AA-1BA/Ag	no	yes (w)	no	na
3AA-1BA-1AA/Ag	no	yes (w)	no	na
2AA-1BA-2AA/Ag	yes	yes (w)	no	na
1AA-1BA-3AA/Ag	no	yes	no	na
1BA-4AA/Ag	yes	yes	yes	na
	SIMS of Fatty Acids on Germanium			
4AA-1BA/Ge	yes (vw)	yes (w)	na	na
3AA-1BA-1AA/Ge	yes (w)	yes (vw)	na	na
2AA-1BA-2AA/Ge	yes	yes (w)	na	na
1AA-1BA-3AA/Ge	yes	yes (w)	na	na
1BA-4AA/Ge	yes	yes	na	na
	SIMS of Fatty Acid Salts on Germanium			
4AA-1BA/Ge	no	no	na	no
3AA-1BA-1AA/Ge	no	no	na	no
2AA-1BA-2AA/Ge	no	no	na	no
1AA-1BA-3AA/Ge	yes (w)	yes (w)	na	no
1BA-4AA/Ge	yes (w)	yes	na	yes (w)

[a] Yes = signal of moderate intensity, w = weak signal, vw = very weak signal, no = no signal, and na = not applicable.

Table 10.3 Results of SIMS sampling depth study on L–B film systems.

layer is at the metal interface, some 125 Å from the surface. However, when acid salts are used and the films are more crystalline, with fewer defects, the surface sensitivity increases and the $(M-H)^-$ is only detectable from the top two positions in the array. The $(M+H)^+$ ion is detectable from all five positions only for the acids prepared on the more inert germanium substrate; it is not detectable from the silver substrate unless present in the top three layers. The molecular ions from cationization processes—whether preformed, as in the $(M-H+Ba^{2+})^+$, or the $(M+Ag)^+$ ion, which requires attachment in the selvedge—are only detectable from analyte layers in the topmost layer. Thus, the substrate metal has a profound effect on interfacial molecular ion formation, as in the silver preventing $(M+H)^+$ formation and the germanium allowing it. The integrity of the layers, greater in the acids on germanium and the acid-salt layers, prevents formation of molecular ions from deeper than two layers. Finally, the state of the molecule (ionized or preformed) is important, as barium does not completely deprotonate the acids in the trough; $(M+H)^+$ is still detectable in the acid-salt arrays from the first two layers.

These studies are the first to explore quantitatively the sampling depth of *molecular ions* in static SIMS and allow a better understanding of sampling depth issues, which have previously focused on atomic ions or comparisons to other techniques.

In more recent work, Hagenhoff et al. have done studies on fatty acid and polymeric L–B films.[37] In these studies, imaging SIMS was used in conjunction with classical depth-profiling studies to examine defects in both types of films. In this study, the distinction was made between the "information depth" and the "sampling depth." Information depth is the thickness of the surface volume from which secondary ions can be emitted due to the sputtering process, while sampling depth refers to the depth from which secondary ions are detected. The sampling depth will usually be greater than the information depth due to such factors as defects in surface films. This study found that the sampling depth was much shallower for polymeric films, and imaging SIMS found that there were far fewer defects in the polymeric films.

It is important to further static SIMS by exploring sampling depth in quantitative terms for molecular ions and for ions from polymer surfaces. The ability of the L–B technique to construct model systems will allow the application of SIMS to thin film layered systems and allow the determination of the presence of species on the surface or in depth.

10.4 Quantitation in SIMS

Development of Quantitation Methods

As mentioned previously, the major disadvantage of static SIMS is the difficulty in obtaining reproducible, accurate quantitative information. Quantitative information with a precision between 2% and 15% relative standard deviation is considered obtainable from only the elemental ions produced in dynamic SIMS.[56] Still, the ability to provide elemental quantitative information through secondary ion yield is complicated by the variation by four orders of magnitude across the periodic table. In addition, matrix effects, instrumental parameters, and ion collection efficiencies influence secondary ion yield and must be considered when attempting to quantitate SIMS data. Obtaining quantitative molecular ion information directly from static SIMS would be a major advancement in the technique. Recently, research from the authors' laboratory introduced a method of obtaining quantitative molecular ion[57] information using thin organic films produced from micropipette deposition and L–B techniques. Biological molecules which are precursors for the biopolymer eumelanin form thin partially polymerized films at the air–water interface. These precursor molecules were studied as monolayer films produced from solution deposition. 1-β-3,4-Dihydroxyphenylalanine (L-DOPA) and various indolic molecules, along with dopamine, were used to investigate how the secondary molecular ion peak intensity varied with concentration and ion dosage. It was noted that the intensity varied with ion dosage, as expected, given the nature of molecular ion formation in SIMS. This obviously limits the ability to obtain quantitative

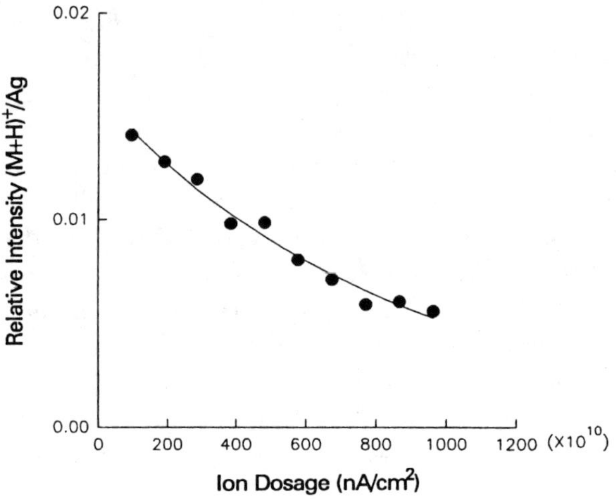

(a)

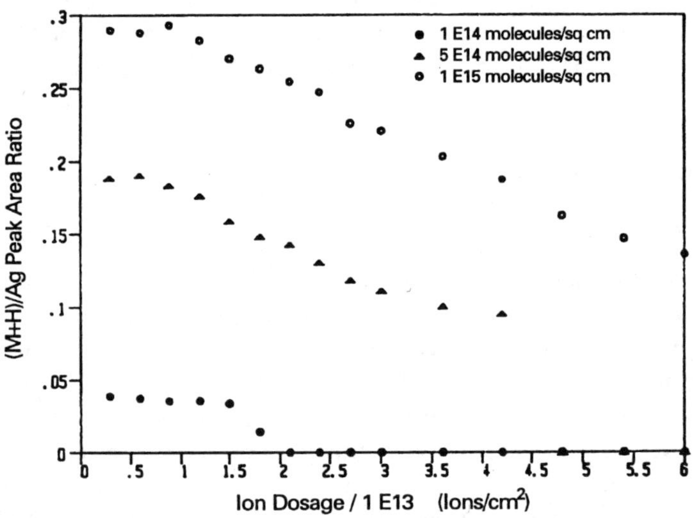

(b)

Figure 10.4 **(*a*) $(M+H)^+$ profile; pH = 4.8. The curve fitting equation used is $y = Ae^{-Bx}$, with $A = 1.602 \times 10^{-2}$ and $B = 1.114 \times 10^{-2}$. (*b*) DHI on Ag.**

information from static SIMS by simply taking relative peak area ratios.[58] The solution proposed was to integrate the signal from the molecular ion (in this case a protonated molecular ion) peak intensity as ratioed to the silver ion intensity by summation over the ion dosage (Figure 10.4). This method was suggested by the approach commonly used to extract concentrations from depth profiles in dynamic SIMS. Ion implantation of internal standards, such as an isotope of the analyte, is used and the known amount implanted is related to the integrated intensity of ion signal over depth. This value is then used to calibrate the signal to concentration relationship for each point in the depth profile.

In the method modified for molecular ions, the assumption is that the total ion signal is a better estimate of concentration than the signal at any one point during ion bombardment, and this integrated amount is related to the concentration. For the solution cast films of known amounts, this method of quantitation yielded a linear relationship, within error limits, with a good correlation coefficient (e.g., 0.998). However, the results still were related to the ion formation probability, in that a calibration curve with a slope related to the formation probability was determined. That meant that the results were still specific to the matrix. In order to overcome this specificity, an "internal standard approach" was developed using mixed L–B films of known concentration.[31] In this study, the ratio of molecular ion intensities should then be directly correlated (i.e., a slope of one and an intercept of zero) to the molar concentration ratio. This model is contingent upon a similar mechanism of molecular ion formation and no chemical interaction between components of interest.[58] Results showed that high correlation coefficients (>0.990) and intercepts within a few percent error of zero were determined over a concentration ratio ranging from 0.2 to 3. These two studies validated the general approach and now give the opportunity to apply the technique to a number of organic thin film problems which require quantitative analysis of mixtures, where SIMS may have unique capabilities, in particular to mixtures of small and trace amounts in the presence of larger amounts of other species.

Application of Quantitative Schemes to Thin Film Chemistry

Cornelio-Clark and Gardella published an analysis of the polymerization kinetics of diacetylenic L–B monolayers as a model for crosslinking at a polymer surface.[32] The amount of residual monomer was quantified by relating the integrated molecular ion signal from various ions. The results showed that reproducible data could be achieved, and that the results were linear with the concentration of monomer. The extent of polymerization was quantified using this technique, and the order and rate of reaction were determined. Figure 10.5 illustrates these results, showing the SIMS spectra before and after irradiation, and the kinetic plots with first order rate constants determined for two analyses. Results generally showed that the early stages of polymerization proceeded at a much faster rate than later stages, and a two-component fit yielded a much higher correlation coefficient. While much work needs to be done to verify the accuracy of these rates, these results bode well for the use of static SIMS to determine such kinetics in this important class of thin film reactions. The technique has also been extended to study the acid–base chemistry of L–B films of alkanoic amines, specifically octadecylamine. Here, the goal was to see if the technique could determine changes in chemistry, rather than changes in the amounts of species. It is well known that during preparation of alkanoic amine monolayer films at different subphase pH values, the isotherms yield

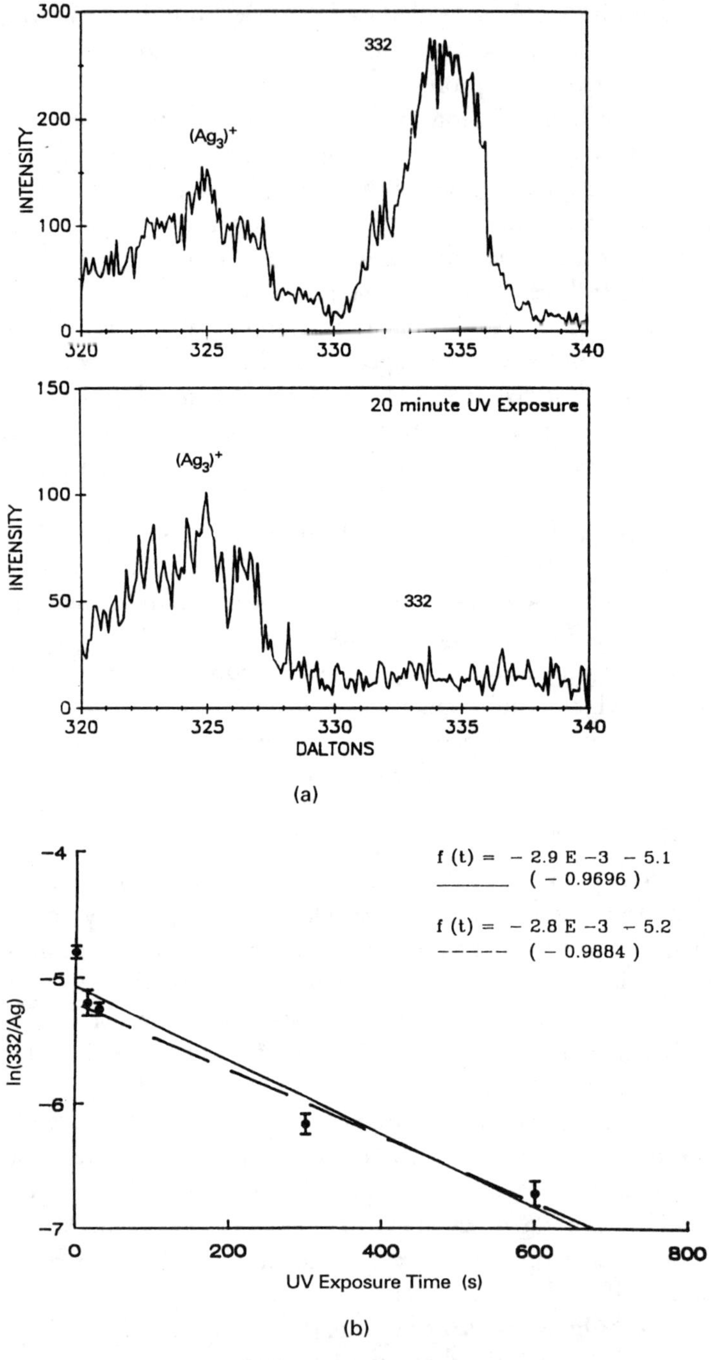

Figure 10.5 **(*a*) SIMS spectra at 0 and 20 min UV exposure. (*b*) Kinetic plots with first order rate constants determined for the two analyses in part (*a*).**

a packing density increase as the pH value increases.[59] The effect of pH changes on octadecylamine L–B films have been studied by Petrov et al.[60] in their study of three-phase contact line motion in the deposition of spread monolayers. The reason for such a dependence is that under acidic conditions, the octadecylamine molecules are protonated, and these functional groups interact strongly by binding the divalent counter anion (e.g., $HPO_4 2-$), so that the area per molecule is relatively larger for lower pH values. As the subphase pH value increases to 7.5, deprotonation of the amine molecules begins. This is due to the equilibrium $RNH_3^+ = RNH_2 + H^+$ being shifted to the left.

Therefore, in basic subphase, the area per molecule is relatively smaller. It is then assumed that this packing density is preserved upon L–B transfer and thus depends strongly on the subphase pH value.

Preliminary results[34] show that the $(M+H)^+$ integrated ion intensity varies as a function of pH, while the $(M+Ag)^+$ signal does not. A plot of the ratio of these two integrated ion intensities yielded a curve analogous to a surface titration curve. Current work involves refining the precision of the method by varying the integration method to provide a firmer value for the area underneath the intensity versus ion dosage curve. Figure 10.4*a* shows the newer approach, fitting the data to an exponential function of the form

$$y = Ae^{-Bt}$$

This equation has the desirable characteristic that the integral can be easily obtained from dividing the coefficients of the exponential terms, A/B.

Thus, by fitting the data and integrating rather than summing the individual points, the method should now, in principle, avoid both the variation in the peak area ratio with increased ion dosage and the nonlinearity between the peak area ratio and the molecular concentration. In the initial work,[31, 34, 57] the integration is carried out over a specific ion dosage. Intuitively, the integration value should be more accurate if carried out over the whole range of the ion dosage until the molecular ion peak disappears. This is not possible if the film is bombarded until there are no more molecular ions emitted, because (1) the creation of defects to the film structure from the ion beam even under static conditions would lead to the attenuated ejection of substrate ions, so that the foundation of using the Ag signal as a reference as the ion dosage becomes relatively large becomes less solid and (2) the inherent noise level in the SIMS instrument would lead to a very large error in the obtained integration value.

Obviously, the development of quantitative techniques in static SIMS should prove useful in determining the composition of mixed SAM systems. Previously, the composition of SAMs that consisted of molecules having varying chain lengths was determined using ellipsometry, contact angle, or ESCA.[48] These techniques are inherently limited, in that their reliability decreases for molecules with small differences in chain length. In contrast, the high resolution of SIMS allows it to distinguish between species of nearly identical chain lengths. In recent studies, we have used the same molecular ion quantitative approach in static SIMS to examine the adsorption of long-chain acids and thiols onto Cu and Au substrates. The results suggest that long-chain species adsorb preferentially over short-chain species, presumably due to an increase in dispersive interactions between the long hydrocarbon chains. The tendency for longer-chain species to adsorb preferentially decreases as the chain length of both species increases.[61]

10.5 Imaging Applications

As mentioned previously, imaging static TOF-SIMS has been utilized by Benninghoven's group[37] to determine the edge defects and contaminants near the edge of multilayers of polymers. Another recent application[61] was the use of imaging TOF-SIMS to demonstrate the selective surface chemistry on lithographically modified polymers using silane self-assembly monolayer chemistry.[62] In this application, the chemically sensitive ion imaging provided the only means to contrast micrometer-level modifications to the surface of a fluoropolymer. The surface modification was accomplished using masking techniques and plasma surface chemistry, followed by direct silane reaction, which was then limited to the lithographically modified regions. No electron microscopic or X-ray methods had the sensitivity to detect these chemical changes, yet neural cell biological model systems were quite sensitive to the spatial limitations. Imaging static SIMS showed that the monolayer surface chemical modification could be directly visualized by imaging unique ion signals from the fluoropolymer and siloxane backbones.

Again, this application of imaging for molecular species is just beginning, but where spatially restricted surface chemistry is necessary, static SIMS promises to provide one of the few chemically sensitive methods to complement scanning tunneling or atomic force microscopy of such films.

10.6 Summary and Prospects

The SIMS experiment has demonstrated capabilities to study the surface chemistry of thin organic films, especially those produced by Langmuir–Blodgett or self-assembly methods. A major advantage is its ability to analyze mixtures of species, especially those of related structures, which can be distinguished by mass spectral analysis (i.e., mass of molecule, fragmentation patterns, etc.). The method has shown the ability to identify degradation pathways in self-asssembled alkanethiolate films on gold.

The use of L–B films as model systems has helped to advance the state of the art of static SIMS and has allowed the further development of qualitative and quantitative surface analysis using the method. With the understanding of sampling depth and the development of new quantitative analysis schemes verified by the application to L–B model systems, it is possible to study the composition of thin films produced by competitive self assembly from solution. From this information, competitive adsorption rates can be determined directly. The effect of small concentrations (in solution) of molecules with potential structural effects can be fully understood by determining the surface composition of the adsorbed film. The linear dynamic range and the ability to distinguish molecules of different mass but with the same molecular functionality allows chemists to induce effective property changes in films with a minimal modification of composition.

A fuller understanding of the role of surface chemistry on the formation of molecular ions is absolutely necessary. However, the development of new instrumentation using time-of-flight and imaging techniques promises to extend the method to a variety of fundamental and practical applications, as demonstrated by the studies reviewed in this chapter.

References

1 A. Benninghoven, F. G. Rudenauer, and H. W. Werner. In *Secondary Ion Mass Spectrometry: Basic Concepts, Instrumental Aspects, Applications and Trends.* (P. J. Elving, J. D. Wineforder, and I. M. Kolthoff, Eds.) Wiley, New York, 1987.

2 P. Steffens, E. Niehuis, D. Friese, D. Greifendorf, and A. Benninghoven. *J. Vac. Sci. Technol. A.* **3** (3), 1322, 1985.

3 A. Benninghoven, E. Niehuis, T. Heller, and H. Feld. *J. Vac. Sci. Tech. A.* **5** (4), 1243–1246, 1987.

4 I. Bletsos, D. Hercules, D. van Leyen, and A. Benninghoven. *Macromolecules.* **20**, 407–413, 1987.

5 D. Fowler, R. Johnson, D. van Leyen, and A. Benninghoven. *Anal. Chem.* **62**, 2088–2092, 1990.

6 A. Benninghoven. *Z. Phys.* **230**, 403, 1970.

7 A. Benninghoven, D. Jaspers, and W. Sichtermann. *Appl. Phys.* **11**, 35, 1976.

8 A. Benninghoven and W. K. Sithermann. *Int. J. Mass Spec. Ion Phys.* **12**, 595, 1977.

9 J. A. Gardella, Jr. and D. M. Hercules. *Anal. Chem.* **53** (12), 1879–1884, 1981.

10 J. A. Gardella, Jr. and D. M. Hercules. *Anal. Chem.* **52** (2), 226–232, 1980.

11 K. J. Hook, T. J. Hook, J. H. Wandass, and J. A. Gardella, Jr. *Appl. Surf. Sci.* **44**, 29–41, 1990.

12 D. Briggs. *SIA, Surf. Interface Anal.* **9**, 391, 1986.

13 D. Briggs. *Org. Mass Spectrom.* **22**, 91, 1987.

14 A. Brown and J. C. Vickerman. *SIA, Surf. Interface Anal.* **8**, 75, 1986.

15 I. V. Bletsos, D. M. Hercules, J. H. Magill, D. van Leyen, E. Niehuis, and A. Benninghoven. *Anal. Chem.* **60**, 938, 1988.

16 A. Benninghoven and W. M. Sichtermann. *Anal. Chem.* **50**, 1180, 1978.

17 A. Benninghoven and W. M. Sichtermann. *Int. J. Mass Spectrom. Ion Phys.* **38**, 351, 1981.

18 P. Sigmund. In *Sputtering by Particle Bombardment I.* (R. Behrisch, Ed.) Springer-Verlag, Berlin, 1981, p. 9.

19 A. Benninghoven. In *Ion Formation From Organic Solids.* (A. Benninghoven, Ed.) Springer, Berlin, 1983, p. 64.

20 G. Falcone, Z. Sroubek, G. Sindona, and N. Uccella. *Int. J. Mass Spectrom. Ion Proc.* **83**, 223, 1988.

21 R. Galera, J. Blais, and G. Bolbach. *Int. J. Mass Spec. Ion Proc.* **107**, 531–543, 1991.

22 B. J. Garrison and N. Winograd. *Science.* **226**, 805, 1982.

23 M. Lombarski and J. Allison. *Int. J. Mass Spec. Ion Pro.* **65**, 31, 1985.

24 J. C. Vickerman. In *Methods of Surface Analysis, Techniques and Applications.* (J. M. Walls, Ed.) Cambridge University Press, Cambridge, 1989.

25 C. Greenlief and J. White. In *Investigations of Surfaces and Interfaces-Part A.* (B. W. Rossiter and R. C. Baetzold, Eds.) Wiley, New York, 1993.

26 N. H. Turner, and R. J. Colton. *Anal. Chem.* **54**, 293R, 1982.

27 J. H. Wandass and J. A. Gardella, Jr. *J. Am. Chem. Soc.* **107**, 6192, 1985.

28 J. H. Wandass, R. L. Schmitt, and J. A. Gardella, Jr. *Appl. Surf. Sci.* **40**, 85–96, 1989.

29 P. Cornelio-Clark and J. A. Gardella, Jr. *Langmuir.* **7**, 2279, 1991.

30 K. J. Hook and J. A. Gardella, Jr. *J. Vac. Sci. Technol. A.* **7** (3), 1795, 1989.

31 P. A. Cornelio and J. A. Gardella, Jr. *J. Vac. Sci. Technol. A.* **8** (3), 2283, 1990.

32 P. A. Cornelio-Clark and J. A. Gardella, Jr. *Langmuir.* **7**, 2454, 1991.

33 R. W. Johnson, Jr., P. A. Cornelio-Clark, and J. A. Gardella, Jr. In *Macromolecular Assemblies in Polymeric Systems.* ACS Symposium Series No. 493. (P. Stroeve and A. Balazs, Eds.) American Chemical Society, Washington, D.C., 1992, p. 113.

34 R. W. Johnson, Jr., P. A. Cornelio-Clark, J.-X. Li, and J. A. Gardella, Jr. In *Secondary Ion Mass Spectrometry (SIMS) VIII.* (A. Benninghoven, C. A. Evans, K. T. F. Janssen, J. Tmpner, and H. Werner, Eds.) Wiley, New York, 1992, pp. 293–296.

35 G. Bolbach, R. Beavis, W. Ens, D. E. Main, B. Schueler, and K. G. Standing. *Abstracts of the 33rd ASMS Meeting.* San Diego, 1985.

36 G. Bolbach, M. Plissonnier, R. Galera, J. Blais, G. Dufour, and H. Roulet. *Thin Solid Films.* **211**, 524, 1992.

37 B. Hagenhoff, M. Deimel, A. Benninghoven, H. Siegmund, and D. Holtkamp. *J. Phys. D: Appl. Phys.* **25**, 818, 1992.

38 L. Laxhuber, H. Mohwald, and M. Hashmi. *Int. J. Mass Spec. Ion Phys.* **51**, 93, 1983.

39 L. Laxhuber, H. Mohwald, and M. Hashmi. *Coll. and Surf.* **10**, 231, 1984.

40 K. Wittmaack, L. Laxhuber, and H. Mohwald. *Nucl. Inst. Meth. Phys. Res.* **B18**, 639, 1987.

41 F. Toyokawa, H. Abe, K. Furuya, and T. Kikuchi. *Surf. Sci. Lett.* **133**, L429–L431, 1983.

42 I. Langmuir and K. B. Blodgett. *Phys. Rev.* **51**, 317, 1937.

43 K. B. Blodgett. *J. Am. Chem. Soc.* **56**, 495, 1934.

44 K. B. Blodgett. *J. Am. Chem. Soc.* **57**, 1007, 1935.

45 G. L. Gaines, Jr. *Insoluble Monolayers at Liquid Gas Interfaces.* Wiley Interscience, New York, 1966.

46 *A Guide to Ultrathin Organic Films.* (A. Ulman, Ed.) Academic, Boston, 1991.

47 M. Tarlov and J. Newman. *Langmuir.* **8**, 1398, 1992.

48 Y. Li, J. Hung, R. McIver, and J. Hemminger. *J. Am. Chem. Soc.* **114**, 2428, 1992.

49 A. Benninghoven, W. Lange, M. Jirikowsky, and D. Holtkamp. *Surf. Sci.* **123**, L721, 1982.

50 P. Hakansson and B. Sundqvist. *Vacuum.* **39** (2–4), 399, 1989.

51 G. Save, P. Hakansson, B. Sundqvist, R. Johnson, E. Soderstrom, S. Lindquist, and J. Berg. *J. Appl. Phys. Lett.* **51**, 1379, 1987.

52 G. Save, P. Hakansson, B. Sundqvist, E. Soderstrom, S. Lindquist, and J. Berg. *Int. J. Mass Spec. Ion Pro.* **78**, 259, 1987.

53 G. Bolbach, S. Della-Negra, C. Deprun, Y. Le Beyec, and K. Standing. *Rapid Comm. Mass Spec.* **1** (2), 22, 1987.

54 G. Bolbach, R. Beavis, S. Della-Negra, C. Deprun, W. Ens, Y. Le Beyec, D. Main, B. Schueler, and K. Standing. *Nucl. Inst. Meth. Phys. Res.* **B30**, 74, 1988.

55 G. Bolbach, A. Viari, R. Galera, A. Brunot, and J. Blais. *Int. J. Mass. Spec. Ion Proc.* **112**, 93, 1992.

56 H. M. Werner. *SIA, Surf. Interface Anal.* **2**, 56, 1980.

57 M. B. Clark and J. A. Gardella, Jr. *Anal. Chem.* **62**, 871, 1990.

58 S. Tamaki, W. K. Sichtermann, and A. Benninghoven. *Jap. J. Appl. Phys.* **23**, 544, 1984.

59 G. L. Gaines, Jr. *Nature.* **298**, 544, 1982.

60 J. G. Petrov, H. Kuhn, and D. Mobius. *J. of Colloid and Interface Science.* **73**, 66, 1980.

61 D. J. Hook, T. G. Vargo, J. A. Gardella, Jr., K. S. Litwiler, F. V. Bright. *Langmuir.* **7**, 142, 1991.

62 T. G. Vargo, P. M. Thompson, L. J. Gerenser, R. F. Valentini, P. Aebischer, D. J. Hook, and J. A. Gardella, Jr. *Langmuir.* **8**, 130, 1992.

11

X-Ray Photoelectron Spectroscopy Of Organic Thin Films

ABRAHAM ULMAN and JAMES F. ELMAN

Contents

11.1 Introduction

The chemical composition of organic surfaces is of great importance, since it helps in the understanding of interfacial phenomena—such as wetting, adhesion, friction, etc.—and how they relate to the physicochemical parameters of functional groups. Furthermore, following the composition of an organic surface or interface as a function of time helps in understanding surface stability, interfacial diffusion, etc. X-ray photoelectron spectroscopy (XPS) is the surface analytical tool of choice when surface composition and its depth profile are under study. The XPS of organic materials has been an active field of study for over twenty years. The maturation of this spectroscopy can be demonstrated by the recent publication of high-quality core and valence band spectra acquired on this truly exotic instrument.[1]

In an XPS experiment, a sample is exposed to X radiation and the properties of inner-shell electrons are probed. If E_0 is the energy of the X ray, and E_j is that of a core electron in the atom (*s*, *p*, etc.), $E_0 - E_j$ will be the energy of the ejected electron. Since the value of E_j is characteristic of a particular electron in a particular atom, the determination of E_j immediately provides atomic identification. Since the number of electrons ejected is proportional to the number of atoms present, composition can also be established quantitatively. Chemical state information can often be extracted from small shifts ("chemical shifts") in E_j. For example, for sulfur 2*s* electrons, one observes an increase of over 5 eV in the ionization energy as the valence state of the sulfur varies from −2 (R–S–R′) to +6 (R–SO_2–R′). This effect was used in analyzing the oxidation of a thioether to a sulfoxide (S^{4+}) in a self-assembled monolayer (SAM).[2] Because of the small depths of photoelectrons, special techniques are used that can emphasize the contribution from atoms in different depth positions in a film. For example, a grazing electron takeoff geometry emphasizes the contribution from surface atoms, and a study of photoelectron spectroscopy as a function of the takeoff angle offers an excellent way to study compositional depth distribution in a film.[3]

The principles of XPS, together with practical examples of use for surfaces, interfaces, and thin films are fully explained in the chapter on XPS in the lead volume of this Series, *Encyclopedia of Materials Characterization.* An extensive review of applications to polymer surfaces can be found in Reference 4.

11.2 Experimental Considerations

To produce respectable polymer spectra today one should use an instrument with the following attributes: a "fast" electron detection system, a monochromatic X-ray source (usually Al kα, but Ag lα sources also are available), and adequate charge compensation (either by sample preparation or by electronic means). Many of the improvements found in today's spectrometers and some of the history of their development have been reviewed.[5] Rapidly acquired monochromatic spectra have inherently finer resolution and show much less radiation-induced damage than older slower instruments. However, use of a monochromatic X ray may not be sufficient to eliminate photon or photoelectron degradation. Cooling the sample with liquid nitrogen will add additional protection, but some materials systems which have particularly high yields of photoelectrons (i.e., Au, Ag, or Cu substrates) are intrinsically damaging to thin organic films.[6] Such situations can be addressed only by carefully monitoring X-ray dosage and continually moving to fresh undamaged areas on the sample. Good charge compensation for insulating samples typically uses an electron "flood" gun, but can also include a conductive screen placed above the sample,[7] a conductive mask that surrounds the analysis area, and/or the appropriate choice of flood electron amperages and voltages (from about 2 eV to hundreds of electronvolts). As will be seen later, even the temperature of insulating thin films can have a measurable effect on sample charging. In addition to the above instrumental approaches, sample preparation can have a profound effect on the ultimate quality of XPS spectra. For polymers that are soluble in the appropriate solvents, spin casting is the preferred method. This technique allows the preparation of extremely smooth thin films that often do not require a flood gun. The effect of film smoothness on peak resolution is dramatically shown in Figure 11.1, where

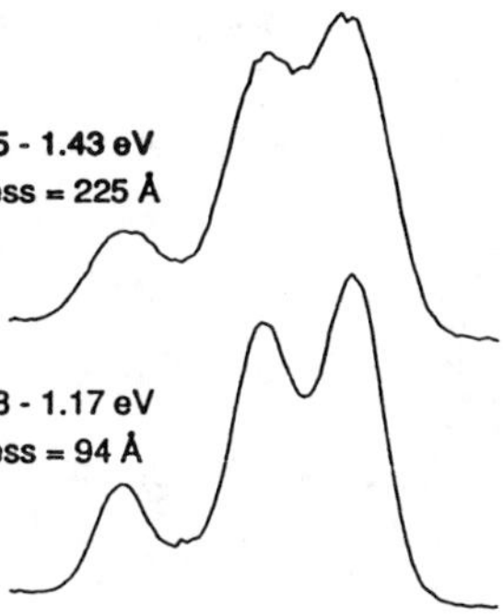

Figure 11.1 The effect of film smoothness on peak resolutions for a C 1*s* XPS spectrum.

the same polymer was prepared with two different levels of roughness, as measured by atomic force microscopy (AFM).[8] Spun-cast film thicknesses of 250–600 Å will not show substrate peaks (i.e., Si 2*s* and 2*p* from the underlying wafer) and should provide excellent spectral resolution.[9] The optimum film thickness may vary from spectrometer to spectrometer. Obviously, self-assembled (SA) and Langmuir–Blodgett (L–B) films are also excellent ways to prepare thin film materials, with surfaces engineered at the molecular level,[10] to provide good-quality XPS data.

11.3 Binding Energy Shifts

The Briggs database has assembled for the first time a true encyclopedia of chemical shifts encountered in polymer XPS.[11, 12] While instrument manufacturers usually provide a database on these binding energy (BE) shifts,[13] their data reflect an archiving of gradually acquired spectra rather than such an intentional systematic study. The Briggs catalogue lists the well-known alpha shifts (e.g., the BE shift in the C 1*s* level for oxygen substitution directly on a carbon atom), but also quantifies the magnitude of several subtle but measurable beta and gamma shifts that previously were controversial as to their existence. As instrument resolution gets better, the experimentalist sees more fine spectral features in organic systems. For instance, the F 2*s* BE is particularly sensitive to mono-, di-, or trifluoro substitution on a carbon atom. While these structural changes have well-known shifts in the C 1*s* region, the corresponding F 2*s* spectral effects have not been reported previously. Recently, Briggs has identified analogous features in the O 2*s* region for oxygen containing polymers, and offered a simple, elegant interpretation based on space group considerations. He points out that rigorous MO calculations are a preferable method for investigating these spectral features, and such studies are underway for the O 2*s*,[14] and F 2*s* orbitals.[15]

11.4 XPS of Molten Films

A convenient method of enhancing spectrum quality (better spectral resolution) is to heat a spun-cast sample to above 100 °C and preferably above either the T_m or

Polymer	Temperature, °C	C 1*s* Peak Position, eV	C 1*s* Peak FWHM, eV
PE	$<T_m$ (40)	several: 286.9 to 285.2	each peak 0.85
PE	$>T_m$ (130)	285.1	0.83
PS_{iso}	$<T_g$ (40)	285.24	0.86
PS_{iso}	$>T_m$ (255)	285.91	0.78

Table 11.1 Spectral features of polyethylene and isotactic polystyrene thin films at different temperatures.

T_g of the material under study. Direct XPS measurement of a polymer melt has several advantages: the removal of any contaminating material, the shifting of peaks to their "conductive" binding energy without the use of an electron flood gun (i.e., close to 284.6 eV for neutral carbon), a significant narrowing of peak widths—thus allowing the study of thicker films while still retaining excellent resolution—and observing atomic percent values very close to stoichiometric. The systems amenable to this experimental variation must not sublime, decompose, or be far from their thermodynamic equilibrium. Most homopolymers above the entanglement molecular weight, and blends or blocks of such materials, would be ideal candidates for melt XPS. SA or L–B films may not have sufficient thermal stability to warrant such treatment. This is unfortunate, as these systems often adsorb a contamination layer which can interfere with an XPS analysis of the film.

As can be seen in Table 11.1, polyethylene (PE) and isotactic polystyrene (PS_{iso}) thin films show improved spectral features when observed at elevated temperatures.† The PE film as spun cast had several discrete thicknesses, each of which (due to differential charging) gave rise to a distinct C 1*s* peak. When this sample was melted in a vacuum, these peaks coalesced into a single sharp peak, indicative of a homogeneous amorphous polymer film. The PS_{iso} was never properly annealed to induce crystallinity and thus may be considered essentially amorphous. It also showed spectral enhancements: binding energies shift about 0.3 eV, and the C 1*s* narrows from 0.86 eV to 0.78 eV. Thicker polymer (1000 Å and above) films have even more pronounced temperature effects, since they show much more sample charging to begin with.

11.5 Angular Dependent XPS

A particularly powerful XPS experiment is to vary electron takeoff angle (ETOA), and in this way nondestructively probe the distribution of chemical species near the

† The polyethylene (PE) was National Bureau of Standards SRM 1482, M_n = 11.4 k, M_w = 13.6 k, and M_w/M_n = 1.19. The polystyrene (PS_{iso}, 90% isotactic) was purchased from Scientific Polymer Products, Inc., M_n = 415 k.

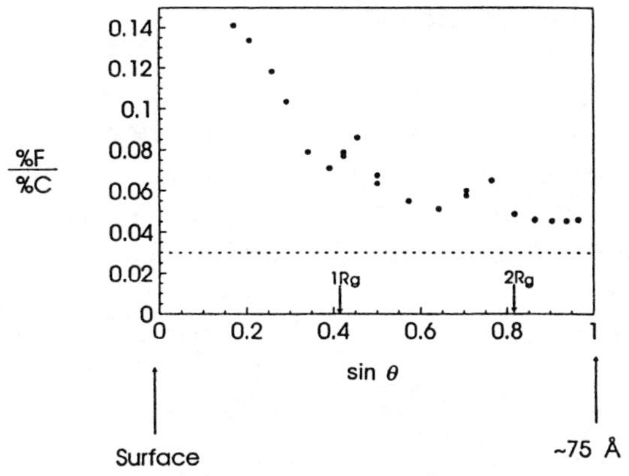

Figure 11.2 **Fluorine concentration in a film of fluorosilane terminated polystyrene versus the sine of the takeoff angle, which can be related to the probing depth.**

surface as a function of depth. (The general approach is described in the *Encyclopedia* and will not be repeated here.) Such studies require, however, mathematical algorithms to transform ETOA to units of angstroms in the film normal direction. This requires a knowledge of the inelastic mean free path (IMF) of the photoelectrons through organic materials. It is known that the IMF is a function of the KE of the photoelectrons, and it is often assumed that it is the same for different organic materials. This may not be the case, and there have been very few recent attempts to actually measure this important parameter. Certainly, though, qualitative information is readily interpretable from these data sets. Angular dependent studies can provide important insights into highly structured films that contain hetero atoms (see below). When the surface of a block copolymer is studied, it is frequently found that the surface is richer with the less-polar component, because of the lower surface free energy of the resulting film. Thus, the surface of a film made of a block copolymer of polystyrene (21 mol %) and poly(ethylene oxide) is richer in the less-abundant polystyrene.[16]

A recent study of fluorosilane terminated polystyrene strongly suggests that experimentalists should acquire angular data on a finer angular scale than is commonly used, particularly if the XPS instrument in question has a small cone of acceptance (less than ±4°).[17] In Figure 11.2, three discrete enhancements of fluorine concentration are encountered as one probes deeper into the film and approaches the ambient stoichiometric fluorine level (θ is the takeoff angle). These fluorine-rich layers correlate with the chain dimensions in this low molecular weight polymer ($M_n = 5000$). Such features would probably have not been noticed if the typical three to five ETOA data points had been acquired. In hindsight, when one examines previous SA studies, there were indications of such fine structure in the angular dependent data.

11.6 ETOA XPS of Self-Assembled Monolayers

XPS can be very useful in understanding the substrate attachment chemistry of SA films and the relative populations for mixed SA layers, and to determine if any post–SA reactions have gone to completion. One caution for XPS analysis is that the vacuum required may compromise the film by sublimation. Liquid-nitrogen cooling will certainly diminish this effect, but it may explain discrepancies found between infrared and XPS results.[18]

XPS has been widely used in the study of monolayers and multilayers.[10] ETOA XPS can be particularly illuminating for L–B and SA systems where one wishes to know the orientation of a molecule relative to the substrate. If the molecule in question has several different heteroatoms in its structure (as is the case below) then ETOA XPS is indeed a powerful tool to determine if the film is aligned and which is the head and which the tail.

The example we discuss here is of SAMs of sulfone-containing alkanethiolates on Au(111) surfaces (Figure 11.3).[19] This study provides a good demonstration of the utility of XPS in the studies of monolayers.

Angle-resolved XPS measurements were carried out on SAMs of **I** and **III**, so that the change in concentration of various elements could be followed as a function of depth. Table 11.2 presents the concentration of each element as a function of the takeoff angle for a sample of **I**. As expected, the concentration of both types of sulfur decreases, while the concentration of gold increases for greater electron takeoff angles (greater depth of measurement). On the other hand, examination of the oxygen-to-sulfur ratio reveals values between 1.5:1 and 1.8:1 for the sulfone group for all angles—close to the expected value of 2 for an SO_2 group. This probably does not result from the known differences in the inelastic mean free path (IMFP) for the S $2p$ and O $1s$ electrons, since this can account for only a small discrepancy (~20%), which in any case would be smoother with angular change. A possible explanation for the observation is that the actual IMFP for O $1s$ electrons may be strongly angular dependent at low angles.

One very important feature noticed during the data analysis was the shape of the sulfur $2p$ peak. A high-resolution scan of the sulfur $2p$ region should resolve the $2p_{1/2}$ and $2p_{3/2}$ doublet, which should have an area ratio of 1:2. The peaks due to the SO_2 sulfur and the thiolate sulfur should also be resolved, with the SO_2 sulfur

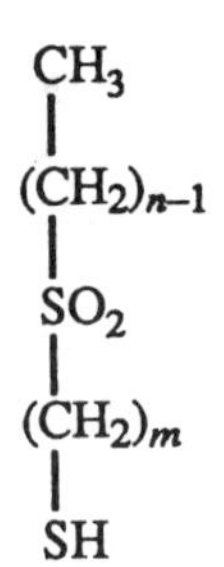

I $n = 6$, $m = 11$
II $n = 8$, $m = 9$
III $n = 11$, $m = 6$

Figure 11.3 Sulfone-containing alkanethiols.

occurring at ~5 eV higher than the thiolate sulfur. This is because of the difference in oxidation state of the two sulfur atoms. *Ab initio* calculations show that the charge on the thiolate sulfur at a hollow site of the Au(111) surface is −0.4 e,[20] thereby making it easier to remove electrons from this sulfur, and causing the thiolate sulfur peaks to shift to a lower binding energy. Using the same rationale, for a sulfur bonded to two electron-withdrawing oxygen atoms, the peaks move to a higher binding energy. Figure 11.4 presents a high-resolution scan of the sulfur $2p$ region for monolayers of **III** and **I**. The two types of sulfur atoms are easily resolved for **III** (Figure 11.4*a*), and the ratio between the $2p_{1/2}$ and $2p_{3/2}$ peaks is 1:2. Figure 11.4*b* is representative of spectra obtained for all the takeoff angles for sample **I**. Note that while the two sulfur doublets are resolved, the lower binding energy doublet does not have the characteristic 1:2 ratio. This peak, in fact, consists of two sets of doublets as the peak-fitting routine shows. The lower-energy doublet belongs to the thiolate sulfur, while the higher-energy doublet is due to SO_2 sulfur that has been reduced. The areas of reduced SO_2 sulfur peaks were added to the area of the highest binding energy sulfur peak to determine the total amount of SO_2 sulfur. Reduction of the SO_2 sulfur was confirmed later when a sample was analyzed for a short time period and then analyzed again overnight. Evidence of the developing additional energy doublet is clearly shown after the overnight analysis. Further support for the reduction of a sulfone group has been seen in SERS experiments.[21] Two possible causes can be suggested for the observed SO_2 reduction: effects from X-ray exposure and effects from secondary electrons originating from the gold substrate.

The results for samples **I** and **III** plotted as a function of takeoff angle versus the ratio of the SO_2 sulfur to the thiolate sulfur (SO_2/S/Au) are shown in Figures 11.5 and 11.6, respectively. Qualitatively, the shapes of both curves are similar. With increasing takeoff angle, the ratio of SO_2 to S/Au should decrease as the analysis depth increases and the SO_2 sulfur becomes less predominant. In both graphs there is a break in the curve occurring at 30° in **I** and 38° in **III**. Notwithstanding the limited data above 40°, these breaks are real and not artifacts. As the takeoff angle is increased, the effective analysis depth also increases. Thus, if we assume a value

	Atomic Percent				
Element	**10°**	**20°**	**30°**	**38°**	**78°**
Carbon	66.2	54.6	44.9	42.9	36.4
Oxygen	7.1	5.8	5.8	5.2	5.0
Sulfur (SO_2)	17.2	4.0	3.8	3.2	2.7
Sulfur (S/Au)	4.2	1.4	1.3	1.3	1.3
Gold	5.4	34.1	47.5	47.4	54.6

Table 11.2 Atomic percent versus takeoff angle for a monolayer of I. (From Reference 19; © American Chemical Society.)

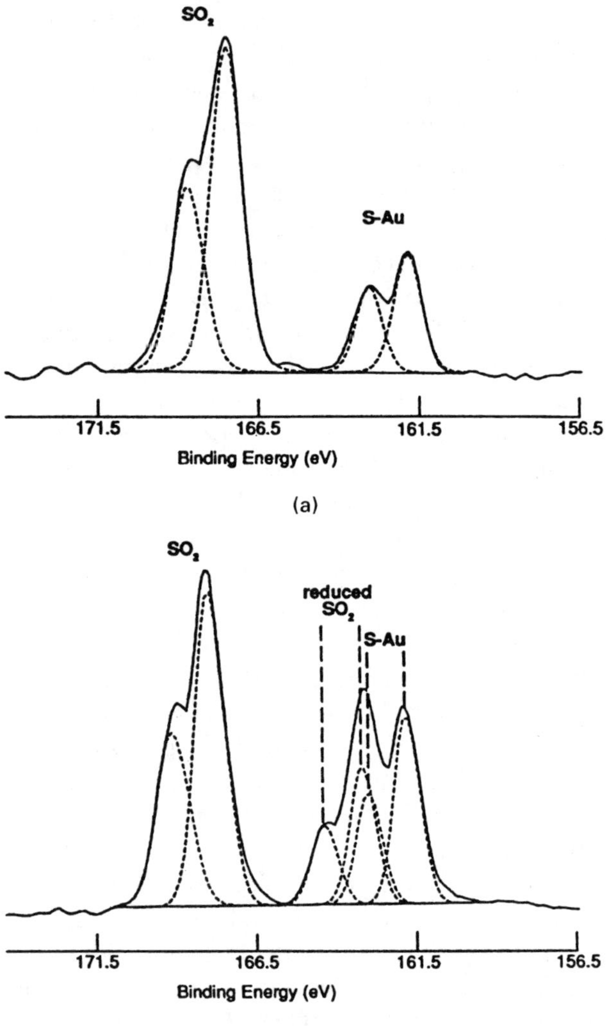

Figure 11.4 **High-resolution XPS spectra of the S 2*p* region for a monolayer of (*a*) III and (*b*) I taken at an electron takeoff angle of 20°. The analysis time was ~2 h for for sample III and ~15 h for sample I. (From Reference 19; © American Chemical Society.)**

of 26 Å for the IMFP of 1300 eV electrons in an organic matrix and work through the appropriate equations converting takeoff angle into depth probed, the data of Figures 11.5 and 11.6 suggest that the SO_2 sulfur in sample **I** is 3 Å closer to the surface than the SO_2 sulfur in sample **III**.[22] If the molecules are in an all-trans conformation, as is apparent from FTIR studies,[19] and the alkyl chains are oriented perpendicular to the surface, one can estimate from space-filling models the SO_2

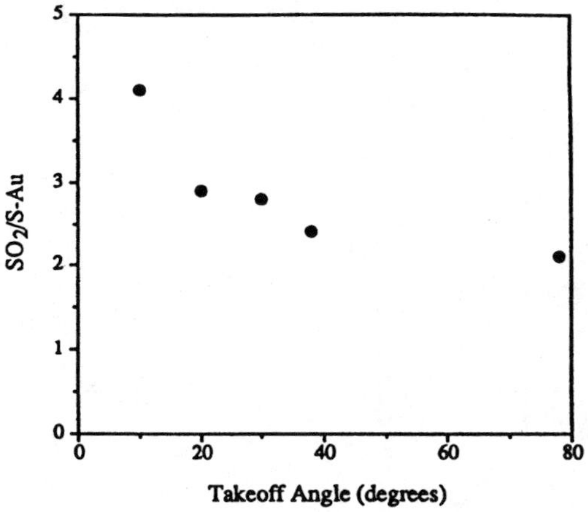

Figure 11.5 **Takeoff angle versus SO_2:S/Au for a monolayer of I. (From Reference 19; © American Chemical Society.)**

sulfur to be 7.75 and 14 Å from the surface in samples **I** and **III**, respectively. However, if the chains are tilted 40° with repect to the surface normal—as evident both from ellipsometry and FTIR studies—then the difference in depth for the two molecules decreases to 5 Å from the space-filling model.

An analysis of the angle dependent data can also provide an estimate of the overlayer thickness and indicate if it is continuous. The method used for the SAM samples was reported previously for surfactant films.[23] An equation for the angle dependent ratio of the overlayer to the substrate was developed by Fadley[24]:

$$R_{\text{overlayer/substrate}} = Ke^{\tau/\sin\theta - 1}$$

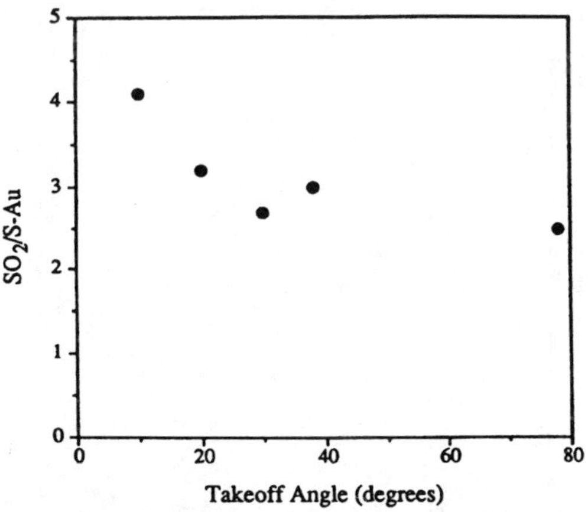

Figure 11.6 **Takeoff angle versus SO_2:S/Au for a monolayer of III. (From Reference 19; © American Chemical Society.)**

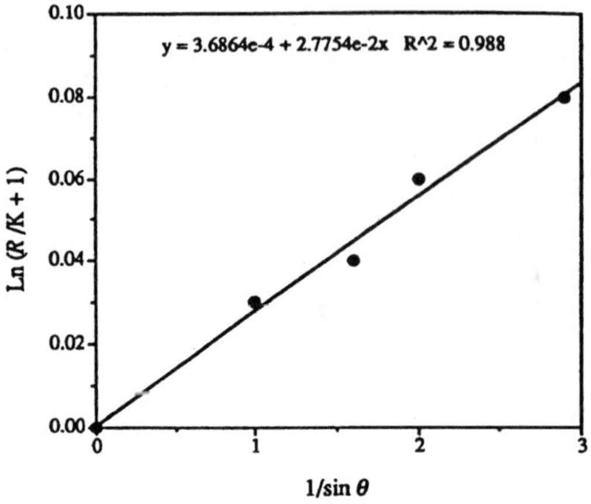

Figure 11.7 **1/sin θ versus Ln($R_{overlayer/substrate}/K + 1$) for a monolayer of I using the thiolate sulfur. (From Reference 19; © American Chemical Society.)**

where t = effective overlayer thickness, θ = electron takeoff angle, and K = a constant. K is dependent on several instrument parameters and the atom density. For simplicity, K was assumed to have the value of 1.0. The concentration of a particular atom in the overlayer and of a different atom in the substrate must be known for the calculation. Since only sulfur was measured in sample **III**, the above calculation will be applied to sample **I** only, where both the sulfur and the gold substrate were monitored.

A plot of $1/\sin\theta$ versus $\ln(R_{overlayer/substrate}/K + 1)$ should give a straight line of slope τ passing through the origin. Figure 11.7 shows the plot obtained for sample

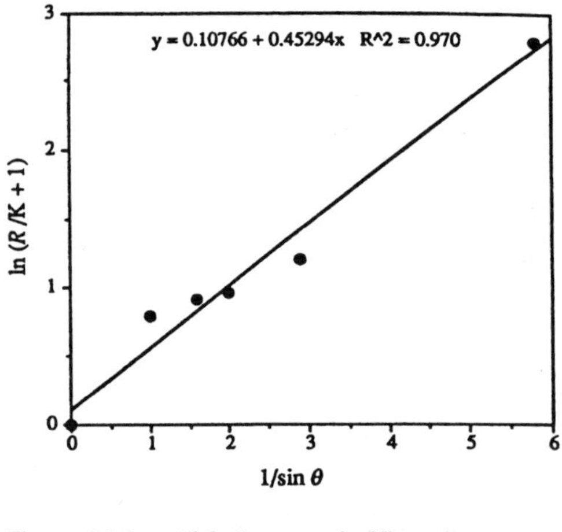

Figure 11.8 **1/sin θ versus Ln($R_{overlayer/substrate}/K + 1$) for a monolayer of I using the SO_2 sulfur. (From Reference 19; © American Chemical Society.)**

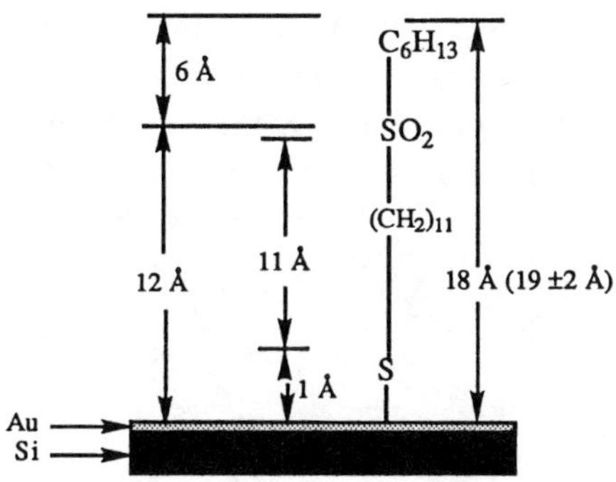

Figure 11.9 **A diagram showing the thickness for monolayer I as determined by XPS. The number in parentheses is the thickness estimated by ellipsometry. (From Reference 19; © American Chemical Society.)**

I, for the thiolate sulfur and the gold. The excellent correlation achieved to a straight-line fit suggests that the monolayer is continuous, which is in agreement with contact angle results. The overlayer thickness can be calculated from the slope of the straight line by using the equation

$$t = \tau / \Lambda_{\text{overlayer}}$$

where τ is the slope, t is the actual overlayer thickness, and Λ is the electron IMFP. With an estimated value of 26 Å for Λ, the overlayer thickness is ~1 Å. However, this is only the thickness of the thiolate sulfur at the gold surface and not the entire overlayer. The value obtained is in reasonable agreement with the covalent radius of sulfur, which is 1.09 Å.

Similar calculations can also be performed for the SO_2 sulfur. The plot obtained is shown in Figure 11.8. Although the correlation is not as good as the previous graph, the data are reasonable. The overlayer thickness calculated from the slope is 12 Å, which includes the $(CH_2)_m$ chain of Figure 11.3 but does not take into account the six-carbon chain above the SO_2 group. The thickness of the lower C_{11} hydrocarbon chain can thus be estimated to be 11 Å (12 – 1 Å), giving a value of 1 Å per methylene unit. This is in an excellent agreement with a value of 1 Å/CH_2 unit estimated from a contribution of 1.25 Å/CH_2 unit for an all-trans alkyl chain perpendicular to the surface, and a tilt angle of 40°. Using this value, the thickness of the six-carbon chain above the SO_2 group should be 6 Å, and the thickness of the entire overlayer 18 Å (Figure 11.9). This compares very well with the 19 ± 2 Å estimated from ellipsometry.

11.7 Conclusions

X-ray photoelectron spectroscopy of organic thin films is probably one of the most important analytical techniques available today. It provides information on the chemical composition of the surfaces, and thus helps in the understanding of interfacial phenomena. It gives a depth profile of chemical composition, and hence helps in understanding surface stability, interfacial diffusion, etc. X-ray damage and reduction by secondary electron may cause problems; thus results obtained under a different exposure time should be carefully evaluated. Good data also can be used for calculations of film thickness, providing values that do not depend upon a selection of refractive index, as is necessary in ellipsometry.

References

1 G. Beamson and D. Briggs. *High Resolution XPS of Organic Polymers: The Scienta ESCA300 Database.* Wiley, Chichester, U.K., 1992.

2 N. Tillman, A. Ulman, and J. Elman. *Langmuir.* **5**, 1020, 1989.

3 N. Tillman, A. Ulman, and J. Elman. *Langmuir.* **6**, 1512, 1990.

4 J. D. Andrade. In *Surface and Interfacial Aspects of Biomedical Polymers,* Vol. 1. (J. D. Andrade, Ed.) Plenum, New York, 1985, chapt. 5.

5 J. A. Gardella. *Anal. Chem.* **61** (9), 589A–600A, 1989.

6 P. E. Laibinis, R. L. Graham, H. A. Biebuyck, and G. M. Whitesides. *Science.* **254**, 981–983, 1991.

7 C. E. Bryson. *Surf. Sci.* **189/190**, 50, 1987.

8 R. Burger and V. DePalma. Personal communication.

9 C. W. Extrand. *Langmuir.* **9**, 475, 1993.

10 A. Ulman. *Introduction to Thin Organic Films: From Langmuir–Blodgett to Self-Assembly.* Academic Press, Boston, 1991.

11 D. Briggs and G. Beamson. *Anal. Chem.* **64**, 1729–1736, 1992.

12 D. Briggs and G. Beamson. *Anal. Chem.* **65**, 1517–1523, 1993.

13 J. F. Moulder, W. F. Stickle, P. E. Sobol, and K. D. Bomben. *Handbook of X-Ray Photoelectron Spectroscopy.* Perkin-Elmer, Eden Prairie, MN, 1992.

14 P. M. A. Sherwood, G. Beamson, and D. Briggs. Manuscript in preparation.

15 R. Baetzold, L. J. Gerenser, J. M. Pochan, and J. F. Elman. Manuscript in preparation.

16 H. R. Thomas and J. J. O'Malley. *Macromolecules.* **12**, 323, 1979.

17 C. Jalbert, J. F. Elman, T. E. Long, and J. T. Koberstein. Manuscript in preparation.

18 T. L. Marshanks, H. K. Jugduth, W. N. Delgass, and E. I. Franses. *Thin Solid Films.* **232**, 126, 1993.

19 S. D. Evans, K. E. Goppert-Berarducci, E. Urankar, L. J. Grenser, A. Ulman, and R. G. Snyder. *Langmuir.* **7**, 2700, 1991.

20 H. Sellers, A. Ulman, Y. Shnidman, and J. E. Eilers. *J. Am. Chem. Soc.* **115**, 9389, 1993.

21 S. D. Evans, E. Urankar, A. Ulman, and N. Ferris. *J. Am. Chem. Soc.* **113**, 4121, 1991.

22 M. P. Seah and W. A. Dench. *SIA, Surf. Interface Anal.* **1**, 2, 1979.

23 L. J. Gerenser, J. M. Pochan, M. G. Mason, and J. F. Elman. *Langmuir.* **1**, 305, 1985.

24 C. S. Fadley. *Prog. Solid State Chem.* **2**, 265, 1976.

12

Molecular Orientation in Thin Films as Probed by Optical Second Harmonic Generation

ROBERT M. CORN and DANIEL A. HIGGINS

Contents

12.1 Introduction

Optical second harmonic generation (SHG) is a nonlinear process that converts two photons of frequency ω to a single photon of frequency 2ω. In the electric dipole approximation, this process requires a noncentrosymmetric medium.[1] For thin films present at the interface between two centrosymmetric media—such as the interface between two liquids, the interface between a liquid and air, or for the surfaces of a variety of insulators and semiconductors—only the molecules which participate in the asymmetry of the interface contribute to the SHG (higher order bulk contributions must also be considered in a rigorous analysis of the surface SHG response). SHG has been employed at a wide variety of surfaces because of this unique surface sensitivity and selectivity,[2–6] and the nearly instantaneous optical response time of the technique has resulted in its application to time-resolved optical measurements of chemical reactions and molecular dynamics at interfaces.[2]

For systems where the SHG from the interface is dominated by molecular contributions to the surface nonlinear susceptibility, the extent of adsorption and the average orientation of the molecules at the interface can be obtained from the optical measurements. The SHG from the interface is found to be proportional to the square of the surface concentration of adsorbed molecules in this instance. In addition, the polarization dependence and phase of the molecular SHG from the surface can be used to determine the average molecular orientation. These attributes have led to the wide application of SHG to studies of the adsorption of dyes, surfactants, and a variety of other small organic molecules at both solid and liquid surfaces.

The preparation of noncentrosymmetric thin organic films for nonlinear optics has been pursued by a number of groups around the world.[2–6] These films have been prepared either by the Langmuir–Blodgett dipping process or by the spontaneous organization of molecular assemblies, and they range in thickness from submonolayer to multilayers. The SHG from these thin films has often been used to monitor the deposition of the monolayers, as well as the extent of molecular orientation and order within the films.

This chapter provides an in depth outline of the procedures necessary for obtaining average molecular orientation information for an adsorbed monolayer or thin film from an analysis of the polarization dependence of the surface second harmonic response. We do not attempt to survey either the large number of applications of molecular SHG to orientation measurements at interfaces, or the application of nonresonant SHG measurements to interfaces; several papers reviewing both subjects have recently appeared.[2–6] Instead, this chapter is directed to the spectroscopist or surface scientist who wishes to make SHG molecular orientation measurements on thin film and monolayer systems. By carefully outlining the required experiments and calculations, and by providing a limited number of examples, we encourage the reader to apply the methodology summarized here to a thin film system of particular interest.

The remainder of this chapter is divided into five sections: (1) a brief description of some experimental requirements for undertaking surface SHG measurements, (2) an examination of the resonant SHG response from molecules, (3) a discussion of the general equations for SHG at surfaces, (4) an analysis of the relationship between the average surface molecular orientation and the SHG polarization measurements, and (5) an analysis of the phase of the surface tensor elements in terms of the absolute molecular orientation.

12.2 Experimental Considerations

The sample or thin film to be studied can be prepared on the substrate of choice by one of numerous methods: the Langmuir–Blodgett technique, spin coating, spreading with a volatile solvent, spontaneous adsorption from solution, or a variety of other techniques. In the examples presented here, monolayer films of *p*-nitrophenol were prepared on an optically flat fused silica substrate by dipping in a dilute ethanolic solution. The fused silica substrates were mounted in a Teflon cell that could either be flushed with nitrogen or filled with an appropriate liquid for liquid–solid experiments. In these experiments, the SHG from the back side of the

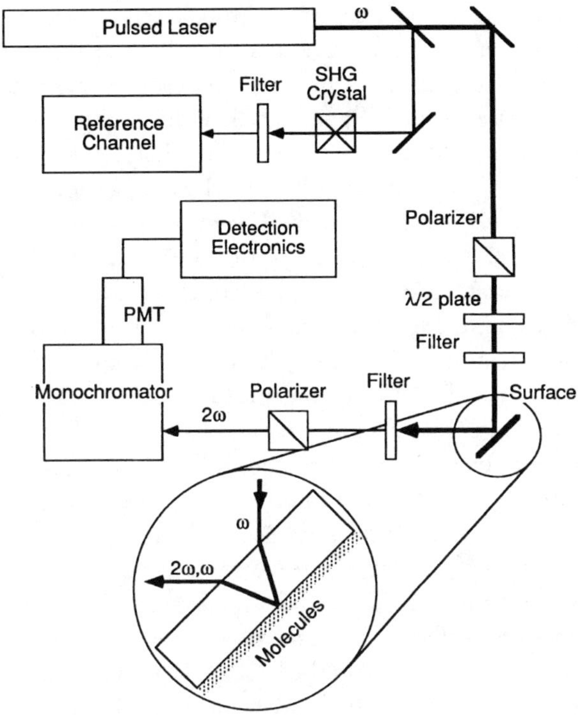

Figure 12.1 **The SHG experimental apparatus. A high-powered pulsed laser is usually used as the light source. A small portion of the laser beam is used in a reference channel to remove laser fluctuations, while the majority of the beam is sent to the sample surface to generate the second harmonic. The fundamental light is usually linearly polarized, focused, and filtered (to remove any light at the second harmonic in the laser beam) prior to incidence on the sample. The sample can consist of thin films adsorbed at a solid–gas, solid–liquid, liquid–gas, or liquid–liquid interface. The inset shows the sample geometry discussed in the text, with the SHG being collected from the molecules on the back face of a fused silica substrate. The SHG from the surface is collected and sent through a polarizer and filters, which remove the residual fundamental while passing the SHG. A small monochromator is used to further remove any fundamental light as well as any fluorescence. The detection electronics typically consist of a gated photon counter or a boxcar averager.**

fused silica substrate was collected (see inset in Figure 12.1). Additional SHG experiments on monolayer films at water–air and water–dichloroethane interfaces are also presented. In the water–air experiments, *p*-nitrophenol monolayers were formed by spontaneous adsorption from solution and *p*-dodecylnitrophenyl ether monolayers were formed by spreading on the water surface with hexane. The experiments at the water–dichloroethane interface involved the spontaneous adsorption of *p*-hexadecylsulfonylaniline to the interface from solution.

A block diagram of a typical experimental apparatus employed in the SHG measurements is shown in Figure 12.1. The SHG apparatus incorporates a high-powered pulsed laser as the light source. The most commonly used lasers are either nanosecond or picosecond Nd:YAG lasers, from which either the 1064 nm fundamental or the doubled fundamental at 532 nm can be used. Frequently, the Nd:YAG laser is used to pump a dye laser to provide for wavelength tunability. In the experiments presented in this chapter, dye lasers pumped with either a nanosecond Nd:YAG (10 ns pulse width, 5 mJ, 10 Hz repetition rate) or a picosecond Nd:YAG (3 ps pulse width, 20 nJ, 4 MHz repetition rate) were used to provide tunable light in the 532–640 nm range. A small portion of the laser beam can be sent through a doubling crystal to generate SHG. This SHG can be detected with a photodiode and used as a reference channel for removing laser fluctuations in the SHG from the sample. The majority of the laser beam, however, is sent to the sample surface to generate SHG from the thin film being studied.

The fundamental laser beam is passed through a polarizer and half-wave plate to set the polarization of the fundamental beam to *p*-polarization (0° with respect to the plane of incidence), *s*-polarization (90°), or some intermediate value. The polarized light can be focused onto the surface, if necessary. A cutoff filter placed in the beam path immediately before the surface removes any second harmonic light generated by the optics prior to the surface. In the SHG experiments presented here, the laser light was focused near the surface at an incident angle of 45° (corresponding to an incident angle of approximately 30° on the back surface of the fused silica and at the water–dichloroethane interface). The power density of the fundamental laser light incident on the surface typically ranges from 0.5 to 300 MW/cm^2. The usable range of power density is limited on the low end by the nonlinear susceptibility of the sample, and on the high end by the damage threshold of the sample. A good method for selecting a safe power level while simultaneously verifying the nonlinear nature of the observed signal is to demonstrate that the observed signal changes with the square of the incident laser intensity.

The second harmonic light from the monolayer is then collected and sent through a second polarizer set to pass either *s*-polarized or *p*-polarized light from the surface. By the selection of the appropriate filter, the second harmonic light can easily be separated from the residual fundamental light reflected from the surface. A small monochromator can be used to further discriminate between SHG, reflected fundamental, and any possible fluorescence. The SHG is detected with a PMT, the output of which can be monitored with either a boxcar averager (for the nanosecond experiments) or a gated photon counter (for the picosecond experiments). Conversion efficiencies in these experiments are normally very small, on the order of 10^{-12}%. Thus, the observed SHG signal from surfaces typically ranges from 5 to 50 000 counts/s, depending on the laser intensity, surface coverage, and nonlinear response of the film.

The phase of the surface second harmonic response can be measured with an interference technique described previously by Kemnitz et al. and depicted in Figure 12.2.[7] These measurements involve the insertion of a thin z-cut quartz crystal

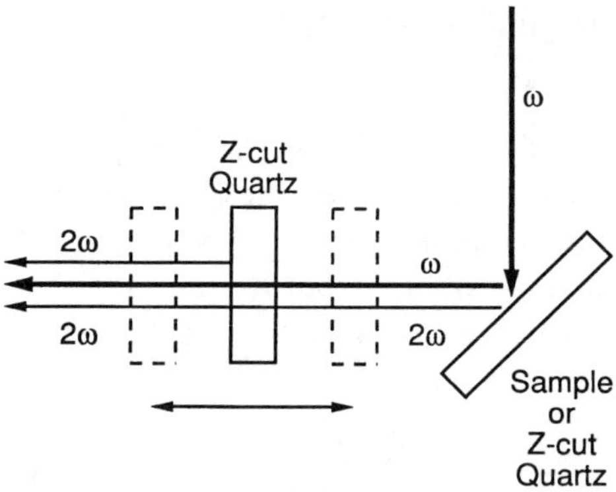

Figure 12.2 **Block diagram of the surface SHG phase measurement. The fundamental light at frequency ω generates light at frequency 2ω at the sample surface. Note that for absolute molecular orientation determination, the SHG must come from the molecules on the front side of the fused silica surface. The fundamental light reflected from the surface propagates along with the SHG to a thin slab of z-cut quartz, from which more SHG is generated. The phase relationship between these two sources of SHG is determined by the position of the z-cut quartz along the beam path. By translating the z-cut quartz parallel to the beam path while monitoring the SHG, an interference pattern is produced that contains information which describes the relative phase between the fundamental light incident on the sample surface and the SHG generated at the surface. This phase information can be extracted from the interference pattern by replacing the sample with a second piece of z-cut quartz of known orientation and then repeating the above interference experiment.**

in the beam reflected from the sample surface. The z-cut quartz generates SHG, which interferes with the SHG from the sample surface. A translation of the z-cut quartz parallel to the beam path changes the phase relationship between the two sources of SHG, thus producing an interference pattern which contains the phase relationship between the SHG from the sample surface and the incident electric field. In order to obtain the phase of the surface SHG relative to the input electric field, it is necessary to calibrate the interference pattern with a standard for which the phase of the SHG is known. This is accomplished by replacing the sample with a second piece of z-cut quartz of known absolute orientation and repeating the aforementioned interference experiment with the same polarizations. Note that in order to obtain meaningful absolute orientation results from phase measurements of this type, only the SHG from the front face (for example, the first air–silica interface or the air–water interface, etc.) of the sample and standard should be collected (see Figure 12.2). In the experiments reported here, the z-cut quartz standard was aligned so that its positive *x*-axis (as determined by the piezoelectric effect[7]) was antiparallel to the laboratory *x*-axis. The incident beam was polarized at 45°, and the *s*-polarized SHG was collected. The results of phase measurements of this type are typically accurate to within ±10°.

12.3 Molecular Nonlinear Polarizability Calculation

In order to obtain average molecular orientation information from the polarization dependence of the surface SHG, knowledge of the complex second harmonic response of an isolated molecule is required. The sign and magnitude of this molecular SHG response is described by the second order molecular nonlinear polarizability tensor, β. Molecular SHG results from the nonlinear polarization, $\alpha^{(2)}(2\omega)$, induced in the molecule by the incident laser light field $E(\omega)$:

$$\alpha^{(2)}(2\omega) = \beta : E(\omega)\, E(\omega) \tag{12.1}$$

Using perturbation theory, the components of β can be expressed in terms of the molecular energies and wavefunctions[8, 9]:

$$\begin{aligned}
\beta_{ijk} + \beta_{ikj} = \frac{-e^3}{4h^2} \sum_{n,n'=g}^{N} & (r^j_{gn'} r^i_{n'n} r^k_{ng} + r^k_{gn'} r^i_{n'n} r^j_{ng}) \\
& \times \left[\frac{1}{\omega_{n'n} - 2\omega + i\Gamma_{n'n}} \left(\frac{1}{\omega_{ng} + \omega - i\Gamma_{ng}} - \frac{1}{\omega_{n'g} - \omega + i\Gamma_{n'g}} \right) \right. \\
& \left. \pm \frac{1}{\omega_{n'n} - 2\omega - i\Gamma_{n'n}} \left(\frac{1}{\omega_{ng} + \omega + i\Gamma_{ng}} - \frac{1}{\omega_{n'g} - \omega - i\Gamma_{n'g}} \right) \right] \\
& + (r^j_{gn'} r^k_{n'n} r^i_{ng} + r^k_{gn'} r^j_{n'n} r^i_{ng}) \\
& \times \left[\frac{1}{(\omega_{n'g} - \omega + i\Gamma_{n'g})(\omega_{ng} - 2\omega + i\Gamma_{ng})} \pm \frac{1}{(\omega_{n'g} - \omega - i\Gamma_{n'g})(\omega_{ng} - 2\omega - i\Gamma_{ng})} \right] \\
& + (r^i_{gn'} r^j_{n'n} r^k_{ng} + r^i_{gn'} r^k_{n'n} r^j_{ng}) \\
& \times \left[\frac{1}{(\omega_{ng} + \omega - i\Gamma_{ng})(\omega_{n'g} + 2\omega - i\Gamma_{n'g})} \pm \frac{1}{(\omega_{ng} + \omega + i\Gamma_{ng})(\omega_{n'g} + 2\omega + i\Gamma_{n'g})} \right]
\end{aligned} \tag{12.2}$$

where β_{ijk} is the tensor element referenced to the molecular coordinate axes i, j, and k; ω is the incident laser frequency; $h\omega_{ng}$ is the energy difference between excited state n and the ground state; Γ_{ng} is the lifetime (approximately the dephasing time) of excited state n; and r^i_{ng} is the transition dipole matrix element between states n and g along the molecular i-axis. Excited state lifetimes (Γ_{nn}) and population feeding terms have been neglected in this equation. Where the notation "±" occurs in

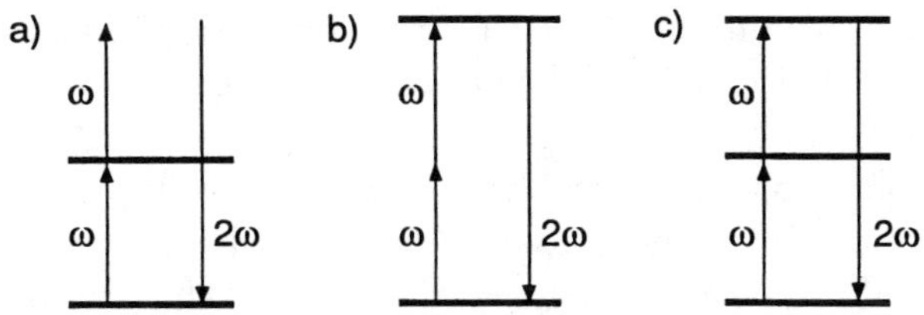

Figure 12.3 **Resonance possibilities for molecular SHG: (*a*) A molecular resonance at the fundamental wavelength. (*b*) A molecular resonance at the second harmonic wavelength, such as for *p*-nitrophenol with 610 nm fundamental light. (*c*) A molecular resonance at both the fundamental and second harmonic, such as for methylene blue with 615 nm fundamental light.**

the equation, the "+" signs are used to obtain the in-phase (real) component of β, and the "–" signs are used to obtain the 90° out-of-phase (imaginary) component.

The complexities involved in calculating β from Equation 12.2 are obvious. Fortunately, the salient features of the molecular second harmonic response can be obtained from a simple two-level model, where one considers only a ground and a single excited state (typically a state responsible for an optical transition near the wavelength of interest), as depicted in Figures 12.3*a* and 12.3*b*. The equations for the elements of the nonlinear polarizability tensor are then given by

$$\beta_{ijk} = \frac{-e^3}{2\hbar^2}\left[\frac{\Delta r_n^i r_{ng}^j r_{ng}^k}{\omega_{ng}^2 - \omega^2} + r_{ng}^i (r_{ng}^j \Delta r_n^k + \Delta r_n^j r_{ng}^k)\frac{(\omega_{ng}^2 + 2\omega^2)}{(\omega_{ng}^2 - 4\omega^2)(\omega_{ng}^2 - \omega^2)}\right] \qquad (12.3)$$

where Δr_n^i is the difference in the permanent dipole moment between excited state *n* and the ground state *g*. Note the following characteristics of the second harmonic response from this two level system:

(1) β has resonances at two wavelengths—once when the laser frequency is equal to ω_{ng}, and a second time when the laser frequency is equal to $\omega_{ng}/2$ (see Figure 12.3). This second condition corresponds to the second harmonic wavelength being in resonance with a transition in the molecule and is particularly useful since resonant enhancement of the SHG can be obtained at a wavelength where there is virtually no absorption of the incident laser power. SHG measurements which solely utilize resonances at twice the laser frequency avoid the problems of photobleaching, burning, and fluorescence. A second advantage of such measurements is that molecules with resonances in the near-UV can be monitored with the use of readily available visible dye lasers. A third possibility for resonance in β (which cannot be described with the above two-level model) arises when the fundamental light is resonant with a transition to one excited state and the second harmonic is resonant with a transition to a different excited state (Figure 12.3*c*). In this case, the molecule is considered to be "doubly resonant enhanced," resulting in a very large molecular SHG response.[10, 11]

(2) The elements of β are proportional to the elements of the change in the permanent dipole moment of the molecule from the ground to excited state, $\Delta \mathbf{r}_n$. Thus, any molecule whose dipole moment does not change upon excitation to a given energy level will not exhibit an optical second harmonic response. Typically, however, the molecules used in most SHG studies to date have been chosen for the presence of large charge transfer absorptions in their near-UV spectra. These molecules usually incorporate a large π-electronic structure, with an electron-donating group at one end of the molecule (an OH, NH_2, or similar group) and an electron-accepting group at the other end (an NO_2, acid, or similar group). Such a configuration leads to relatively low-energy excited states which involve the transfer of electron density from the donating group to the accepting group and hence a large $\Delta \mathbf{r}_n$ and a large nonlinear response.[12–14]

(3) β is a tensor formed by the direct product of the vectors $\mathbf{r}_{ng}$ and $\Delta \mathbf{r}_n$. Thus, any component of β that involves the axis perpendicular to the plane defined by $\mathbf{r}_{ng}$ and $\Delta \mathbf{r}_n$ will be zero. If $\mathbf{r}_{ng}$ and $\Delta \mathbf{r}_n$ are collinear, then only one tensor element, defined as β_{zzz}, will be nonzero. Most of the previous papers that determined the average molecular orientation from SHG measurements assumed that the nonlinear response of the adsorbed molecules could be described by a single dominant molecular nonlinear polarizability tensor element.[15, 16] In this chapter, that assumption is relaxed to allow for the contribution of a second tensor element.

The simple two-state model provides for a clear description of the nonlinear polarizability for small molecules (such as *p*-nitrophenol) in spectral regions far from resonance. However, for large dye molecules with many low-energy electronic transitions, and even for small molecules in spectral regions near resonance, this model breaks down, and a more complete calculation utilizing Equation 12.2 is required. In these more complicated systems, the calculation can be simplified for most aromatic molecules by assuming that only the π-electronic structure contributes to the molecular nonlinear polarizability in the UV-visible region of the electromagnetic spectrum. For these molecules, a π-electron calculation, such as the method of Pople–Pariser–Parr (PPP), can be used to obtain the energies and wavefunctions for the molecule.[17, 18] The elements of β can then be calculated from Equation 12.2.[8] Using this methodology, the principle β tensor elements for *p*-nitrophenol have been calculated and are plotted in Figure 12.4. For *p*-nitrophenol it is found that two tensor elements, β_{zzz} and β_{zxx}, dominate the molecular nonlinear polarizability around 600 nm.[8, 19] Both the real and imaginary components of these two elements are plotted as a function of input photon energy in the figure. For this calculation, the values of Γ_{ng}, $\Gamma_{n'g}$, and $\Gamma_{n'n}$ were all set to 2000 cm^{-1}, the experimentally determined half width at half maximum for the first electronic absorption of *p*-nitrophenol.

As seen in Figure 12.4, a resonance is observed at half the energy of the first excited state, which is at approximately 4.3 eV (287 nm). This resonance involves an intramolecular charge transfer of electron density from the OH to the NO_2 of

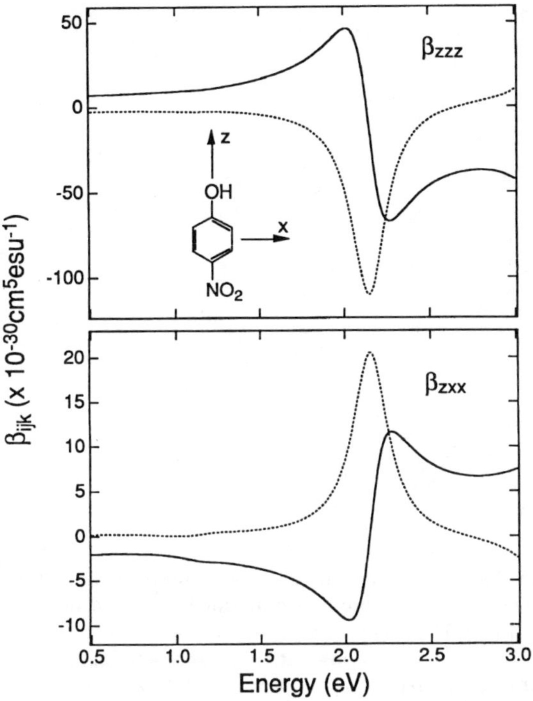

Figure 12.4 **The π-electron molecular nonlinear polarizability calculation for *p*-nitrophenol. The tensor elements β_{zzz} and β_{zxx} were found to be the dominant elements and are defined with respect to the molecular axes depicted in the figure. Both real (—) and imaginary (---) components are shown for β_{zzz} and β_{zxx}.[8]**

p-nitrophenol. In general, the position of such charge transfer bands are highly sensitive to the condensed phase environment of the molecule and can be drastically different from the calculated position. For example, in Figure 12.5 the UV-visible absorption spectrum for a *p*-nitrophenol monolayer adsorbed onto a silica substrate exhibits an absorbance maximum at 320 nm. With incident light at 610 nm such that the second harmonic at 305 nm is in resonance with this transition, the tensor element β_{zzz} is theoretically found to dominate the second harmonic response (where the molecular *z*-axis corresponds to the long axis of the molecule, as shown in Figure 12.4), with the tensor element β_{zxx} having the next most important contribution. By inspection of the calculated results from Equation 12.2, it can be determined that the dominance of the β_{zzz} element arises from fact that the electronic transition to the first excited state involves both a large transition dipole (r_{ng}) and a large change in the permanent dipole moment (due to the charge transfer effects described above), both of which occur in the *z*-direction. The nonzero contribution of β_{zxx} is due to the presence of a higher energy transition at 5.5 eV that possesses a large *x*-transition dipole as well as a large *x*-transition dipole to the first excited state ($r_{n'n}$), thus leading to a large contribution from the term $r^x_{gn'}\, r^x_{n'n}\, r^z_{ng}$ in

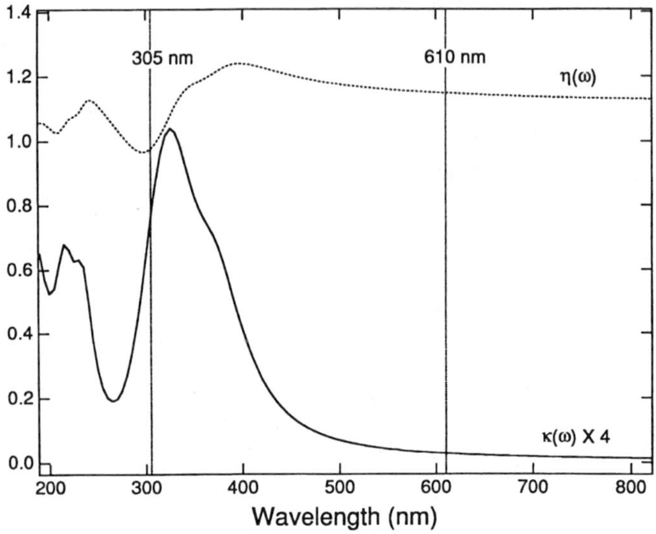

Figure 12.5 **Optical absorption spectrum for a monolayer of *p*-nitrophenol adsorbed to a fused silica surface. The spectrum is plotted as the imaginary component of the dielectric constant, κ × 4 (—). The absorption maximum at 320 nm corresponds to a absorbance of 1×10^{-2}. SHG is observed with incident laser light at 610 nm due to the resonance at the second harmonic wavelength of 305 nm (see diagram). Also plotted is the real component of the dielectric constant, η (---), as calculated by a Kramers–Kronig analysis of the absorption spectrum. The dielectric constant is given by $(\eta + i\kappa)^2$.[8]**

Equation 12.2. Note that the contribution of this second term would be neglected in the two-state approximation.

The nonlinear polarizability results shown in Figure 12.4 could theoretically be used to determine the relative contributions of these two tensor elements; however, due to the approximate nature of these semi-empirical calculations, this ratio cannot be taken from the nonlinear polarizability calculation, but instead can be determined experimentally from the polarization dependence of the SHG signal.[8, 19] In addition to the identification of the primary molecular tensor elements for the orientation calculation, the phase characteristics of the tensor elements obtained from nonlinear polarizability calculations are required for the determination of the absolute molecular orientation at the interface from the surface SHG phase measurements, as is discussed below.

12.4 Measurements of the Surface Nonlinear Susceptibility

When adsorbed to a surface, a monolayer of molecules collectively responds to an incident laser light field to create an induced nonlinear polarization on the surface at frequency 2ω, $P^{(2)}(2\omega)$:[1]

$$P^{(2)}(2\omega) = \sum \alpha^{(2)}(2\omega) = \chi^{(2)} : E(\omega)\ E(\omega) \tag{12.4}$$

where $\chi^{(2)}$ is defined as the surface nonlinear susceptibility tensor. The surface nonlinear polarization leads to the generation of light at frequency 2ω at the surface. The observed intensity at the detector, $I(2\omega)$, is related to $\chi^{(2)}$ by[4, 20]

$$I(2\omega) = \frac{32\pi^3\omega^2\sec^2\theta_{2\omega}}{c^3}\left|\mathbf{e}(2\omega) \cdot \chi^{(2)} : \mathbf{e}(\omega)\ \mathbf{e}(\omega)\right|^2 I(\omega)^2 \tag{12.5}$$

where $I(\omega)$ is the intensity of the incident laser light, $\theta_{2\omega}$ is the angle from the surface normal at which the SHG signal occurs, the vectors $\mathbf{e}(\omega)$ and $\mathbf{e}(2\omega)$ describe the polarization and Fresnel factors for the fundamental and second harmonic light fields at the surface, and all other symbols have their usual meanings.

The components of $\mathbf{e}(\omega)$ and $\mathbf{e}(2\omega)$ are given in Equation 12.6 for the example system presented in this paper, that of a monolayer adsorbed on the back side of a fused silica substrate surrounded by air. Explicit equations for $\mathbf{e}(\omega)$ and $\mathbf{e}(2\omega)$ for other experimental geometries can be found in papers by Heinz[4] and by Sipe.[20]

$$\begin{aligned}
e_x(\omega) &= t_1^{\parallel}(\omega)\ \left[1 - r_2^{\parallel}(\omega)\right]\cos\theta(\omega) \\
e_y(\omega) &= t_1^{\perp}(\omega)\ \left[1 + r_2^{\perp}(\omega)\right] \\
e_z(\omega) &= t_1^{\parallel}(\omega)\ \left[1 + r_2^{\parallel}(\omega)\right]\sin\theta(\omega) \\
e_x(2\omega) &= -t_1^{\parallel}(2\omega)\ t_2^{\parallel}(2\omega)\cos\theta(2\omega) \\
e_y(2\omega) &= t_1^{\perp}(2\omega)\ t_2^{\perp}(2\omega) \\
e_z(2\omega) &= t_1^{\parallel}(2\omega)\ t_2^{\parallel}(2\omega)\sin\theta(2\omega)
\end{aligned} \tag{12.6}$$

In Equation 12.6, $\theta(\omega)$ is the angle of incidence for the fundamental laser light as measured inside the silica, and $\theta(2\omega)$ is the angle at which the second harmonic is generated in air. The Fresnel coefficients $t_1^{\parallel}$ and $t_1^{\perp}$ are for transmission of light (p-polarized and s-polarized, respectively) at the appropriate frequency through the first interface (air–silica), while $r_2^{\parallel}$ and $r_2^{\perp}$ are for reflection of light from the second interface (silica–air). Equation 12.6 was derived following the method of Sipe, assuming that the driving fields are in the fused silica and the SHG is generated in air.[20] For an s-polarized input light vector (i.e., with the electric field perpendicular to the plane of incidence), e_y is the only nonzero component of $\mathbf{e}(\omega)$, and for a p-polarized input light vector (i.e., one with the electric field parallel to the plane of incidence) e_x and e_z are the only nonzero components. The Fresnel coefficients are usually calculated from the tabulated optical constants for the bulk media, assuming that any corrections due to the presence of the adsorbed monolayer can be neglected.

If the SHG from the adsorbed monolayer is invariant during rotation of the surface about the surface normal (the Z axis), then the molecules within the laser spot are randomly oriented about the surface normal, and $\chi^{(2)}$ has only three unique elements: χ_{XXZ}, χ_{ZXX}, and χ_{ZZZ}. The intensity of the s-polarized and p-polarized SHG signals ($I_s(2\omega)$ and $I_p(2\omega)$, respectively) can then be directly related to these elements:

$$I_s(2\omega) \propto |a_1 \chi_{XXZ} \sin^2\gamma|^2 I(\omega)^2 \tag{12.7}$$

$$I_p(2\omega) \propto |(a_2 \chi_{XXZ} + a_3 \chi_{ZXX} + a_4 \chi_{ZZZ})\cos^2\gamma + a_5 \chi_{ZXX} \sin^2\gamma|^2 I(\omega)^2 \tag{12.8}$$

where γ is the polarization angle of the incident light ($\gamma = 0°$ for p-polarized light and $\gamma = 90°$ for s-polarized light), and the a_i terms describe the electric fields in the monolayer and thus include the dielectric constants of the monolayer, $\varepsilon'(\omega)$ and $\varepsilon'(2\omega)$. Guyot-Sionnest et al. have stressed the need for an independent measurement of $\varepsilon'(\omega)$ and $\varepsilon'(2\omega)$ in order to obtain a true molecular orientation measurement.[21] The a_i terms for the example system presented here are given by

$$\begin{aligned}
a_1 &= e_y(\omega)e_z(\omega)e_y(2\omega)[\varepsilon_s(\omega)/\varepsilon'(\omega)] \\
a_2 &= 2e_x(\omega)e_z(\omega)e_x(2\omega)[\varepsilon_s(\omega)/\varepsilon'(\omega)] \\
a_3 &= |e_x(\omega)|^2 e_z(2\omega)[1/\varepsilon'(2\omega)] \\
a_4 &= |e_z(\omega)|^2 e_z(2\omega)[\varepsilon_s(\omega)^2/\{\varepsilon'(\omega)^2 \varepsilon'(2\omega)\}] \\
a_5 &= |e_y(\omega)|^2 e_z(2\omega)[1/\varepsilon'(2\omega)]
\end{aligned} \tag{12.9}$$

where $\varepsilon_s(\omega)$ is the dielectric constant of the substrate at the fundamental frequency. An estimate of the monolayer dielectric constants can be determined separately through a Kramers–Kronig analysis of the optical absorption spectrum.[8, 11] The calculated real and imaginary components of the dielectric constant for p-nitrophenol on fused silica are shown in Figure 12.5. From this calculation, the approximate dielectric constants for the monolayer are found to be 1.3 for $\varepsilon'(\omega)$ at 610 nm and 1.0 for $\varepsilon'(2\omega)$ at 305 nm. These values are only approximate, since the thickness of the monolayer is required for the calculation. Furthermore, the absorption spectrum outside of the UV-visible region is required for an accurate Kramers–Kronig calculation in this region; however, such spectra are not available.

Curves of $I_p(2\omega)$ and $I_s(2\omega)$ as a function of γ for a monolayer of p-nitrophenol adsorbed onto a silica surface and for a monolayer of HDSA molecules adsorbed at a water–dichloroethane interface are shown in Figure 12.6. From this data, the relative magnitudes of the various $\chi^{(2)}$ elements for these two monolayers can be obtained by using Equations 12.7 and 12.8. The observation that $I_p(2\omega)$ goes to zero at an input polarization angle between 0° and 90° implies that the two terms in Equation 12.8 involving $(a_2 \chi_{XXZ} + a_3 \chi_{ZXX} + a_4 \chi_{ZZZ})$ and $a_5 \chi_{ZXX}$ are 180° out of phase. A more complete measurement of the phases of the tensor elements with respect to the fundamental driving fields can be obtained with the interference methods described above, and is discussed further in Section 12.6.

12.5 Molecular Orientation Calculation

Once the elements of $\chi^{(2)}$ have been determined, they can be related to the elements of the molecular nonlinear polarizability, β, by the molecular orientation distribution function. This mathematical relationship can be described as the average of the product of three direction cosines, R_{Xx}, between the laboratory and molecular coordinate systems:

$$\chi_{IJK} = N_s \sum \langle R_{Ii}\, R_{Jj}\, R_{Kk} \rangle \beta_{ijk} = N_s \sum \langle F_{IJKijk}(\phi,\theta,\alpha) \rangle \beta_{ijk} \tag{12.10}$$

where N_s is the surface concentration of adsorbed molecules. The product of the direction cosines that define the molecular orientation on the surface in Equation 12.10 can be expressed as the molecular orientation distribution function $F_{IJKijk}(\phi,\theta,\alpha)$, where the molecular angles φ, θ, and α are defined relative to the space-fixed coordinates, as shown in Figure 12.7. As mentioned above, we only consider cases in which the SHG from the surface does not depend upon the azimuthal angle φ,

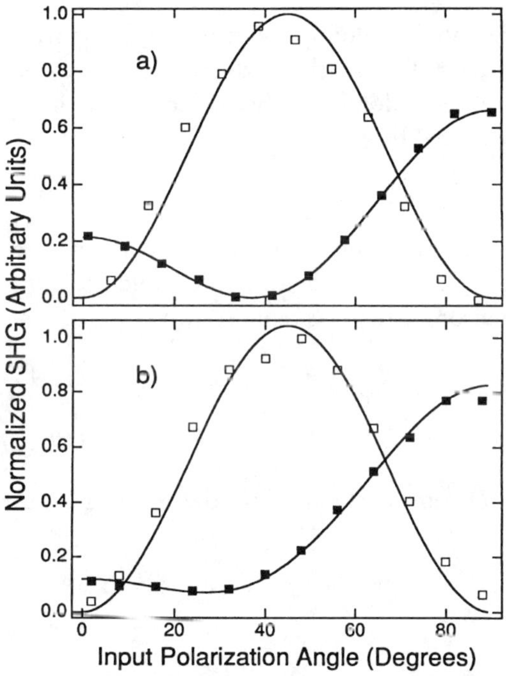

Figure 12.6 **Polarization dependence of the resonant SHG from (*a*) a *p*-nitrophenol monolayer on fused silica at 305 nm and (*b*) HDSA at the water/1,2-dichloroethane interface at 266 nm. The *p*-polarized (filled squares) and *s*-polarized (open squares) SHG from the monolayers are plotted as functions of input polarization, where 0° corresponds to *p*-polarized fundamental light and 90° to *s*-polarized fundamental light. The solid lines are the theoretical fits used to determine the values of χ_{ZXX}, χ_{XXZ}, and χ_{ZZZ} for the surface SHG.[8, 30]**

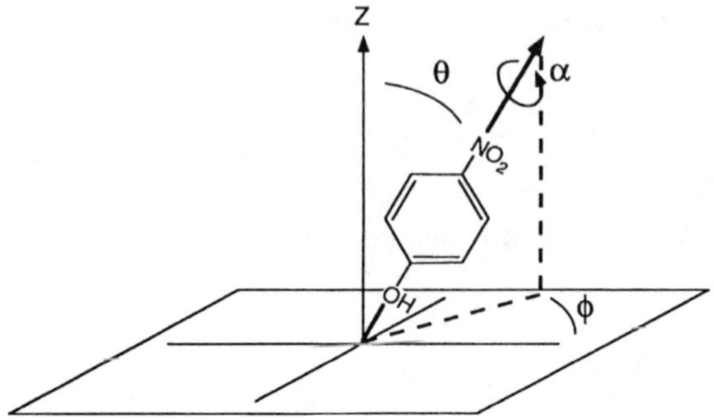

Figure 12.7 **Definition of α, ϕ, and θ. The surface normal is defined as the laboratory *Z*-axis. The molecular orientation calculation usually involves assumptions as to the orientation distribution with respect to the angles α and ϕ. The distribution in ϕ is usually assumed to be random as can be shown by demonstrating the isotropic nature of the SHG from the surface with respect to rotation about the surface normal. A variety of assumptions have been made as to the distribution in ϕ, as discussed in the text.**

which implies that the molecules are oriented randomly about the surface normal and therefore F_{IJKijk} can be integrated over all ϕ. Further reduction of F_{IJKijk} depends upon which molecular tensor elements dominate the molecular nonlinear polarizability. The separate cases are presented below.

Case 1:* β_{ZZZ} *only

For the case where β_{zzz} is the only nonzero tensor element, the molecular orientation distribution function (see Equation 12.10) is greatly simplified, and the number of independent surface nonlinear tensor elements is reduced to two:

$$\chi_{ZXX} = \chi_{XXZ} = \tfrac{1}{2} N_s \langle \cos\theta \, \sin^2\theta \rangle \beta_{zzz} \tag{12.11}$$

$$\chi_{ZZZ} = N_s \langle \cos^3\theta \rangle \beta_{zzz} \tag{12.12}$$

The molecular orientation parameter D can be obtained by the ratio of tensor elements given in Equation 12.13:

$$D = \frac{\langle \cos^3\theta \rangle}{\langle \cos\theta \rangle} = \frac{\chi_{ZZZ}}{\chi_{ZZZ} + 2\chi_{XXZ}} \tag{12.13}$$

If the distribution $\langle\theta\rangle$ is sufficiently narrow, an average molecular angle $\langle\theta\rangle = \mathrm{Cos}^{-1}(D^{1/2})$ can be determined from the SHG measurements. For example, in the case of HDSA at a water–dichloroethane interface (see Figure 12.6*b*), a D of 0.6 is observed, which corresponds to a $\langle\theta\rangle$ of 39°. The assumption of a single nonzero β_{zzz} tensor element for HDSA requires that values of 2.52 and 4.75 be assumed for $\varepsilon'(\omega)$ and $\varepsilon'(2\omega)$, respectively, in order to satisfy the requirement that $\chi_{XXZ} = \chi_{ZXX}$.

This methodology for the incorporation of the monolayer dielectric constants into the orientation calculation has been discussed by Zhang et al.[22]

This case of a single dominant β_{ZZZ} molecular nonlinear polarizability tensor element is the one most commonly assumed; it was first proposed by Heinz et al.[16] in the SHG study of *p*-nitrobenzoic acid adsorbed at an ethanol–silica interface. In that paper, the authors employed an alternative polarization nulling method at three different wavelengths to measure a $\langle\theta\rangle$ of $40 \pm 5°$. The assumption of a single nonzero tensor element in the orientation calculation has been subsequently used for many molecules at a wide variety of interfaces.

Case 2: β_{ZXX} *only*

A second possibility which has been described in the literature arises when β_{ZXX} can be assumed to be the single nonzero molecular nonlinear polarizability tensor element. This has been suggested to be the case for dye molecules such as rhodamine 6G and rhodamine B.[15, 23] In these instances, the dependence of F_{IJKijk} on the angle α must be considered. In the SHG measurements on rhodamine 6G, it was assumed that the adsorbed molecules had no preferential orientation of their molecular planes about their symmetry axes. With that assumption, F_{IJKijk} can be integrated over α and the orientation parameter D is given by

$$D = \frac{\langle\cos^3\theta\rangle}{\langle\cos\theta\rangle} = \frac{2\chi_{ZXX} - \chi_{ZZZ}}{2\chi_{ZXX} + \chi_{ZZZ}} \tag{12.14}$$

For rhodamine 6G at the silica air interface, an average angle $\langle\theta\rangle$ of 34° was obtained.[15] An alternate analysis was used by Peterson et al. for the case of rhodamine B. They demonstrated through trigonometric relations that combinations of ϕ and α could be related to γ, the angle that the long axis of the dye molecule made with the surface normal, and that the average angle $\langle\gamma\rangle$ could be determined from the surface tensor elements.[23]

Case 3: β_{XXZ} (= β_{XZX}) *only*

A third possibility, that of a dominant β_{xxz} element, has been suggested for the rhodamine dyes in experiments performed with a 1064 nm laser.[24, 25] Once again, the molecular distribution in α can be assumed to be random. In this case, the molecular orientation parameter can be obtained from the experimentally determined surface nonlinear susceptibility elements as follows:

$$D = \frac{\langle\cos^3\theta\rangle}{\langle\cos\theta\rangle} = \frac{2\chi_{XXZ}}{\chi_{ZZZ} + 2\chi_{XXZ}} \tag{12.15}$$

Case 4: β_{ZZZ} *and* β_{ZXX}

For the case of a dominant β_{ZZZ} with a non-negligible contribution from β_{ZXX} (as is the case with *p*-nitrophenol in resonant or near-resonant experiments), the relative magnitude of these two tensor elements can be obtained from the experimentally determined values of the surface nonlinear susceptibility by Equation 12.16:

$$\frac{\beta_{zxx}}{\beta_{zzz}} = 2\frac{\chi_{ZXX} - \chi_{XXZ}}{\chi_{ZZZ} + 2\chi_{XXZ}} \qquad (12.16)$$

It is apparent from this equation that, as mentioned above, χ_{XXZ} and χ_{ZXX} must be equal for β_{zzz} to be the only contributing β tensor element. Using the experimental SHG results shown in Figure 12.6*a*, the ratio for the two dominant β elements, β_{zxx}/β_{zzz} was found to be −0.31 for *p*-nitrophenol adsorbed to fused silica.[8] Note that the two tensor elements are observed to be 180° out of phase, as is predicted in the nonlinear polarizability calculations (Figure 12.4).

For the determination of the average molecular orientation, a second relationship can be derived from Equation 12.10, which provides a value for the orientation parameter, *D*, directly from the measured $\chi^{(2)}$ values (assuming a random distribution in α):

$$D = \frac{\langle \cos^3\theta \rangle}{\langle \cos\theta \rangle} = \frac{\chi_{ZZZ} - \chi_{ZXX} + \chi_{XXZ}}{\chi_{ZZZ} + 3\chi_{XXZ} - \chi_{ZXX}} \qquad (12.17)$$

If $\beta_{ZXX} = 0$, this equation reduces to Equation 12.13. For *p*-nitrophenol at the silica–air interface, *D* was measured to be 0.33 ± 0.07.[8] For comparison, if the *p*-nitrophenol molecules in the monolayer were distributed randomly with angle θ ranging between 0° and 90°, *D* would be equal to ⅔. If all the *p*-nitrophenol molecules are aligned on the surface with exactly the same orientation, a *D* of 0.33 corresponds to an angle θ of 55°. This assumption of a single molecular orientation on the surface is somewhat unrealistic; an alternative interpretation of the orientation parameter in terms of a Gaussian distribution of the orientation angle θ has been suggested elsewhere.[8]

Case 5: β_{ZXX} *and* β_{XXZ} (= β_{XZX})

A final case has been observed—of a dominant β_{ZXX} tensor element with a non-negligible contribution from β_{XXZ}—in the analysis of the doubly resonant SHG from methylene blue molecules adsorbed to platinum electrodes and silica surfaces.[10, 11] As in the previous case, the relative magnitudes of the surface nonlinear susceptibility tensor elements can be determined from the experimental polarization curves, which can then be used to determine the relative contributions of β_{ZXX} and β_{XXZ} from the following equation:

$$\frac{\beta_{xxz}}{\beta_{zxx}} = \frac{\chi_{ZZZ} + 2\chi_{XXZ}}{\chi_{ZZZ} + 2\chi_{ZXX}} \qquad (12.18)$$

The ratio β_{xxz}/β_{zxx} was found to be 0.75 for 615 nm light.[11]

For the determination of the orientation parameter, D, the following equation can be derived using the assumption of a random distribution in α:

$$D = \frac{\langle \cos^3\theta \rangle}{\langle \cos\theta \rangle} = \frac{2\chi_{ZXX} + \chi_{ZZZ} + 4\chi_{XXZ}}{2\chi_{ZXX} + 3\chi_{ZZZ} + 4\chi_{XXZ}} \tag{12.19}$$

From this equation, D was found to be 0.51, which corresponds to a $\langle\theta\rangle$ of 44° (the angle between the short axis of methylene blue and the surface normal) if all of the methylene blue molecules are adsorbed in the same orientation. As in case 2, these results can also be interpreted following the methodology of Peterson et al. to include the angle α.[23] In this instance, the angle between the dominant transition dipole (the long axis of methylene blue) and the surface normal is found to be 60.5°.[11]

12.6 Absolute Molecular Orientation Measurements

Although the average molecular orientation angle can be obtained directly from the polarization dependence of the surface SHG, further SHG measurements can help deduce the absolute direction of the adsorbed molecules on the surface.[7] These experiments entail the measurement of the phase of the χ_{XXZ} surface susceptibility tensor element through the interference method described in the experimental section of this chapter. As a representative example, the interference curve for *p*-nitrophenol at the air–water interface at 610 nm is shown in Figure 12.8. Upon comparison of the measured phase of this tensor element with that expected from the molecular nonlinear polarizability calculation, the absolute orientation with respect to the surface (which in the case of *p*-nitrophenol at the air–water interface means NO_2 up or down) can be ascertained. The utility of these measurements has been demonstrated for phenol at the water–air interface and for water itself by Eisenthal's research group.[7, 26]

The expected phase of χ_{XXZ} can be calculated from the molecular nonlinear polarizability by Equation 12.10. The results of this calculation for *p*-nitrophenol when the molecule has its nitro group up are plotted in Figure 12.9. As seen in the figure, the phase of χ_{XXZ} is calculated to be 180° below resonance, 90° on resonance, and 0° above resonance. In contrast, if the molecule were oriented with its nitro group down, the phase of χ_{XXZ} would range between 0° and 180°, with a value of −90° on resonance. For the *p*-nitrophenol molecule at the air–water interface, the phase of χ_{XXZ} was found to be 80°, 70°, and 60 ± 10° at 640, 610, and 585 nm, respectively.[8, 27] These results suggest that the nitro group is up (i.e., pointing out of the water), a result consistent with previous studies.[7]

Also shown in Figure 12.8 is the phase curve for *p*-nitrophenyl ether at 610 nm at the air–water interface. When comparing the phase of χ_{XXZ} for two slightly different molecules at the same interface, or the same molecule at two different interfaces, it is important to remember that the observed phases will change due to a change in the wavelength of resonance. For example, the phase of χ_{XXZ} is −10°

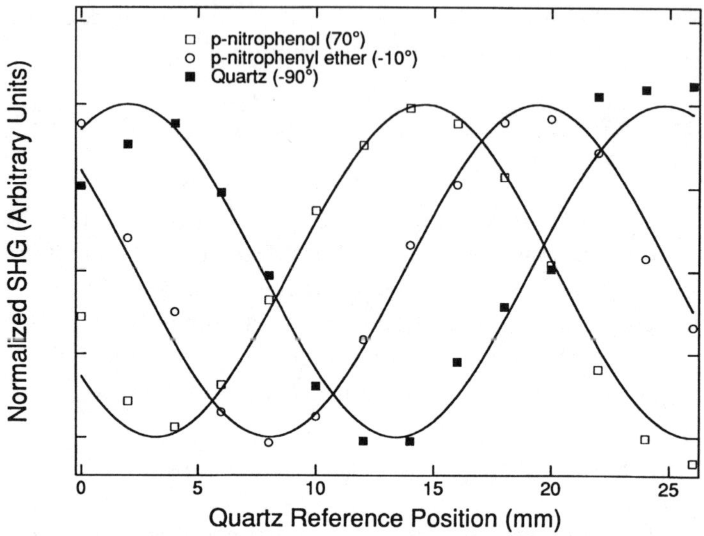

Figure 12.8 **Interference measurements of the phase of the resonant SHG signal at 610 nm used to determine the absolute orientation of the *p*-nitrophenol and *p*-nitrophenyl ether molecules at the air–water interface. The filled squares are from the z-cut quartz standard and correspond to a phase of 270° relative to the fundamental electric fields. The open squares are from the *p*-nitrophenol monolayer, and the open circles are from the *p*-nitrophenyl ether monolayer. These results exhibit phases of 70° for *p*-nitrophenol and −10° for *p*-nitrophenyl ether, indicating that the NO_2 group is up for *p*-nitrophenol and down for *p*-nitrophenyl ether (see Figure 10.9).[8, 27]**

for *p*-nitrophenyl ether at 610 nm and −50° at 585 nm.[27] This suggests that for *p*-nitrophenyl ether, the nitro group is down at the air–water interface, and the position of resonance has been blue-shifted relative to that for *p*-nitrophenol. Performing the phase measurement at a number of wavelengths both on and slightly off resonance helps to insure that the absolute orientation determination is correct.

In addition to the phase measurements made with a quartz reference, an interferometric method involving two adsorbed monolayers on opposite faces of a fused silica substrate has been described by Sato et al.[28, 29] The advantage of this method is that the reference is another adsorbed layer of the molecule, or a related molecule, so that the phase characteristics of the two monolayers can be matched very closely, leading to a simple interpretation of the data.

12.7 Summary and Conclusions

In this chapter, we have demonstrated how molecular orientation can be monitored in adsorbed monolayers and thin films with surface SHG experiments. In particular, we have shown that the average molecular orientation of adsorbed molecules

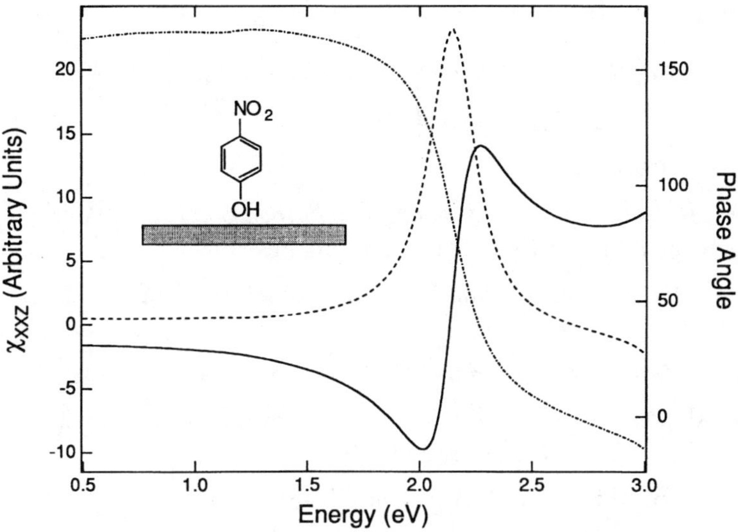

Figure 12.9 The real (—) and imaginary (---) components and the expected phase (-··-) of the χ_{xxz} surface nonlinear susceptibility element as a function of fundamental photon energy. These curves are obtained through Equation 12.10 from the results of Figure 12.4 and the measured orientation of *p*-nitrophenol on silica. The calculation shown here is for the case in which the NO_2 group points up (out of the water–air interface).[8]

can be determined from near-resonant and resonant SHG measurements in conjunction with π-electron molecular nonlinear polarizability calculations. Using the polarization dependence of the surface SHG, the average molecular orientation is described by an orientation parameter $D = \langle\cos^3\theta\rangle/\langle\cos\theta\rangle$. In some instances, a single molecular nonlinear polarizability tensor element dominates the molecular nonlinear optical response, and the analysis can be simplified. In addition, measurements of the phase of the surface SHG both on and off resonance can be compared to the phase expected from the molecular nonlinear polarizability calculations to determine the absolute orientation of the molecules adsorbed at the interface. Knowledge of the linear dielectric constants at the fundamental and second harmonic wavelengths are crucial to the orientational analysis. In the future, we hope to see SHG orientational analysis including the PPP nonlinear polarizability calculation employed for thin films of many varieties, including those specifically designed for their nonlinear optical characteristics.

Acknowledgments

The authors wish to thank the National Science Foundation for support in the preparation of this manuscript.

References

1. Y. R. Shen. *The Principles of Nonlinear Optics.* Wiley, New York, 1984.
2. K. B. Eisenthal. *Annu. Rev. Phys. Chem.* **43**, 627, 1992.
3. R. M. Corn. *Anal. Chem.* **63**, A285, 1991.
4. T. F. Heinz. In *Nonlinear Surface Electromagnetic Phenomena.* (H. E. Ponath and G. I. Stegeman, Ed.) North-Holland, Amsterdam, 1991.
5. G. L. Richmond, J. M. Robinson, and V. L. Shannon. *Prog. Surf. Sci.* **28**, 1, 1988.
6. Y. R. Shen. *Annu. Rev. Phys. Chem.* **40**, 327, 1989.
7. K. Kemnitz, K. Bhattacharyya, J. M. Hicks, G. R. Pinto, K. B. Eisenthal, and T. F. Heinz. *Chem. Phys. Lett.* **131**, 285, 1986.
8. D. A. Higgins, M. B. Abrams, S. B. Byerly, and R. M. Corn. *Langmuir.* **8**, 1994, 1992.
9. J. F. Ward. *Rev. Mod. Phys.* **37**, 1, 1965.
10. D. J. Campbell, D. A. Higgins, and R. M. Corn. *J. Phys. Chem.* **94**, 3681, 1990.
11. D. A. Higgins, S. K. Byerly, M. B. Abrams, and R. M. Corn. *J. Phys. Chem.* **95**, 6984, 1991.
12. J. Zyss and D. S. Chemla. In *Nonlinear Optical Properties of Organic Molecules and Crystals,* Vol. 1, 1987.
13. A. Ulman. *J. Phys. Chem.* **92**, 2385, 1988.
14. B. F. Levine. *Chem. Phys. Lett.* **37**, 516, 1976.
15. T. F. Heinz, C. K. Chen, D. Ricard, and Y. R. Shen. *Phys. Rev. Lett.* **48**, 478, 1982.
16. T. F. Heinz, H. W. K. Tom, and Y. R. Shen. *Phys. Rev. A.* **28**, 1883, 1983.
17. R. Pariser and R. G. Parr. *J. Chem. Phys.* **21**, 466, 1953.
18. J. A. Pople. *Trans. Faraday Soc.* **49**, 1375, 1953.
19. A. J. Bell, J. G. Frey, and T. J. VanderNoot. *J. Chem. Soc. Faraday Trans.* **88**, 2027, 1992.
20. V. Mizrahi and J. E. Sipe. *J. Opt. Soc. Am.* **B5**, 660, 1988.
21. P. Guyot-Sionnest, Y. R. Shen, and T. F. Heinz. *Appl. Phys. B.* **42**, 237, 1987.
22. T. G. Zhang, C. H. Zhang, and G. K. Wong. *J. Opt. Soc. Am. B.* **7**, 902, 1990.
23. E. S. Peterson and C. B. Harris. *J. Chem. Phys.* **91**, 2683, 1989.
24. P. Di Lazzaro, P. Mataloni, and F. DeMartini. *Chem. Phys. Lett.* **114**, 103, 1985.

25 R. M. Corn. In *Frontiers in Electrochemistry: Adsorption of Molecules at Metal Electrodes.* (J. Lipowski and P. N. Ross, Eds.) VCH Publishers, New York, 1992.

26 M. C. Goh, J. M. Hicks, K. Kemnitz, G. R. Pinto, K. Bhattacharyya, K. B. Eisenthal, and T. F. Heinz. *J. Phys. Chem.* **92**, 5074, 1988.

27 D. A. Higgins and R. M. Corn. Unpublished results.

28 O. Sato, R. Baba, K. Hashimoto, and A. Fujishima. *J. Phys. Chem.* **95**, 9636, 1991.

29 O. Sato, R. Baba, K. Hashimoto, and A. Fujishima. *J. Electroanal. Chem.* **306**, 291, 1991.

30 D. A. Higgins and R. M. Corn. Manuscript in preparation.

Appendixes: Techniques Summaries

The technique summaries on the following pages of this appendix are reprinted from the lead volume of this series, *Encyclopedia of Materials Characterization*, by C. Richard Brundle, Charles A. Evans, Jr., and Shaun Wilson; they are summaries of full-length articles appearing there.

Auger Electron Spectroscopy (AES) 1

Auger electron spectroscopy (AES) uses a focused electron beam to create secondary electrons near the surface of a solid sample. Some of these (the Auger electrons) have energies characteristic of the elements and, in many cases, of the chemical bonding of the atoms from which they are released. Because of their characteristic energies and the shallow depth from which they escape without energy loss, Auger electrons are able to characterize the elemental composition and, at times, the chemistry of the surfaces of samples. When used in combination with ion sputtering to gradually remove the surface, Auger spectroscopy can similarly characterize the sample in depth. The high spacial resolution of the electron beam and the process allows microanalysis of three-dimensional regions of solid samples. AES has the attributes of high lateral resolution, relatively high sensitivity, standardless semiquantitative analysis, and chemical bonding information in some cases.

Range of elements	All except hydrogen and helium
Destructive	No, except to electron beam-sensitive materials and during depth profiling
Elemental Analysis	Yes, semiquantitative without standards; quantitative with standards
Absolute sensitivity	100 ppm for most elements, depending on the matrix
Chemical state information	Yes, in many materials
Depth probed	5–100 Å
Depth profiling	Yes, in combination with ion-beam sputtering
Lateral resolution	300 Å for Auger analysis
Imaging/mapping	Yes, called scanning Auger microscopy (SAM)
Sample requirements	Vacuum-compatible materials
Main use	Elemental composition of inorganic materials
Instrument cost	$100,000–$800,000
Size	10 ft × 15 ft

Dynamic Secondary Ion Mass Spectrometry (Dynamic SIMS) 2

In secondary ion mass spectrometry (SIMS), a solid specimen, placed in a vacuum, is bombarded with a narrow beam of ions, called primary ions, that are sufficiently energetic to cause ejection (sputtering) of atoms and small clusters of atoms from the bombarded region. Some of the atoms and atomic clusters are ejected as ions, called secondary ions. The secondary ions are subsequently accelerated into a mass spectrometer, where they are separated according to their mass-to-charge ratio and counted. The relative quantities of the measured secondary ions are converted to concentrations, by comparison with standards, to reveal the composition and trace impurity content of the specimen as a function of sputtering time (depth).

Range of elements	hydrogen to uranium; all isotopes
Destructive	Yes, material removed during sputtering
Chemical bonding information	In rare cases, from molecular clusters, but see Static SIMS
Quantification	Standards usually needed
Accuracy	2% to factor of 2 for concentrations
Detection limits	10^{12}–10^{16} atoms/cm^3 (ppb–ppm)
Depth probed	2 nm to 100 μm (depends on sputter rate and data collection time)
Depth profiling	Yes, by the sputtering process; resolution 2–30 nm
Lateral resolution	50 nm to 2 μm; 10 nm in special cases
Imaging/mapping	Yes
Sample requirements	Solid conductors and insulators, typically ≤ 2.5 cm in diameter, ≤ 6 mm thick, vacuum compatible
Main use	Measurement of composition and of trace-level impurities in solid materials a function of depth, excellent detection limits, good depth resolution
Instrument cost	\$500,000–\$1,500,000
Size	10 ft × 15 ft

Fourier Transform Infrared Spectroscopy (FTIR) 3

The vibrational motions of the chemically bound constituents of matter have frequencies in the infrared regime. The oscillations induced by certain vibrational modes provide a means for matter to couple with an impinging beam of infrared electromagnetic radiation and to exchange energy with it when the frequencies are in resonance. In the infrared experiment, the intensity of a beam of infrared radiation is measured before (I_0) and after (I) it interacts with the sample as a function of light frequency, $\{w_i\}$. A plot of I/I_0 versus frequency is the "infrared spectrum." The identities, surrounding environments, and concentrations of the chemical bonds that are present can be determined.

Information	Vibrational frequencies of chemical bonds
Element range	All, but not element specific
Destructive	No
Chemical bonding information	Yes, identification of functional groups
Depth profiling	No, not under standard conditions
Depth probed	Sample dependent, from μm's to 10 nm
Detection limits	Ranges from undetectable to $< 10^{13}$ bonds/cc. Sub-monolayer sometimes
Quantification	Standards usually needed
Reproducibility	0.1% variation over months
Lateral resolution	0.5 cm to 20 μm
Imaging/mapping	Available, but not routinely used
Sample requirements	Solid, liquid, or gas in all forms; vacuum not required
Main use	Qualitative and quantitative determination of chemical species, both trace and bulk, for solids and thin films. Stress, structural inhomogeneity
Instrument cost	\$50,000–\$150,000 for FTIR; \$20,000 or more for non-FT spectrophotometers
Instrument size	Ranges from desktop to (2 × 2 m)

High-Resolution Electron Energy Loss Spectroscopy (HREELS) 4

In high-resolution electron energy loss spectroscopy (HREELS), a highly mono-energetic beam of low energy (1–10 eV) electrons is focused onto a sample's surface, and the scattered electrons are analyzed with high resolution of the scattering energy and angle. Some of the scattered electrons suffer small characteristic energy losses due to vibrational excitation of surface atoms and molecules. A vibrational spectrum can be obtained by counting the number of electrons versus the electron energy loss relative to the elastically scattered (no energy loss) electron beam. This spectrum is used mainly to identify chemical species (functional groups) in the first layer of the surface. Often this layer contains adsorbed species on a solid.

Information	Molecular vibrational frequencies
Main use	Nondestructive identification of the molecular functional groups present at surfaces
Range of elements	Not element specific
Bonding	Any chemical bonds that have vibrations in the range 50–4000 cm^{-1}
Detection limits	0.1% monolayer for strong vibrational bands
Quantification	Difficult, possible with standards
Depth probed	2 nm
Lateral resolution	1 mm^2
Sample requirements	Single-crystal samples of conductors best; other solid samples are suitable, including polycrystalline metals, polymeric materials, semiconductors, and insulators, ultrahigh vacuum compatible; typically ≥ 5 mm diameter, 1–3 mm thick
Instrument cost	$100,000 plus associated techniques and vacuum system
Size	Attaches to vacuum chamber by 8–14 in. diameter flange.

Low-Energy Electron Diffraction (LEED) 5

In low-energy electron diffraction (LEED) a collimated monoenergetic beam of electrons in the energy range 10–1000 eV ($\lambda \approx 0.4–4.0$ Å) is diffracted by a specimen surface. In this energy range, the mean free path of electrons is only a few Å, leading to surface sensitivity. The diffraction pattern can be analyzed for the existence of a clean surface or an ordered overlayer structure. Intensities of diffracted beams can be analyzed to determine the positions of surface atoms relative to each other and to underlying layers. The shapes of diffracted beams in angle can be analyzed to provide information about surface disorder. Various phenomena related to surface crystallography and microstructure can be investigated. This technique requires a vacuum.

Range of elements	All elements, but not element specific
Destructive	No, except in special cases of electron-beam damage
Depth probed	4–20 Å
Detection limits	0.1 monolayer; any ordered phase can be detected; atomic positions to 0.1 Å; step heights to 0.1 Å; surface disorder down to ~10% of surface sites
Resolving power	Maximum resolvable distance for detecting disorder: typically 200 Å; best systems, 5 μm
Lateral resolution	Typical beam sizes, 0.1 mm; best systems, ~10 μm
Imaging capability	Typically, no; with specialized instruments (e.g., low-energy electron microscopy), 150 Å
Sample requirements	Single crystals of conductors and semiconductors; insulators and polycrystalline samples under special circumstances; 0.25 cm^2 or larger, smaller with special effort
Main uses	Analysis of surface crystallography and microstructure; surface cleanliness
Cost	≤$75,000; can be home built cheaply
Size	Generally part of other systems; if self-standing,~8 m^2

Raman Spectroscopy 6

Raman spectroscopy is the measurement, as a function of wavenumber, of the inelastic light scattering that results from the excitation of vibrations in molecular and crystalline materials. The excitation source is a single line of a continuous gas laser, which permits optical microscope optics to be used for measurement of samples down to a few mm. Raman spectroscopy is sensitive to molecular and crystal structure; applications include chemical fingerprinting, examination of single grains in ceramics and rocks, single-crystal measurements, speciation of aqueous solutions, identification of compounds in bubbles and fluid inclusions, investigations of structure and strain states in polycrystalline ceramics, glasses, fibers, gels, and thin and thick films.

Information	Vibrational frequencies of chemical bonds
Element range	All, but not element specific
Destructive	No, unless sample is susceptible to laser damage
Lateral resolution	1 μm with microfocus instruments
Depth profiling	Limited to transparent materials
Depth probed	Few μm to mm, depending on material
Detection limits	1000 Å normally, submonolayer in special cases
Quantitative	With difficulty; usually qualitative only
Imaging	Usually no, although imaging instruments have been built
Sample requirements	Very flexible: liquids, gases, crystals, polycrystalline solids, powders, and thin films
Main use	Identification of unknown compounds in solutions, liquids, and crystalline materials; characterization of structural order, and phase transitions
Instrument cost	\$150,000–\$250,000
Size	1.5 m × 2.5 m

Scanning Electron Microscopy (SEM) 7

The scanning electron microscope (SEM) is often the first analytical instrument used when a "quick look" at a material is required and the light microscope no longer provides adequate resolution. In the SEM an electron beam is focused into a fine probe and subsequently raster scanned over a small rectangular area. As the beam interacts with the sample it creates various signals (secondary electrons, internal currents, photon emission, etc.), all of which can be appropriately detected. These signals are highly localized to the area directly under the beam. By using these signals to modulate the brightness of a cathode ray tube, which is raster scanned in synchronism with the electron beam, an image is formed on the screen. This image is highly magnified and usually has the "look" of a traditional microscopic image but with a much greater depth of field. With ancillary detectors, the instrument is capable of elemental analysis.

Main use	High magnification imaging and composition (elemental) mapping
Destructive	No, some electron beam damage
Magnification range	10×–300,000×; 5000×–100,000× is the typical operating range
Beam energy range	500 eV to 50 keV; typically, 20–30 keV
Sample requirements	Minimal, occasionally must be coated with a conducting film; must be vacuum compatible
Sample size	Less than 0.1 mm, up to 10 cm or more
Lateral resolution	1–50 nm in secondary electron mode
Depth sampled	Varies from a few nm to a few μm, depending upon the accelerating voltage and the mode of analysis
Bonding information	No
Depth profiling capabilities	Only indirect
Instrument cost	$100,000–$300,000 is typical
Size	Electronics console 3 ft × 5 ft; electron beam column 3 ft × 3 ft

Scanning Tunneling Microscopy and Scanning Force Microscopy (STM and SFM) 8

In scanning tunneling microscopy (STM) or scanning force microscopy (SFM), a solid specimen in air, liquid or vacuum is scanned by a sharp tip located within a few Å of the surface. In STM, a quantum-mechanical tunneling current flows between atoms on the surface and those on the tip. In SFM, also known as atomic force microscopy (AFM), interatomic forces between the atoms on the surface and those on the tip cause the deflection of a microfabricated cantilever. Because the magnitude of the tunneling current or cantilever deflection depends strongly upon the separation between the surface and tip atoms, they can be used to map out surface topography with atomic resolution in all three dimensions. The tunneling current in STM is also a function of local electronic structure so that atomic-scale spectroscopy is possible. Both STM and SFM are unsurpassed as high-resolution, three-dimensional profilometers.

Parameters measured	Surface topography (SFM and STM); local electronic structure (STM)
Destructive	No
Vertical resolution	STM, 0.01 Å; SFM, 0.1 Å
Lateral resolution	STM, atomic; SFM, atomic to 1 nm
Quantification	Yes; three-dimensional
Accuracy	Better than 10% in distance
Imaging/mapping	Yes
Field of view	From atoms to > 250 μm
Sample requirements	STM—solid conductors and semiconductors, conductive coating required for insulators; SFM—solid conductors, semiconductors and insulators
Main uses	Real-space three-dimensional imaging in air, vacuum, or solution with unsurpassed resolution; high-resolution profilometry; imaging of nonconductors (SFM).
Instrument cost	$65,000 (ambient) to $200,000 (ultrahigh vacuum)
Size	Table-top (ambient), 2.27–12 in. bolt-on flange (ultrahigh vacuum)

Static Secondary Ion Mass Spectrometry (Static SIMS) 9

Static secondary ion mass spectrometry (SIMS) involves the bombardment of a sample with an energetic (typically 1–10 keV) beam of particles, which may be either ions or neutrals. As a result of the interaction of these primary particles with the sample, species are ejected that have become ionized. These ejected species, known as secondary ions, are the analytical signal in SIMS.

In static SIMS, the use of a low dose of incident particles (typically less than 5×10^{12} atoms/cm^2) is critical to maintain the chemical integrity of the sample surface during analysis. A mass spectrometer sorts the secondary ions with respect to their specific charge-to-mass ratio, thereby providing a mass spectrum composed of fragment ions of the various functional groups or compounds on the sample surface. The interpretation of these characteristic fragmentation patterns results in a chemical analysis of the outer few monolayers. The ability to obtain surface chemical information is the key feature distinguishing static SIMS from dynamic SIMS, which profiles rapidly into the sample, destroying the chemical integrity of the sample.

Range of elements	Hydrogen to uranium; all isotopes
Destructive	Yes, if sputtered long enough
Chemical bonding information	Yes
Depth probed	Outer 1 or 2 monolayers
Lateral resolution	Down to ~100 μm
Imaging/mapping	Yes
Quantification	Possible with appropriate standards
Mass range	Typically, up to 1000 amu (quadrupole), or up to 10,000 amu (time of flight)
Sample requirements	Solids, liquids (dispersed or evaporated on a substrate), or powders; must be vacuum compatible
Main use	Surface chemical analysis, particularly organics, polymers
Instrument cost	\$500,000–\$750,000
Size	4 ft × 8 ft

Transmission Electron Microscopy (TEM) 10

In transmission electron microscopy (TEM) a thin solid specimen (≤ 200 nm thick) is bombarded in vacuum with a highly-focused, monoenergetic beam of electrons. The beam is of sufficient energy to propagate through the specimen. A series of electromagnetic lenses then magnifies this transmitted electron signal. Diffracted electrons are observed in the form of a diffraction pattern beneath the specimen. This information is used to determine the atomic structure of the material in the sample. Transmitted electrons form images from small regions of sample that contain contrast, due to several scattering mechanisms associated with interactions between electrons and the atomic constituents of the sample. Analysis of transmitted electron images yields information both about atomic structure and about defects present in the material.

Range of elements	TEM does not specifically identify elements measured
Destructive	Yes, during specimen preparation
Chemical bonding information	Sometimes, indirectly from diffraction and image simulation
Quantification	Yes, atomic structures by diffraction; defect characterization by systematic image analysis
Accuracy	Lattice parameters to four significant figures using convergent beam diffraction
Detection limits	One monolayer for relatively high-Z materials
Depth resolution	None, except there are techniques that measure sample thickness
Lateral resolution	Better than 0.2 nm on some instruments
Imaging/mapping	Yes
Sample requirements	Solid conductors and coated insulators. Typically 3-mm diameter, < 200-nm thick in the center
Main uses	Atomic structure and microstructural analysis of solid materials, providing high lateral resolution
Instrument cost	\$300,000–\$1,500,000
Size	100 ft^2 to a major lab

Variable-Angle Spectroscopic Ellipsometry (VASE) 11

In variable-angle spectroscopic ellipsometry (VASE), polarized light strikes a surface and the polarization of the reflected light is analyzed using a second polarizer. The light beam is highly collimated and monochromatic, and is incident on the material at an oblique angle. For each angle of incidence and wavelength, the reflected light intensity is measured as a function of polarization angle, allowing the important ellipsometric parameter to be determined. An optimum set of angle of incidence and wavelength combinations is used to maximize measurement sensitivity and information obtained. Physical quantities derivable from the measured parameter include the optical constants of bulk or filmed media, the thicknesses of films (from 1 to a few hundred nm), and the microstructural composition of a multiconstituent thin film. In general only materials with parallel interfaces, and with structural or chemical inhomogeneities on a scale less than about one-tenth the wavelength of the incident light, can be studied by ellipsometry.

Main use	Film thicknesses, microstructure, and optical properties
Optical range	Near ultraviolet to mid infrared
Sample requirements	Planar materials and interfaces
Destructive	No, operation in any transparent ambient, including vacuum, gases, liquids, and air
Depth probed	Light penetration of the material (tens of nm to μm)
Lateral resolution	mm normally, 100 μm under special conditions
Image/mapping	No
Instrument cost	\$50,000–\$150,000
Size	0.5 m × 1 m

X-Ray Diffraction (XRD) 12

In X-ray diffraction (XRD) a collimated beam of X rays, with wavelength $\lambda \approx$ 0.5–2 Å, is incident on a specimen and is diffracted by the crystalline phases in the specimen according to Bragg's law ($\lambda = 2d \sin \theta$, where d is the spacing between atomic planes in the crystalline phase). The intensity of the diffracted X rays is measured as a function of the diffraction angle 2θ and the specimen's orientation. This diffraction pattern is used to identify the specimen's crystalline phases and to measure its structural properties, including strain (which is measured with great accuracy), epitaxy, and the size and orientation of crystallites (small crystalline regions). XRD can also determine concentration profiles, film thicknesses, and atomic arrangements in amorphous materials and multilayers. It also can characterize defects. To obtain this structural and physical information from thin films, XRD instruments and techniques are designed to maximize the diffracted X-ray intensities, since the diffracting power of thin films is small.

Range of elements	All, but not element specific. Low-Z elements may be difficult to detect
Probing depth	Typically a few μm but material dependent; monolayer sensitivity with synchrotron radiation
Detection limits	Material dependent, but ~3% in a two phase mixture; with synchrotron radiation can be ~0.1%
Destructive	No, for most materials
Depth profiling	Normally no; but this can be achieved.
Sample requirements	Any material, greater than ~0.5 cm, although smaller with microfocus
Lateral resolution	Normally none; although ~10 μm with microfocus
Main use	Identification of crystalline phases; determination of strain, and crystallite orientation and size; accurate determination of atomic arrangements
Specialized uses	Defect imaging and characterization; atomic arrangements in amorphous materials and multilayers; concentration profiles with depth; film thickness measurements
Instrument cost	\$70,000–\$200,000
Size	Varies with instrument, greater than ~70 ft^2

X-Ray Fluorescence (XRF) 13

In X-ray fluorescence (XRF), an X-ray beam is used to irradiate a specimen, and the emitted fluorescent X rays are analyzed with a crystal spectrometer and scintillation or proportional counter. The fluorescent radiation normally is diffracted by a crystal at different angles to separate the X-ray wavelengths and therefore to identify the elements; concentrations are determined from the peak intensities. For thin films XRF intensity–composition–thickness equations derived from first principles are used for the precision determination of composition and thickness. This can be done also for each individual layer of multiple-layer films.

Range of elements	All but low-Z elements: hydrogen, helium, and lithium
Accuracy	±1% for composition, ±3% for thickness
Destructive	No
Depth sampled	Normally in the 10-μm range, but can be a few tens of Å in the total-reflection range
Depth profiling	Normally no, but possible using variable-incidence X rays
Detection limits	Normally 0.1% in concentration.
Sensitivity	10–10^5 Å in thickness can be examined
Lateral resolution	Normally none, but down to 10 μm using a microbeam
Chemical bond information	Normally no, but can be obtained from soft X-ray spectra
Sample requirements	≤5.0 cm in diameter
Main use	Identification of elements; determination of composition and thickness
Instrument cost	\$50,000–\$300,000
Size	5 ft × 8 ft

X-Ray Photoelectron Spectroscopy (XPS) 14

In X-ray photoelectron spectroscopy (XPS), monoenergetic soft X rays bombard a sample material, causing electrons to be ejected. Identification of the elements present in the sample can be made directly from the kinetic energies of these ejected photoelectrons. On a finer scale it is also possible to identify the chemical state of the elements present from small variations in the determined kinetic energies. The relative concentrations of elements can be determined from the measured photoelectron intensities. For a solid, XPS probes 2–20 atomic layers deep, depending on the material, the energy of the photoelectron concerned, and the angle (with respect to the surface) of the measurement. The particular strengths of XPS are semiquantitative elemental analysis of surfaces without standards, and chemical state analysis, for materials as diverse as biological to metallurgical. XPS also is known as electron spectroscopy for chemical analysis (ESCA).

Range of elements	All except hydrogen and helium
Destructive	No, some beam damage to X-ray sensitive materials
Elemental analysis	Yes, semiquantitative without standards; quantitative with standards. Not a trace element method.
Chemical state information	Yes
Depth probed	5–50 Å
Depth profiling	Yes, over the top 50 Å; greater depths require sputter profiling
Depth resolution	A few to several tens of Å, depending on conditions
Lateral resolution	5 mm to 75 μm; down to 5 μm in special instruments
Sample requirements	All vacuum-compatible materials; flat samples best; size accepted depends on particular instrument
Main uses	Determinations of elemental and chemical state compositions in the top 30 Å
Instrument cost	\$200,000–\$1,000,000, depending on capabilities
Size	10 ft × 12 ft

Index